WEALTH FROM DIVERSITY

Economics of Science, Technology and Innovation

VOLUME 9

The titles published in this series are listed at the end of this volume.

WEALTH FROM DIVERSITY

Innovation, Structural Change and
Finance for Regional Development
in Europe

Edited by

XAVIER VENCE-DEZA
University of Santiago de Compostela

and

J. STANLEY METCALFE
University of Manchester

**UNIVERSIDADE DE
SANTIAGO DE COMPOSTELA**

KLUWER ACADEMIC PUBLISHERS
DORDRECHT / BOSTON / LONDON

Library of Congress Cataloging-in-Publication Data

```
Wealth from diversity : innovation, structural change and finance for
  regional development in Europe / edited by Xavier Vence-Deza, J.
  Stan Metcalfe.
      p.   cm. -- (Economics of science, technology, and innovation ;
  v. 9)
    Selected papers presented at the International Congress "European
  Periphery Facing the New Century" held in Oct. 1993 in Santiago de
  Compostela, Galicia.
    Includes bibliographical references and index.
    ISBN 0-7923-4115-5 (alk. paper)
    1. Regional planning--Europe--Congresses.  2. Structural
  adjustment (Economic policy)--Europe--Congresses.  3. Technological
  innovations--Economic aspects--Europe--Congresses.   I. Vence Deza,
  Xavier.   II. Metcalfe, J. S. (J. Stanley)  III. International
  Congress "European Periphery Facing the New Century"  (1993 :
  Santiago de Compostela (Spain))  IV. Series.
  HT395.E8W43   1996
  338.94--dc20                                              96-8870
```

ISBN 0-7923-4115-5

CONTENTS

LIST OF CONTRIBUTORS

Amendola, Mario. Professor of Economics. University La Sapienza. Roma, Italy.

Ashcroft, Brian. The Fraser of Allander Institute. University of Strathclyde. Glasgow, Scotland.

Bruno, Sergio. Professor of Economics. University La Sapienza. Roma, Italy.

Charbit, Claire. Economics Department. E.N.S. Télécomunications. Paris, France.

Cowan, Robin. University of Western. Ontario, US.

Dow, Sheila C. Department of Economics. University of Stirling, Scotland.

Foray, Dominique. IRIS/CNRS/University ParisIX-Dauphine. Paris, France.

Gaffard, Jean Luc. Professor of Economics. LATAPSES and University of Nice, France.

Gordon, Richard. Center for the Study of Global Transformations. University of California, Santa Cruz, US.

Justman, Moshe. Industrial Development Policy Group. Jerusalem Institute for Israel Studies. Jerusalem. Israel

Messori, Marcello. Professor of Economics. University of Cassino, Italy.

Metcalfe, John Stan. Professor of Economics. PREST and Faculty of Economics. University of Manchester, England.

Patel, Parimal. SPRU. University of Sussex at Brighton, England.

Pavitt, Keith. SPRU. University of Sussex at Brighton, England.

Quéré, Michel. LATAPSES. University of Nice, France.

Quévit, Michel. RIDER. Université Catholique de Louvain, Louvain-la-Neuve, Walonie.

Stöhr, Walter B. Professor Emeritus. Interdisciplinary Institute for Urban and Regional Studies. University of Economics and Business Administration, Vienna.

Teubal, Morris. Industrial Development Policy Group. Jerusalem Institute for Israel Studies. Jerusalem, Israel.

Vence-Deza, Xavier. Professor of Applied Economics. IDEGA. University of Santiago de Compostela, Galicia.

ACKNOWLEDGEMENTS

This book is a selection of the papers presented at the International Congress on "European Periphery facing the New Century" held in Santiago de Compostela in October 93, organised by IDEGA and University of Santiago de Compostela with the financial support of the Conselleria de Cultura of the Xunta de Galicia and the SA de Xestión Xacobeo 93. We wish to express our gratitude to all persons and institutions that helped us in the whole undertaking. A special mention should be paid to the Vice-rector of Culture, Justo Beramendi, for his personal support in attaining the funding to organize the Congress.

We also thank Marga Bures, Luis Castañón, Isabel Diéguez, Ana Dopico and Xulia Guntin for their invaluable assistance in handling organizational matters. We would also like to thank the seventy participants in the Congress and particularly the members of the Scientific Commitee for their advice, their time and their expertise; a special mention should be paid to professors Mario Amendola, Juan Ramón Cuadrado Roura and Miren Etxezarreta for their generous help in designing the Congress.

Finally, in producing this volume we thank all authors for their excellent contributions and for their care in revising the papers for publication.

LIST OF FIGURES

LIST OF TABLES

1 INTRODUCTION

X. Vence-Deza & J.S. Metcalfe

U. of Santiago de Compostela & U. of Manchester

This book contains selected papers presented at the International Congress "European Periphery Facing the New Century" held in Santiago de Compostela, Galicia, at the end of 1993. The general aim of this congress was to rethink the great economic and social changes in Europe during the last decade from a critical view, specially focused on peripheral regions and the conditions for an enduring process of development. Both economic, social and political changes affect the characteristic diversity of Europe and they have a special impact on the countries and regions that were traditionally known as the periphery within Western Europe. The list of concerns is long: regions with different levels of development compete within a new single market; the free movement of productive factors modifies the traditional pattern of industrial location, increasing tendencies to regional concentration shift the balance of income generation; new competitors enter traditional markets; information technology creates new possibilities of industrial organization and cooperation; competitivity is based on different capabilities to innovate and to promote structural change, and these capabilities differ among regions; traditional regional policies fail in the present even more than in the past. Hence the central concern of this volume, to explore the links between diversity and regional development.

These aspects have great importance in evaluating the real tendencies towards regional cohesion and the possibilities of convergence. The process of integration in Europe opens the barriers for action according to the principles of the open market, accentuating uneven tendencies between the countries and above all between regions. The action of Community Policies tends to reproduce this inequality instead of correcting it, except in the case of specific regional policies; moreover, this corrective effect of regional policy, focussed on the endowment of infrastructures, is of limited efficiency. Therefore, despite the inclusion of cohesion principles in the Maastricht Treaty, it will be quite difficult for the limited policies of the European Comission to counteract the strong tendency towards geographical concentration whose impulse is strengthened by the liberalization of the flow of both goods and production factors.

X. Vence-Deza and J. S. Metcalfe (eds.), Wealth from Diversity, 1–10.
© 1996 *Kluwer Academic Publishers. Printed in the Netherlands.*

REGIONAL DISPARITIES IN THE EUROPEAN UNION

The existence of strong contrasts in the levels of development in European regions is a known reality in any research undertaken on the subject. The historical process has formed a great area of development in Europe, the hard core being the triangle-London-Paris-Hamburg; in the sixties and seventies this zone extended towards the south encompassing the South of Germany, North of Italy and, more recently, also the French and Catalonian Mediterranian. This is what constitutes the dynamic core, referred to by some authors as the "Great Dorsal" or "Great Banana", which continues to show the highest growth rate and attract the largest flows of investment in Europe. Of course, by defining the dynamic core we simultaneously define the periphery.

Taking the data of the GDP per capita for the years 1989-1991 in the 179 regions of the EU it was found that 119 displayed values under the average, 43 of which were more than 25% below the average (so called Objective 1 regions in the Community classification). These 43 regions alone represent 22% of the Community population in 1991 (EC, 1994). The differences in these aggregated indicators coincide with analogous differences in other variables such as infrastructure endowment (transport, energy, telecommunications ...), human capital formation, research and development activities and high value added productive activities. The relative importance of these variables in each region constitutes an explanatory factor of the disparities. Anyhow, in no way can we explain the existing inequality nor the dynamics through time on the basis of static endowments of the productive factors: we must include other elements and other aspects that determine the dynamic processes involved in economic development. The other elements have been emphasised for example by Michael Porter in his recent book **The Competitive Advantage of Nations**. Thus, investments in new industries tended to be located in the established industrial centres (NEI, 1993), attracted by the proximity of the markets and a complex industrial system with a network of suppliers, subcontractors and advanced services which generate important external economies and agglomeration economies.

The observation of the regional trends based on the data of the basic macroeconomic indicators reveals a significant increase in the regional disparities in Europe during the eighties. The disparities of GDP and of income per capita increased and at the same time a concentration of employment growth and investments was observed in a reduced number of metropolitan areas. Taking as an index of disparities, the standard deviation of the GDP per capita weighted by the size of the population of each region, this was observed to increase from 26.6% in 1980 up to 27.9% in 1991 for the regions (NUTS 2) that form the European Union. A revealing fact comes from the comparison between the averages of the richest 25 regions and the poorest 25 regions of the EU. Whilst the former were to be found

2

at a level of 135.3% above the average in 1980 increasing to 137.0% in 1991, the latter experienced a contrary trend descending from 54.9% to 54.4%. This widening of the gap throughout the decade between the most developed regions and those less developed, shows uneven performance increasing in the first half of the decade coinciding with the phase of slow growth and slowly reducing in the second half coinciding with the phase of economic recovery.

The evolution of employment and unemployment also shows a markedly different pattern among the European regions. Employment growth during the period from 1981 to 1991 was localised in a reduced number of regions of intermediate industrial tradition and scarcely grew or even decreased in agricultural regions and in those characterized by mature and declining industries. On the other hand, the rates of unemployment also reflect an accentuation of the disparities from the mid-seventies up to today (except for a short lapse of time between 1988 and 1990), increasing at the same time as the volume of unemployment). This image of disparities also offers a comparison of the rates of unemployment which show that 71 regions, representing 43% of the population, have rates of unemployment above the average (greater than 10%) whilst 34 other regions have unemployment rates lower than 5%; in fact the weighted standard deviation with respect to the EUR average reaches 50.4%.

Several policy instruments to reduce these differences in the EU have been adopted by Community Policy makers, such as so called Structural Funds, investment loans and Cohesion Funds. Even if financial effort is important, a strong question remains: to what extent can these policy instruments really help the relatively poor regions of Europe Union to catch up?. Certainly, answering this question requires knowing which forces drive regional growth differences. The prevailing idea in the European Commission seems to be based on the idea that income and productivity depend on convergence in certain macroeconomic indicators (inflation, public sector deficit, external account, interest rate etc). The theoretical basis of convergence criteria defined by EC seems to be a combination of standard macroeconomic theory and the neoclasical theory of economic growth. A central assumption is that if markets are allowed to work, poor regions should be expected to catch up towards rich regions. Convergence will allow them to share the same level of income and grow at the same rate. The papers in this volume do not support this optimistic view and in this they are drawing upon different traditions of thought.

THE LIMITS OF THE NEOCLASSICAL EXPLANATION OF THE DISPARITIES

Given that important disparities exist across countries and across regions we need a convincing explanation for them. Even more, we need an explanation of

3

trends of these disparities. Most explanations are an extension of international growth rate differences analysis and their statements are well known. Here we are concerned with disparities across regions but, of course, at European level; that implies considering differences between regions and, at the same time, differences across countries.

Some preliminary questions should be posed. The first one to be considered is the concept of region itself. Whilst explaining existing disparities and the tendencies in the concentration of economic activity and that of employment, it is necessary to specify the degree of internal coherence of the regions being compared and analyse which are the dynamic factors that feed them.

In fact, regions are not internally homogeneous spaces and therefore the aggregate regional disparities conceal other types of internal differences within each of these regions (metropolitan areas, agricultural areas, mountain areas etc). Moreover, the evolution of the disparities is not displayed in an identical way in all economic and social variables but, on the contrary, relatively divergent patterns among them can be observed; for example, the evolution of income and employment can show asymmetry and the same can occur if we consider the evolution of human capital or infrastructure endowment. This is particularly relevant if we take into account the fact that each element of a productive system gives rise to dynamic and cumulative effects which are of very different importance in the long term. In this sense, the relative evolution of the different categories of employment can be more relevant in the medium and long term than that of aggregate employment; likewise, the importance of the disparities of investments in infrastructures does not have the same effect on the creation of wealth for the future as disparities in, for example, technology generation or industrial investment.

With respect to the explanation of the disparities, several theoretical approaches exist. As is well known, in the neoclassical approach, the spacial inequalities are the result of uneven endowment of factors and will tend to disappear spontaneously as a consequence of the mobility of the factors. Factor holders are driven towards the regions where they perceive higher remuneration. The workers of poor regions with a lower capital intensity, will direct themselves towards regions with high capital intensity that offer higher salaries which will tend to even out relative income levels, by decreasing demographic pressure in the former and by pushing down the salaries in the labour market of the latter. As for capitals, they will flow in the opposite direction towards the regions with lower salaries promoting the development of backwardness areas. In this way, factor proportions and relative remuneration rates will tend to even out in the different regions. Recent empirical works in this area (Sala i Martin, 1994) show that convergence takes place, although at a rate too slow to be consistent with the neoclassical growth model (Solow).

4

Beyond the theoretical limitations of the basis of the neoclassic models (assumption of full employment of resources, perfect competition, perfect information, perfect flexibility of factor prices...) there is an essential limitation derived from the fact that these models leave hidden important determining factors in the growth process. In a dynamic model we can consider a variation in the growth rate of the labour force and technical progress in the regions which alters the conditions of the equilibrium aforementioned. For example, an increase in the population in low-wage regions can be an obstacle to income growth; likewise, upward displacement of the marginal productivity function of capital in high-wage regions, owing to, for example a technological change, can attract capital to such regions. Moreover, other important aspects exist that question and weaken the neoclassical approach. First, labour mobility cannot be explained only by the wage gap between regions and nor is it demonstrated that emigration plays a balanced role, on the contrary it seems to accentuate the disparities (brain drain, demographic dynamism etc). Second, the regions are considered as simple economic units and the spatial dimension only enters into play as a consequence of a difference in the remuneration of factors in the different regions (proximity, externalities etc, are not taken into consideration). Third, the assumption of technology as a global public good is not consistent with recent empirical data; on the contrary, technology appears as a localized process that implies tacit knoledge, embedded in firms, clusters of firms and regions. Fourth, the assumption of constant returns to scale and the ignorance of external economies do not take into account the positive spillovers that investments in education or R&D may have. Fifth, the only active agents are microeconomic (workers and entrepreneurs), without the capacity to influence markets; the existence of cities, agglomerations and regional institutions is not considered but, in fact, these agents can promote the development capability of each region in different ways and with uneven efficiency. In fact we can observe a very different rationale for the implementation of policies in the fields of, for example, training, technology, infrastructure, trade and tax.

ALTERNATIVE THEORIES OF UNEVEN REGIONAL DEVELOPMENT: DISPARITIES AND DIVERSITY OF REGIONS

In contradistinction to the neoclassical approach of convergent, uniform development, there are alternative theories which try to explain uneven development according to which divergence in regional growth rates tend to increase. One of these is the structural imbalance theory which states that the capitalist development process is, in itself, a process of concentration of the productive capacity from the point of view of the ownership as well as from the territorial viewpoint. In these approaches the development of some areas is founded in the underdevelopment of others, thus establishing a hierarchical and dependent relationship (core-periphery model), with each specialising in a different function in the global process (spacial division of work). The polarisation is a result of the fact that capital concentration creates agglomeration economies and at the same time allows hierarchical decentralisation of certain activities in order to reduce production costs, exploiting the specific advantages of inter-regional division of the production processes.

In the cumulative circular causation theory (Myrdal, Hirschman...), the differences in income tend to increase more than decrease, above all as a result of the effect of technical progress differentials (with different production functions in a static sense or different innovative capacities in a dynamic sense). This can contribute to a reversal in the direction of the flow of the production factors contrary to that stated in the neoclassic model. This can occur in a particularly clear way in the case of capital, since an initial advantage derived from a higher level of technical progress can allow a higher rate of profit in a central region in spite of the fact that the capital factor could be greater. In this way, knowledge and capital as much as labour will tend to flow in the same direction towards the higher developed regions. As a result, the flow of workers has an impoverishing effect on backward regions to the extent that emigration is selective and above all it will be the young and the qualified who move. The backward regions support the costs of the upbringing and education of these people but it will be the developed regions who take advantage of their productive and creative capacity. The net result is a transfer of resources, costly for the poor regions,which contributes to a widening of the gap between the regions. This "brain drain" constitutes, in the Schumpeterian approach, one explanation for the low entrepreneurial and investing capacity of the backward regions. These difficulties are accentuated by the role of the banking system in facilitating the drain of savings and financial resources in the direction of the dominant markets of capital and credit. Lastly, we must recognise a wide range of dynamic factors which increase economic divergence between the regions: development itself increases demand which generates new investment stimulating income generation and yet more demand and new investment, making development a process with a strong propensity to be self-supporting, in a cumulative dynamic.

The introduction of demand in the regional growth model, in a Keynesian approach, begs the consideration of the market size and exports as a growth motor - export based theory. In its basic formulation, external demand is the determinant of the growth of so-called "base sectors" and at the same time, the income generated by these is the variable that drives the growth of " non-base sectors". Obviously, the greater the local consumer satisfaction is with local production, the greater the multiplier will be.

It was Kaldor who established the connection between export growth, induced income growth and the increase of competitivity of the base sector. In Kaldor's approach, cumulative causation derives from the existence of increasing returns to scale, which derive not mainly from production scale economies but more significantly from the advantages of growth in the whole industry (external scale economies) and subsequent new knowledge, improved diffusion of information and know-how, increase in specialisation and division of work. The main factor which determines the economic growth rate (of productivity) in a region, is the export growth rate via the effect that this has, through the accelerator, on the growth rate of industrial capacity. In turn, exports depend on exogenous factors (increase in world demand) and endogenous factors, such as the so called "efficiency wage": the lower the relationship between the monetary wage and productivity the higher will be the competitiveness of a region in relation to the other regions. Moreover, if the monetary wage tends to increase at the same rate in all regions, the efficiency wage will tend to reduce in those regions where productivity growth is faster and this will occur,in accordance with Verdoorn's Law,in regions with higher output growth rate. This is precisely the manifestation of cumulative causation.

This model displays an explanatory capacity notably higher than the neoclassic model, however it also shows an important number of limitations. Many critics have paid much attention to the practical aspects of the model (the difficulty in establishing the distinction between base sector and non-base sector, the stability of the relationship between both sectors, the role of region size in the determination of the multiplier...); more fundamental limits exist concerning the relevance attributed to external factors and the subsequent lack of consideration of the internal characteristiques of regional space. To consider these internal factors it is necessary to explain the causes of export and competitivity factors which can permit greater market shares to be won. The true force of a regional economy derives from its capacity to adapt to economic change and from this long-term perspective elements such as accumulated investment, human capital, innovative capacity, institutional system, financial system, etc, prove to be essential.

These limits gave rise to approaches that emphasise endogenous factors of regional development. In these approaches territory is conceptualised in a totally different way and acquires an active role which is the result of the innovative and dynamized activity of local agents (private and public). A good number of reasons exist which explain the emergence of this type of approach in the eighties: the introduction of new technologies provoke a structural change in the worldwide productive system and modifies links between production plants and space, modifies competitive advantages of the regions and facilitates the emergence of a new organisational model of firms. All of this occurs in a context in which economic processes are globalised and competition between regions to attract new investment and develop new activities is increased. The role of innovation in this context acquires pre-eminence as an element in raising the competitiveness of firms as well as in the development of new industries and the modernization of existing industry. The innovative capacity of a territory is the result of the learning processes from both the past and the present which crystalize in entrepreneural structures, human resourses and institutions. Moreover, the efficiency of innovative system depends on the capacity to coordinate these elements in a dynamic process in which are generated external economies, agglomeration economies, interactive economies, cooperation networks, etc. which together form an "innovative environment".

ABOUT THIS VOLUME

Many of these themes are explored in detail in the papers which follow. Amendola and Gaffard focus upon the central role of diversity in a process of regional development in which the sequence of development stages flow one from the other in path-dependent fashion, emphasizing the "out of equilibrium" nature of this process. Metcalfe too explores the role of diversity, linking this to necessary returns and drawing upon fundamental evolutionary principles which relate the rate and direction of economic change to measured variety in economic behaviour. Several papers explore the role of learning mechanisms on the generation of variety. Pavitt and Patel focus on the differences between large firms in the rate and direction of learning and establish the clear empirical result that one dominant learning mechanism, the technological activities of large firms, tends to be located in the home countries of the firms not their overseas affiliates. Cowan and Foray consider how the development of information technology systems will change accepted patterns of learning and thus patterns of catching-up in the development process. An important conclusion is the relevance of supporting activities which are not at the current technological frontier but which may constitute technologies for future recombination. Any discussion of learning mechanisms leads naturally to the concepts of industrial districts and innovative agglomerations of firms and other supporting institutions. This theme is taken up in the paper by Gordon, who develops a typology of regional innovation systems while emphasising the importance of cross regional linkages, Gaffard and Quéré, by comparison build their taxonomy around different complementarities and patterns of structural change, while, Vence-Deza, considers a range of policies to strengthen local innovation learning systems including those which emphasise the demand for innovation in order to cope with specific problems of each regional production system, particularly in peripheral regions. An important theme in all these papers is the open-ended historical nature of regional development, it is not a transition between equilibrium positions but rather the cardinal discovery of new pathways for economic activity in a self generating and cumulative fashion. These themes come together quite naturally in the discussion of innovation systems and technology policy infrastructure which is the concern of the paper by Justman and Teubal. Technology policy is given two roles, building capabilities and building markets which in turn imply problems related to co-ordination, uncertainty and the attainment of critical mass. Particularly important is their treatment of the limitations of market failure analysis and the need for technology policy experiments.

It follows from the above that one desirable characteristic of the regional development process, from a policy viewpoint, is that it too should be self sustaining and this is the focus of the paper by Stöhr who makes a distinction between 'top-down' and 'bottom-up' regional policies and outlines the characteristics of the portfolio of policy stances which support endogenous development. A good many of these will inevitably concern the create and growth of SMEs and this is the theme explored by Charbit who is particularly concerned with the role of international co-operative arrangements with large firms being an element in the development of regional capabilities, specifically in peripheral regions.

The 'core'-periphery theme is taken up in detail by Ashcroft who is concerned with the external control of economic activities in peripheral areas, via investments and acquisitions and their different effects on regional economic performance in terms of competitiveness, innovation and the creation and growth of firms. Since the benefits for individual firms may differ from the broader regional benefits he argues that inward investment is to be preferred to acquisition. Clearly a policy is required to manage a peripheral region's external relations. The shifting balance of regional activity depend not only on the diverse behaviours of individual firms but also on the selection environment which revolves these behaviours into patterns of concentration and catching up. A central feature of the European environment will be the arrangements made for monetary integration and this is the central theme of the paper by Dow. Developing the links between credit creation and regional development, Dow explores the likely impacts of competition on the banking sector, the prospects for financial instability and risk assessment and the implied financial constraints on regional development. The possible adverse consequences for peripheral regions include the elimination of independent local banks and adverse risk assessments by core banks which inhibit the promotion of regional activity. In a paper on banking and finance in the Mezzogiorno, Messori amplifies these concerns, pointing to the effects of an unfavourable supply of credit on local development. In turn this limitation on credit is linked to default risks, peculiarities of balance sheets of local banks, and the segrementation of the local credit market, creating a vicious circle embedding credit supply as a binding constraint on development. The final two papers by Quévit and by Bruno turn to the wider question of the completion of the internal market. Quévit reviews the debate on the costs of not completing the single market and the implications of the unified market for lagging regions, drawing attention to the likely diversity of regional responses to the new trading opportunities. Bruno turns explicit to the policy implications of the single market particularly those relating to longer view questions of growth, capital accumulation, expectations, and competition. While drawing attention to the need for an appropriate use of policies he suggests that it is also important to ensure the 'selective preservation of diversities' rather than design policies solely as instruments of convergence.

In drawing to a close this introduction we must emphasise the challenge to Europe presented by our theme of wealth in diversity. It is a central feature of competition that "success" has lain in the ability to accommodate to radically new concepts of economic activity, or new combinations of activity to use Schumpeter's characterization, and that this process has continually shifted the pattern of location of economic activity. In so doing the basis of diversity is continually eroded yet the continuation of progress depends upon sustaining diversity: that fact which fuels the competitive process. In the context of strengthening European integration the regional basis for the maintenance of diversity will surely remain at the centre of policy making concerns. Progress depends on diversity in innovation but progress must not be allowed to undermine the conditions which recreate diversity and their wider social and economic implications. It is perhaps more important than ever to remember Frank Knight's warning that competitive systems cannot be judged solely in terms of their ability to satisfy economic wants but only by appraising them in terms of a higher standard of values or social ideals. In the modern competitive world of necessary structural change this will be no easy task.

REFERENCES

Comisión Europea (1994), **Competitividad y cohesión: las tendencias de las regiones. Quinto informe periodico sobre la situación y la evolución socioeconómica de las regiones de la Comunidad**, Bruselas, CE.

Fagerberg, Jan and Verspagen, Bart (1995), "Convergence or Divergence in the European Union. Theory and Facts" -Draft-, Maastricht, MERIT.

NEI (1993), **New location factors for mobile investments in Europe**, Luxemburg, CCE, Regional Development Studies, n.6.

Sala i Martin , Xavier (1994), "Regional Cohesion: Evidence and Theories of Regional Growth and Convergence", **CEPR Discussion paper**, n.1075, London, CEPR.

2 CHANGES IN STRUCTURE AND RELATIONS BETWEEN PRODUCTIVE SYSTEMS

Mario Amendola and Jean Luc Gaffard
University La Sapienza & University of Nice

The capacity of the economies to grow is usually made to depend on the presence of some particular requisites (resources, skills, technology,...) that are generally considered as essential for assuring growth. Only the economies that get to possess these requisites are thus able to grow; on the other hand, the very possession of the requisites themselves assures growth. This is the view underlying the standard growth theory in its traditional steady growth version or in its more sophisticated endogenous growth revisitation.

The model is general: it concerns all economies. In this perspective diversity is a curse; it means not to correspond to the standard defined by the possession of the required elements.

The equilibrium approach behind this framework focuses on the definition of a given analytical configuration: an "equilibrium". The existence of the elements which are essential to this definition automatically implies the given configuration. Comparison of alternative some such configurations corresponding to different values of these essential elements are the way in which the analysis, including dynamic analysis, is carried out in this approach.

In the standard growth theory the dynamic configuration of the economy is defined in terms of appurtenance parameters –mainly technologic and demographic parameters– that the theory itself treats as exogenous. As is well known, the steady-state associated with given values of these parameters is approached asymptotically for any arbitrary initial conditions prevailing in the economy: convergence is assured. Thus, if we believe that all countries have access to the same improving technology, convergence to the same steady growth rate not only in different countries belonging to a particular group but in all countries should also be expected, at least in the very long run.

But differences do exist. The reinterpretation of growth theory that goes under the name of "endogenous growth" tries to see to that, that is, to account for

11

X. Vence-Deza and J. S. Metcalfe (eds.), Wealth from Diversity, 11–18.
© 1996 *Kluwer Academic Publishers. Printed in the Netherlands.*

persistence of difference instead of convergence. The endogenous determination of the long run growth rate of the economy through aggregative increasing returns to scale induced mainly by the external effects of human capital accumulation makes it possible to explain diversity of performances not only among economies endowed with different resources but also for those endowed with the same resources. Different preference parameters –referring to the utility function and the rate of time preference– determine the accumulation of human capital and hence influence the growth rate. Thus countries similar in any respect when their development proccesses started may end up growing very differently, better reproducing the observed reality than old Solow–type models.

When applied, e.g., to two countries and two goods, the most simple case of international trade, this analytical set up helps to show how trade may lock in different growth rates that reflect different dynamic learning processes prompted by different initial specialization patterns. Differences, and their maintenance, are thus not only accounted for but strengthened by international trade.

Standard growth models, including endogenous growth models, are equilibrium models where the appropriate choice of the relevant conditions (preference parameters...) guaranteed by perfect foresight, makes the economy jump inmediately to its steady growth rate. As in all equilibrium models no transitional dynamics towards the long run growth rate is, nor can be, contemplated.

Now, in the real world, neither it is possible to jump inmediately on given growth path nor is one condemned to given trajectories once and for all started off by certain initial conditions. Changes take place –also changes of growth rates– which imply an altogether different analytical approach to be dealt with.

A change in the growth rate is a process which, in order to be understood as such, must be considered in its sequential building up and cannot be taken as a whole, with reference to its point of arrival in the ex post perspective that characterizes equilibrium analytical frameworks and comparative analysis. This implies in fact a qualitative change, that is, some sort of structural modification that most likely takes the economy out of equilibrium. It is therefore in the nature of an *out of equilibrium process* which –due to the intertemporal complementarities of production and of the decision processes– is likely to become cumulative and acquire a recurrent character rather than converge to a new equilibrium. The viability of this process –its being able to bring about the change in the productive structure required to sustain a different growth rate– becomes then the crucial analytical issue. The main point that we intend to stress in this paper is that the viability of this out of equilibrium process requires different kinds of external interventions interacting over time, and that relations between different production systems may provide a coherent interaction framework which transforms diversity into an essential element of international growth.

Equilibrium analysis relies on an essentially timeless representation, not in the naive sense that it does not make reference to time but in that it overlooks the sequential nature of the phenomenon of production. The main analytical implications of this kind of representation is that, even when differently dated, inputs and ouputs are analytically, and from an accounting viewpoint, contemporaneous, so that there are always proceeds against which costs can be set and a "current" productive activity out of which they can be financed. This aspect of the timelessness of the production process is what really legitimates equilibrium analysis and its use in (a certain kind of) dynamic economics. It implies in fact a *given relation* between the basic magnitudes of the process itself (output, capital, employment...) which draws the attention to the functioning of the given productive capacity that brings about this regular behaviour (steady–states of the economy), and not to changes in this capacity implying a modification of the given relation (structural modifications).

Now the essence of qualitative change is to be a creative destruction process, the main aspect of which are the repeated changes in technology, preferences and industrial structures that result in a dissociation of costs and proceeds over time, that is, in the appearance of sunk costs. And this is what does not make possible to analyze such a process within an equilibrium framework based on an ex post representation of production.

Focus on the time dimension of production allows to bring to light that commodities are not produced by commodities but by processes and that these processes take place through a sequence of periods that make up a phase of construction and, following it, a phase of utilization of productive capacity. Whereas in a stationary or steady state the attention can be confined to the production of commodities (the "utilization" moment of a production process) this is no longer the case in any non stationary economic system where the production activity is aimed at two distinct objectives –to produce *goods* and to produce *processes*– and where the latter activity must come before the former.

Thus the first requirement of a proper analysis of non stationary states is a representation of production activity as production of processes articulated over time. This makes it possible to investigate the consequences of a breaking of an equilibrium configuration: in the first place a distortion in the age structure of productive capacity which implies dissociating in time inputs from output and costs from proceeds. More generally, the immediate result of the intended change is a distortion of productive capacity which implies a true disruption of economic capacity, that is, the scrapping of some production processes not planned beforehand, the abrupt desappearence of a part of the existing productive capacity, and hence a sudden mismatch between supply and demand. Construction and utilization,

13

investment and consumption, and supply and demand, are no longer harmonized over time.

A first general meaning of being "out of equilibrium" is thus that a change in the balance of processes of production in the different stages of their life (that is, a change in the balance between "construction" processes and "utilization" processes) is under way; which also implies that there is no longer a given and stable relation between the relevant economic magnitudes.

The initial distortion, we have mentioned, stirs an out of equilibrium process which, via expectations, may become cumulative, so that its very viability may be hampered. The problem, then, is to bring to light the conditions for viability, that is, what is required to reestablish the consistency over time of the relevant interacting magnitudes of the economy —which, in the context envisaged, means not only to figure out *what* is required, but also *when* this is required. However the analysis of qualitative change needs more than simply stressing the time articulation of the construction and the utilization phases of production processes with focus on construction and the costs associated with it. It needs to pass from an ex post to an ex ante view of production, which implies in particular interpreting construction not as adoption or realization but as *creation anew*. This creation concerns not only technology and productive options but the very resources involved in the process.

In the standard theory the elementary analytical particle is the firm, identified with the production process as defined by the technology, which also determines the "boundaries" of the firm itself within a given context. This micro representation carries easily over to the macro level by making the elementary process (multiplied by a scalar) stand for a whole sector or for the entire productive system.

This is no longer so when the focus is on the organization and implementation of a process by which new resources, new productive options, new technology, and hence a new and different context, are brought about. This process, in fact, also affects the firm, by changing its configuration and articulation within a context continuously modified and redefined. The picture, then, is no longer that of the firm facing a given market. Aggregates of elements that exhibit different relations among them get structured into different aggregations: this is the essence of irreversible and hence truly dynamic processes. The focus is then on aggregations that render the innovation process viable; and in this light must be stressed in the first place the interaction of organizational design and market behaviour in structuring coherent innovation systems, that is, their nature of *complements* in a process of creation of resources.

A viability problem, we have just seen, arises because, as a result of a

14

structural disturbance, productive activity is no longer harmonized over time. The viability of the intended change thus depends on being able to reestablish a consistent time pattern of construction processes and utilization processes, which, as already mentioned, also implies the consistency over time of investment and consumption and of supply and final demand. However, we have also seen that in the sequencial context considered the built in reaction mechanisms to the initial distortion of productive capacity associated with the attempt to accelerate growth are likely to cause stronger and stronger fluctuations in output, prices and investment that go rather in the direction of increasing the unbalance over time between construction and utilization. External interventions aimed at regulating the working of the existing mechanisms and/or at providing other compensating mechanisms are then the only way to reestablish the harmony required for viability.

We need in fact an adequate mix of different interventions, a sort of "fine tuning" which is not made once and for all but must be continuously modified as the interventions are interventions over time which must deal with perturbations that are recurrent but take on different shape and intensity in time. This clearly means that there is not a unique regulation system good to deal with a given structural disturbance and hence that no system can be preferred ex ante to another one.

We have stressed that external interventions are essential for rendering processes of qualitative change viable. Some of these interventions concern regulation mechanisms which reflect a change in the organizational and institutional set–up. Others, however, are only made possible by the existence of other (different) economic systems, External flows of final demand are required to absorb the excesses of productive capacity at given moments ot time during the process of speeding up the growth rate. Transfers of human resources aimed at sustaining a more intense construction activity during the same process are also needed. Borrowing and lending between countries should also be added; this not only enables transfers of financial resources, but also allows intertemporal consumption smoothing and hence affect the time patterns of demand and prices.

Particular time profiles of these interventions are required, which can only be the outcome of the interacting behaviours of different economic systems. The coexistence of different systems –that is, of different institutional set–up that imply a different bias in the relation between construction and utilization activities and hence require different regulation mechanisms– is in fact the only way to make available at particular moments of time what a given system is not able to provide itself.

However, diversity is not a sufficient condiction. It causes interactions but not necessarily the right kind of interactions over time. Coherent interaction patterns are needed: which means establishing complementary intertemporal feedback such as to smooth the distortions of productive activity in the different systems rather than to exarcerbate them.

It is viability that matters when dealing with out of equilibrium processes. And viability, as we have seen, needs diversity; as opposed to the homogeneity evoked by the uniqueness of optimal solutions in equilibrium contexts, where instead efficiency matters. And it is again with reference to viability that must be considered the coherence of behaviours in a world characterized by diversity.

Finally, it must be stressed that the dynamic analysis sketched out is the expression of a change of perspective which, by leading to interpret production no longer as allocatioin of given resources but as creation of altogether new resources, also implies a change both of the modelling of the creation process involved and of the use of the different type of model considered.

This concerns in the first place the traditional interpretation of the terms "exogenous" and "endogenous". In a model there are variables and parameters: the parameters (given magnitudes and coefficients) reflect the existing constraints. Exogenous, in the standard analysis, are the constraints which exist outside and above the economy and which determine its behaviour. But once we recognize that the time over which creation of resources takes place is a continuing and irreversible process, we have to consider as a parameter, and hence as exogenous, not some given element chosen beforehand in reason of its nature or characteristics, but whatever, at a given moment of time, is inherited from the past. What appears as a parameter at a given moment of time is therefore itself the result of processes which have taken place within the economy: processes during which everything –including resources and the environment, as well as technology– undergoes a transformation and hence is made endogenous to the change undergone by the economy. Thus, while the standard approach focuses on the right place to draw the line between what should be taken as exogenous and what should be considered instead as endogenous in economic modelling –a line that moves according to what we want to be explained by the model– in the different perspective proposed the question is no longer that of drawing a line here or there but rather one of time perspective adopted. Everything can be considered as given at a certain moment of time, while everything becomes endogenous over time: which implies a process in the nature of a sequence "constraints–decisions–constraints".

This process, we have seen, may become cumulative, so that its very viability may be hampered. The relevant problem is then "viability", and this problem is not one calling for general analytical solutions as in the usual perspective of formal (allocative) models. A creation process that builds up along the way can be explored, and its viability ascertained, working out the evolution of the economy along the sequence of periods through which the intertemporal complementarities

16

and the interactions that characterize the functioning of the economy trace out the effects of the initial structural modification. The use of simulations makes this possible. In particular numerical calculations help to bring to light to what extent financial (and human resources) constraints affect the age structure of productive capacity and hence what interventions are required for making the innovation process viable.

REFERENCES

AMENDOLA, M.; GAFFARD, J.–L. (1988), **The Innovative Choice**, Oxford, Basil Blackwell.

AMENDOLA, M.; GAFFARD, J.–L. (1992), "Towards and 'Out of Equilibrium' Theory of the Firm", **Metroeconomica**, 43, pp. 267–288.

BAUMOL, W.J. (1991), "On Formal Dynamics: From Lundberg to Chaos Analysis" in L. Johnung (ed.) **The Sockolm School of Economics Revisited**, New York, Cambridge University Press.

HICKS, J.R. (1978), **Capital and Time**, Oxford, Clarendon Press.

GEORGESCU–ROEGEN, N. (1976), **Energy and Economic Myths**, Oxford, Pergamon Press.

LUCAS, R.E. (1988), "On the Mechanics of Economic Development", **Journal of Monetary Economics**, 22, pp. 3–42.

SOLOW, R.M. (1992), "Siena Lectures on Endogenous Growth Theory", **Collana del Dipartamento di Economia Politica**, Universita degli Sudi di Siena.

3 ECONOMIC DYNAMICS AND REGIONAL DIVERSITY - SOME EVOLUTIONARY IDEAS

J.S. Metcalfe
University of Manchester
PREST and Department of Economics

I. INTRODUCTION

My purpose in this brief note is to sketch some evolutionary perspectives on the process of regional development. Our starting point is the central stylized fact of modern economic growth, namely its uneven incidence across space and time, and I shall suggest that this diversity of growth experience is both a consequence of and a cause of evolutionary change. Increasing returns and cumulative phenomena play an important part in this story and we shall see how increasing returns work out in an evolutionary model. In this way we can identify the relation between familiar Kaldor/Verdoorn mechanism, which has been widely discussed in the regional context, and what I shall call the Fisher mechanism, which has its roots in evolutionary dynamics. My second subsidiary theme will be that while private firms are at the leading edge of the development process, they do not operate in isolation but rather in terms of a wider institutional matrix of supporting activities. This is particularly so with respect to those activities where private behaviours are imperfect and among these the behaviours relating to the accumulation of human and intellectual capital are of prominent importance. Indeed this is a theme extensively explored in the "new growth" theory [Romer (1986, 1990), Lucas (1988)] but it was not unknown in the "old growth" theory either, as the briefest acquaintance with the Marshallian concept of external economies demonstrates.

We shall insist on an institutional perspective on this problem to identify how externalities arise and the particular mechanisms of accumulation with respect to knowledge and skills [Parrinello (1993)]. The types of institutions which are relevant, their objectives and decision making processes, and their connectivity with firms and other institutions become key elements in defining the effectiveness of a region as an engine of growth. Hence, there is a sense in which regions are competing to attract market creating activities, and their institutional bases for the generation of growth enhancing externalities are an important element in this picture. However progress is not irreversible and regions which are unable to adapt their institutions to cope with new opportunities may pass into relative if not absolute

19

X. Vence-Deza and J. S. Metcalfe (eds.), Wealth from Diversity, 19–38.
© 1996 Kluwer Academic Publishers. Printed in the Netherlands.

decline [Arthur (1989, 1990)]. Recovery, if it is to happen at all, may then require a major innovation to open up compensating opportunities for development. A comparison of the Lancashire cotton industry with the Californian micro electronics industry shows exactly what is entailed by this diversity of experience. Success fuels success but not forever and so the capacity of a region to react to changes in its external environment plays an important factor in the maintenance of growth [Steiner (1990)]. The enemy of progress in this context is inertia of institutions and behaviours, and this in itself creates an interesting conundrum since some degree of inertia is necessary for evolution to be possible [Hannan and Freeman (1989), Mokyr (1992), Robertson and Langlois (1992)]. The final point to emphasise here is the connection between competition and regional development: not competition in its static, equilibrium setting, rather competition as a dynamic process contingent on the differential behaviour of firms and supporting institutions. Behind this differential behaviour lie the fundamental phenomena of innovation and entrepreneurship and it is the ability to foster these engines of growth which is crucial to a regional development policy over the longer term.

II. THE EVOLUTIONARY PERSPECTIVE ON REGIONAL CHANGE

The traditional theory of regional economic differences is very much connected with the factor endowments approach to the characterisation of economic equilibrium. In essence this identifies differences in economic structure, and per capita output, with differences in endowments of the productive factors, on the assumption that regions are identical with respect to the preferences of the population and technology. It is the latter assumption of identity of technology which is the most problematic in modern conditions and it is the importance of differences in technology which the evolutionary perspective is particularly concerned to address. However, the significance of differences in technology is not to be found in terms of economic equilibrium but rather in the dynamic processes in which different technologies are absorbed into economic structures [Amendola and Gaffard (1990)]. Indeed the economic resource base of a region is very much contingent on the availability of technologies to exploit those resources.

In this context the proposition that economic change and growth are closely connected with increasing returns is one of the oldest in economic theory. That the development process involves structural change in the presence of positive feedback, is a conjunction of ideas which leads directly to the concept of cumulative causation and the path dependence of economic processes [Arthur (1989), David (1986)]. There can be little doubt, therefore, that increasing returns phenomena are central to the elaboration of an evolutionary theory of economic change at regional level. Evolutionary theories of the kind discussed here are not simply theories of change but theories in which variety of behaviour is the essential element in producing

20

change. Neither the notion of representative behaviour nor the notion of static equilibrium have any role to play and this is one reason why evolutionary competition is a process of change not a state of equilibrium; a process in which the central problem is to explain the origin and consequences of diversity in behaviour. In the presence of increasing returns the competitive process takes on a new dimension, not only does competition select across a given distribution of behaviour, it also changes that distribution *endogenously* such that the relation between selection and variety generation is symbiotic: variety drives selection and selection creates variety. In this sense the roots of the argument are a mix of Marshallian and Schumpeterian ideas: increasing returns far from being incompatible with competition are central to it.

Competition between rival firms or technologies in the presence of constant returns to selection has been widely discussed in the evolutionary economics literature from the perspectives of theory and simulation[1]. Implicit in this view are the four central themes of the evolutionary perspective: that it is differences in behaviour across regimes which drive the evolutionary process; that these differences are evaluated economically within a population of competing behaviours; that this evaluation generates selective pressure to change the relative importance of each distinct form of behaviour in the population; and that these behaviours are subject to inertia, changing slowly relative to the changes imposed by selection. In short variety in behaviour stimulates change at the population level but the behaviours themselves must not change so quickly that they swamp selective forces. Only with this assumption can selection dominate the process of economic change but clearly such an assumption can be problematic unless we assume constant returns to selection. The effect of constant returns is precisely to make the relevant behaviours independent of the selection process and to ensure that selection dominates. Change at the population level reflects change in the relative importance of the competing behaviours, not change in the competing behaviours themselves. With increasing returns to selection this is no longer automatically the case, selection also changes the pattern of behaviour in the population and this may work in the same or the opposite direction to the forces of selection.

Perhaps the key questions to be asked about such dynamic competitive processes relate to their "progressiveness". Does average behaviour within the population improve over time?. Is it the case that economically superior behaviours increase in relative importance over time under the pressure of selection?. In the case of constant returns to selection the answer to both questions is affirmative and

---

[1] Nelson and Winter (1982), Silverberg, Dosi and Orsenigo (1990), Metcalfe and Gibbons (1989), Andersen (1992). Downie (1955) can rightly be claimed to have anticipated much of this work. Smith (1991); Silverberg *et al.* are representative of the complex simulation work which can be undertaken in otherwise simple evolutionary models. Witt (1993a) and Arthur (1989) have been instrumental in raising the question of selection in the presence of increasing returns.

reflects the operation of Fisher's principle. The nature of the argument is worth brief comment.

III. THE FISHER'S PRINCIPLE

In modern evolutionary theory the concept of selection plays a fundamental role in explaining the changing relative importance over time of different patterns of behaviour within a population. Understanding of these processes was considerably sharpened by the work of R.A. Fisher, the eminent evolutionary biologist (1930).

At this point it is useful to distinguish what we shall call Fisher's principle from various forms of his fundamental theorem of natural selection. These have played an important and controversial role in evolutionary biology. The principle is general and states that, in the context of a population of diverse behaviours across which selection is taking place in a constant environment, the rate of change of mean behaviour is a function of the degree of variety in behaviour across the population. In this form it is no more than a general and powerful statement of the fundamental dynamics of selective change within a population of competing behaviours: patterns of change are premised on patterns of differential behaviour. By making more specific assumptions the principle can be transformed into a more precise statement which also defines the exact way in which variety in behaviour is to be measured. When these behaviours are one dimensional, the principle is transformed into the fundamental theorem of natural selection, that the rate of improvement of mean behaviour is proportional to the variance in behaviour in the population. It is this which gives progress its direction for the variance is a statistic of unambiguous sign. Moreover, it is also the case that a secondary but weaker theorem applies, that the rate of change of the mean of a behavioural characteristic is proportional to the covariance between that characteristic and the rates of expansion in the entities which embody the characteristic. Of course, we have nothing in our economic context to correspond to the genetic foundations of Fisher's original argument: nor are any needed, for the content of the theorem is quite independent of genetic considerations. This we must stress, what we have called Fisher's principle is quite free from any basis in genetics, and reflects a general method of analysis[2].

What the Fisher principle does for us is to set an agenda for an evolutionary framework in which the principle issues become: the meaning of variety in a given

---

[2] On the theorem see Sober (1985) chapter 6; Crow and Kimura (1970) chapter 5. Papers by Price (1972), Ewans (1989), Nagylaki and Crow (1974), Kimura (1958) and Turner (1969) cover most of the relevant biological discussion. Crow (1990) and Wynne Edwards (1990) provide interesting discussion of R.A. Fisher's work. Scott Findlay (1990, 1992) provides an extension into biocultural models of selection. Mani (1991) contains an excellent summary of modern evolutionary theory.

context; the appropriate measurement of variety; and the relation of measured variety to change at the population level. The resolution of each of these issues depends on the particular theoretical framework in which one works.

The fundamental theorem also has a further strong implication, that selection destroys the measure of variety on which it depends, so that for example the variance of behaviour is driven to zero by the selection process; a form of 'heat death' equilibrium in which change is absent because everything is uniform. In fact in specific cases one can go further and show that the rate of change in the variance of behaviour is proportional to the skewness of the population distribution, the third moment about the mean behaviour. Thus there is a logical chain of effects in which moments of the behaviour distribution at one level change in proportion to moments at the next higher level.

It is important to stress at this point that the dynamic processes underlying Fisher's principle are quite different from those typically employed in economic theory. There the principal method is to specify an equilibrium, moving or stationary and identify the dynamics in terms of perturbations around that equilibrium. It is a familiar and powerful method but one which from our viewpoint has significant drawbacks. First there is the problem of multiplicity of equilibria which are likely to arise in the presence of increasing returns. Second, is the problem that the response to perturbations close to equilibrium may be qualitatively different from perturbations which take the system far from equilibrium. Finally, there is the problem that equilibrium theory cannot provide a rationalization for out of equilibrium behaviour which must inevitably be modelled in a more or less *ad hoc* fashion. This is perhaps one reason for the appeal of rational expectations models which are *ex hypothesi* in equilibrium continuously apart from (small) random shocks.

The dynamic processes employed here, by contrast, make no reference to equilibrium positions or states of rest of any kind. Rather the dynamic behaviour is governed by replicator equations in which motions are related to comparisons between actual behaviour and average behaviour in the population. They may or may not approach an attractor but irrespective of this the mechanism generating the motions is clear. In this sense the replicator principle provides an important alternative basis for dynamics, a dynamics which exploits the variety of behaviour in evolutionary systems. The principle rationalises a dynamics of discovery, the unfolding of economic structures as a consequence of variety in behaviour.

IV. UNITS OF SELECTION

In order to provide the simplest treatment possible in what follows we focus on a region consisting of a number of business units, firms producing the same commodity. Each firm differs in three dimensions which are the bases of an evolutionary process. Firms differ in efficiency, that is, unit cost of production, they differ in the propensity to grow as measured by the relationship between their growth rate of output and their profitability; and, they differ in their creativity, the ability to enhance their efficiency over time by technological or organisational innovation. Each of these dimensions of firm behaviour is contingent upon the wider properties of the region: its natural and human resources, the operation of the capital market in providing finance for growth, and, of educational and research institutions which directly or indirectly support innovation. As our index of the performance of the region we take its average efficiency, measured across the firms in the region. More specifically efficiency is measured by a firm's unit costs of production which are aggregated into a regional average using the share of each firm in aggregate production. Let us see how the economic performance of the region will evolve under constant returns.

V. SELECTION UNDER CONSTANT RETURNS

To illustrate these points we can briefly outline a simple canonical model of selection[3]. Consider a region defined by a population of competing firms, producing an identical product for sale in a perfect world market and let each firm have a different unit cost level, h_i, which is independent of its scale of output. Unit cost is the only dimension in which the firms differ. Each firm's unit cost level is a reflection of its technology and behavioural rules and these are given for the moment and will not concern us further. The set of firms is fixed. Given the prevailing product price, P, a subset of the firms will be profitable, these are the dynamic firms with the capability to expand by investing profits in capacity expansion. Firms with unit costs in excess of the price are assumed to be declining in absolute terms[4]. Hence we have a selection process with respect to a single competitive characteristic, the level of unit cost. For a given population of firms we can define a cost or rent gradient ranging from worst to best practice[5]. If all firms invest the same proportion of their profits in capacity expansion, have a uniform propensity to accumulate, f, the dynamics of population change follow immediately since the rate of growth of each firm is given by the relation $g_i = f(p - h_i)$. In

---

[3] cf., Metcalfe (1992) for further elaboration.

[4] Alternatively one can assume that all firms with unit cost greater than the prevailing price are out of the market. These firms which just break even would play a passive role in the dynamics of population change. It is simpler for present purposes to treat all firms symmetrically. It is not at all difficult to introduce an imperfect market with an endogenous distribution of prices as in Metcalfe (1989).

[5] Kaldor (1985), p. 44; Eliasson (1991), Dosi (1984) and Metcalfe (1989).

principle one is interested in a variety of measures of the changing relative importance of different behaviours in the population. Typically the moments around the mean provide convenient summary statistics of the pace of evolutionary change, and we shall here focus on one such moment the behaviour of the population mean unit cost level.

Let s_i be the share of firm i in total output (capacity) then $\bar{h} = \Sigma\, s_i h_i$ is the average unit cost level defined across the entire population of firms. Let g_i be the growth rate of the i th firm, $g = \Sigma\, s_i g_i$ is the average growth rate and $ds_i/dt = s_i(g_i - g)$, from which it follows that:

$$\frac{ds_i}{dt} = f s_i\,(\bar{h} - h_i) \tag{1}$$

which is the distance from mean principle that underlies all the selection processes considered here[6]. At each point in time, three categories of firm co−exist in the region: contracting firms for which $h_i > p$; growing firms which are also loosing market share for which $p > h_i > \bar{h}$; and growing firms which are increasing their market share and for which $p > \bar{h} > h_i$

Taking the entire population we find that:

$$\frac{d}{dt}\,(\bar{h}) = \Sigma_i \frac{ds_i}{dt}\, h_i = C_s(h_i, g_i) \tag{2}$$

which, is the appropriate form of the secondary theorem of selection. The rate of change of the mean is proportional to the (share weighted) covariance between unit costs and rates of growth at the firm level[7]. However, given the uniform propensity to accumulate, this covariance is proportional to the variance in unit costs so that (2) becomes:

which is the analogue to Fisher's fundamental theorem of selection, the rate of reduction of the population mean is proportional to the (share weighted) variance

---

[6] Another way of writing **(1)** is to let S represent the market share, economic weight, of all firms other than the i th. Then $ds_i/dt = -f s(1-s)\partial \bar{h}/\partial s_i$. This captures neatly the progressive element in selection. The weight of the i th firm only increases if an increase in its weight contributes to an improvement (ie a reduction) in mean unit cost. We should emphasise that at each point in time, the market price is determined, auction market fashion, by demand in relation to relevant capacity. The dynamics of shifting demand and investment in capacity drive the pattern of change in the price over time.

[7] In this form the direction of the change in $\bar{h}$ is ambiguous since the covariance can be positive or negative.

$$\frac{d}{dt}(\bar{h}) = -fV_s(h_i) \tag{3}$$

within the population. Comparing (2) and (3) we see immediately that $C_s(h_i g_i)$ is negative, lower cost firms have higher growth rates. Similarly, we find that the rate of change of the variance is:

$$\frac{d}{dt} V_s(h) = f\sum_i s_i(h_i - \bar{h})^3 = fK(h_i) \tag{4}$$

when $K(h_i)$ the third moment about the mean, a measure of the skewness of the unit cost distribution.

Since the variance is positive, the selection process is progressive in the sense that it operates to continually reduce average practice unit cost: and it does this not by changing the individual h_i but by increasing the market share of the firms with below average practice unit costs. Ultimately output is concentrated in the best practice firm and the variance becomes zero[8]. Consider now the strict analogue to Fisher's fundamental theorem expressed in terms of the growth rates (fitness) of the individual firms. From the definition of the average growth rate we have:

$$\frac{d}{dt} g = \sum_i \frac{ds_i}{dt} g_i + \sum_i s_i \frac{dg_i}{dt}$$
$$= V_s(g) + f\frac{dp}{dt}$$

The first term is precisely the economic analogue to the Fisher Theorem, while the second term reflects the effect of a change in the environment, a change in the market price, on the fitness of each firm. The variance of growth rates, however, is a secondary or derived variance which in turn is related to the underlying primary variance in unit costs across the firms. Taking account of this it follows that:

$$\frac{d}{dt} g = f^2V_s(h) + f\frac{dp}{dt}$$

Selection operates to increase the average growth rate, just as Fisher argued. An adverse change in the environment (a reduction in price) operates to reduce it.

---

[8] For proof, consult Hofbauer and Sigmund, pp. 84–86.

Notice finally that if we further constrain the growth of total output to equal an exogenous growth rate of market demand, g_D, we would then determine the market price according to the relation $p' = g_D/f + \bar{h}$. For this case, Fisher's law obviously cannot apply to the growth rate, instead it applies to the rate of reduction in the market price, so that $dp/dt = -fV_s(h)$. In short, we have two distant kinds of situations in the market. In the first situation, one of balanced expansion, the growth rate of output equals the exogenous growth rate of the market, and the price satisfies the relation $p = p'$. In the second situation the two growth rates differ and are related to the actual market price p and the balanced market price p' by $g - g_D = f[p - p']$. The second situation would be relevant, for example, in the face of a unexpected upward shift in demand for the regions output. In what follows we restrict our argument to the balanced expansion situation.

All this is by now well known but leads to two important observations. First the fundamental and secondary theorems are special cases of the more general Fisher principle and the important issue to address is the robustness of the principle outside of the canonical case. For example, introducing two behavioural characteristics for each firm, product quality and unit cost, keeps the principle intact, in that the rate of change in the mean of either behavioural character is a function of measured variety, but does not reproduce the relation (3) between changes in population mean and population variance[9]. Secondly, the dynamics of change are such that the rate of change of each firm's weight in the population, its market share, is proportional to the distance between own unit cost and population average unit cost. This distance from mean principle is reflected in a wide range of replicator processes common to evolutionary selection and evolutionary game theory[10], and, in this regard, Fisher's principle is simply a shorthand summary of certain kinds of replicator process. However, it does have also a remarkable implication. Namely that of all the possible patterns of change in the market shares, s_i, it is those governed by the distance from mean principle which maximize the rate of improvement in the population average[11]. In this sense, competition is an optimizing process not with respect to individual behaviours but with respect to the collective outcome of differences in behaviour[12]. It will also be seen that the replicator dynamics of

---

[9] The analysis of this case is contained in Metcalfe (1989). The secondary theorem (1) continues to hold but is not particularly informative.

[10] In terms of the analysis in Hofbauer and Sigmund (1988), this constant returns model is a zero order replicator process. See also Silverberg (1988), Cressman (1990), Hansen and Samuelson (1988), Robson (1990), Friedman (1991) and Banerjee and Weihall (1992) contain up to date discussion of replicator processes in the context of evolutionary games.

[11] An elementary proof is contained in the Appendix.

[12] The relation with Hayek's principle of spontaneous order deserves the attention which it cannot be given here. On this see Witt (1993b).

27

selection are quite different from the dynamic adjustment processes typically assumed in discussions of economic dynamics where change is governed by deviations from a given equilibrium position. In the replicator approach the dynamics are defined in relation to actual not potential behaviours and a central issue then becomes the ability of the process to discover an attractor, an outcome where all relevant variety in behaviour has been eliminated.

VI. INCREASING RETURNS AND SELECTION

Illuminating though this process of competition is with constant returns, it leaves open the issue of what happens to the dynamics of evolutionary change when firm behaviours are changing over time and, more to the point, changing endogenously in response to the selection process. The balance between inertia and selection then becomes problematic for behaviours also change as a consequence of the changing relative importance of the different firms. Within the economic literature, endogenous changes in behaviours have traditionally been associated with the theories of innovation and the theories of increasing returns[13]. It is to the latter that we turn our attention here. In the regional context this is a particularly important issue which has been critical to the work of Kaldor and others on regional dynamics [Kaldor (1966, 1970), Dixon and Thirlwall (1975)].

Three kinds of increasing returns are to be distinguished. Static internal economies which depend upon the scale of output of the individual firm. Static external economies which depend on the output of the industry, population of firms as a whole, although the degree of access to such economies may depend on the relative size of the competing firms. We should not take it for granted that all firms have equal access to these external economies. External economies do have a public good dimension but this does not imply that they are freely available. In this regard the simplest hypothesis is that relative access depends on the relative size of firms. Dynamic economies related to the accumulated experience of the firm which may be internal, or, to the extent that there are knowledge spill–overs, may also be external[14]. In the following we deal only with internal dynamic economies. The significance of the various forms of increasing returns is that they make the degree of variety of behaviour depend endogenously on the structure of the population: as structure changes so does the degree of variety and the distribution of selective pressure. To this extent, the Fisher principle now operates at two interacting levels: selecting for given variety, and changing the pattern of variety on which selection

---

[13] We also draw attention as Penrose did (1961), to the possibility of a relation between the unit cost levels and the growth rate of the firm. This adds a further factor slowing down the selection process but it does not change the fundamentals of selection. The point is not pursued further here.

[14] The simulation of such spill–overs is a central concern of Silverberg *et al.* (1990).

operates. Moreover, in the presence of increasing returns the Fisher principle interacts with the Kaldor/Verdoorn principle relating the growth rate of the region to the generation of scale economies.

We explore this further by assuming the following behaviour function defining the unit cost level in each firm.

$$h_i = h_i(x_i, E_i, L_i) \qquad\qquad (5)$$

when x_i is the firm's current output level, E_i is the extent to which it enjoys external economies, and L_i is the extent to which it has accumulated dynamic internal economies. Unit cost decreases in each of these arguments. Let $E_i = \phi_i(s_i)X$, when X is the aggregate output of the competing firms, and $\phi(s_i)$ is the transfer function indicating the extent to which a firm benefits from those economies. For pure external economies, we would have $\phi(s_i)=1$ for all s_i. It is natural to assume that $\phi(s_i)$ is non–decreasing in s_i. Similarly define L_i as $\psi(\int_0^t x_i(t)dt)$ an increasing function of the entire past history of the firm's production experience. We shall treat each of these economies as reversible, subject to attrition should the firm begin to decline.

It follows that:

$$\frac{dh_i}{dt} = \frac{\partial h_i}{\partial x_i}\frac{dx_i}{dt} + \frac{\partial h_i}{\partial E_i}\frac{dE_i}{dt} + \frac{\partial h_i}{\partial L_i}\frac{dL_i}{dt} \qquad\qquad (6)$$

Now define the following elasticities:

$$\beta_i = \frac{-\partial h_i}{\partial x_i}\frac{x_i}{h_i}, \epsilon_i = \frac{-\partial h_i}{\partial E_i}\frac{E_i}{h_i} \text{ and, } \eta_i = \frac{\partial \phi_i}{\partial s_i}\frac{s_i}{\phi_i}. \text{ Let } \frac{-\partial h_i}{\partial L_i} = \gamma_i.$$

The only restriction we place on these elasticities is that $\eta_i \le 1$, although normally $\beta_i < 1$ if marginal costs of expansion are positive.

Using these definitions we can write (6) as:

$$-\frac{dh_i}{dt} = \beta_i h_i g_i + \epsilon_i h_i (\eta_i(g_i - g) + g) + \gamma_i x_i \tag{7}$$

We can see immediately that the different forms of increasing returns are generated by quite different mechanisms. Learning economies accumulate as a function of the scale of current regional output, while static internal economies accumulate in proportion to the growth rate of a firm's output. External economies are more complex. In the pure case, $\eta_i=0$, they accumulate in proportion to the growth rate of the region. In the impure cases, $\eta_i>0$, their accumulation depends additionally on the distance of the firm's growth rate from the mean industry growth rate. Taking (7) as our summary of the endogenous changes in behaviour, we can now combine it with the traditional selective forces. To do this initially we set the elasticities $\beta_i, \epsilon_i, \eta_i$ and the coefficient γ_i the same across all firms. As before we focus on the evolution of the average level of unit cost in the population of firms

$$\frac{d\overline{h}}{dt} = \sum_i \frac{ds_i}{dt} h_i + \sum_i s_i \frac{dh_i}{dt} \tag{8}$$

where the summation is again over all the firms. The first term is the pure selection effort while the second term averages across all the sources of endogenous change in the single competitive characteristic. This term, of course, is absent in the constant returns case[15]. Carrying out the necessary substitutions we have:

$$\frac{-d\overline{h}}{dt} = (\beta + \epsilon\eta - 1)C_s(h_i g_i) + (\beta + \epsilon)\overline{h}g_D + \gamma HX \tag{9}$$

Consider each of these terms in reverse order. The last one indicates that the average rate of accumulation of dynamic learning economies depends on the product of the region's output rate and the Herfindall index of concentration within the region $H=\sum s_i^2$, which we interpret as the weighted average market share. Clearly, the more concentrated the industry the greater the significance of these dynamic economies for the movement of average unit cost. The second term is the Kaldor–Verdoorn term, it captures the relation between the growth rate of the region, and the rate of reduction of average practice unit costs. The link, of course, is provided by the internal and external economy of scale elasticities. Clearly, it is only with constant returns to scale that the Fisher principle accounts for all the change in the population mean.

---

[15] More generally, the second term would also reflect factors such as fluctuations in the economic environment, say a change in relative factor prices, which change the distribution of unit costs across firms. Similarly it would also incorporate the effects of innovations.

30

The final term is the Fisher principle term, the link between diversity of behaviour and the rate of improvement in average behaviour, and it is in this term that we are most interested. But in this case the principle applies at two levels: to the conventional process of selection across given unit cost levels; and to the endogenous rate of change in those unit cost levels in response to the accumulation of static internal and external economies of scale. That the evolutionary mechanism applies to the rate of generation of economic variety as well as to competitive selection across variety is, we believe, a fundamental characteristic of the dynamics of modern capitalist economies –exactly as Schumpeter would have insisted. Selection not only consumes variety it regenerates it endogenously whenever returns to scale are not constant. With constant returns only selection is involved and it is sufficient that the covariance $C_s(h_i,g_i)$ be negative to produce a progressive competitive process. With increasing returns variety in the population also influences the pattern of changes in behaviour and the outcome depends on the elasticity sum $\beta + \epsilon\eta$. Eliminating the covariance from (9) we can write it as:

$$\frac{-d\bar{h}}{dt} = f(1 - \beta - \epsilon\eta)V_s(h_i) + (\beta+\epsilon)\bar{h}g_D + \gamma HX \qquad (10)$$

which may be compared directly with the constant returns case (3). The overall effect of increasing returns on the evolution of the population mean is of course progressive, it could scarcely be otherwise. Both selection and scale economics lend to a reduction in average unit cost. However, with respect to the static internal and external economies a distinction has to be made between the Kaldor/Verdoorn effect and the Fisher effect. Now while the Kaldor/Verdoorn effect, $(\beta + \epsilon)\bar{h}g_D$, is always progressive, as is the dynamic learning effect, in the presence of strong increasing returns, the Fisher effect need not be. The presence of increasing returns has the effect of weakening the connection between the variety of behaviour and the rate of improvement of average practice costs.

Thus for variety to be progressive in the sense of Fisher's fundamental theorem, it is necessary that the combined efforts of internal and external scale economies are not too strong, that is, $\beta + \epsilon\eta \prec 1$. In short there must be sufficient inertia for the forces of endogenous change not to swamp the effects of selection. Conversely if scale effects are overly strong, $\beta + \epsilon\eta \succ 1$, the direction of the link between selection and progress is reversed, selection is regressive and may outweigh the progressive Verdoorn and learning effects. This is obvious when the market growth rate is zero. A positive growth rate provides an offset to a regressive Fisher effect, such that the greater the growth rate the more likely is the overall outcome to be progressive. The boundary case when $\beta + \epsilon\eta = 1$ is also instructive for then $\bar{h}$ is independent of Fisher effects, even though there is variety in behaviour.

By contrast diminishing internal and external economies $(\beta < 0, \epsilon < 0)$ reinforce the operation of Fisher's fundamental theorem but reverse the operation of the Kaldor/Verdoorn principle. As is so often the case, it is increasing returns which provides the interesting situations of conflict between economic forces.

VII. INSTITUTIONS AND SUPPORT SYSTEMS

Finally some brief remarks on the role of institutional support systems in the generation of external economies and the effect more generally on the ability of firms to innovate. Indeed the idea of industrial districts as a generation of external economies has a distinguished history in the literature. Such a district is an institution in this case a formal and informal network of interactions between firms and their customers to share knowledge to the collective benefit of the district. The ability of a region to generate and sustain such a complex is surely a key factor in explaining the different dynamic histories of different regions. From which follows an important dimension of regional policy. Much regional policy, certainly in the recent history of the UK, has been about channelling investment resources to regions, thus influencing the link between efficiency and growth. What we are identifying here is the importance of policy in terms of the creativity of a region. In particular, the available set of research and training institutions and how they are bridged to firms is surely a central element in the innovation performance of a region. As Steiner (1991) has indicated regions need to be distinguished according to their ability to achieve efficiency through adaptation and their ability to be creative through innovation. The institutions and behaviours to achieve these outcomes are often in conflict. But this is properly the topic of another paper.

VIII. CONCLUDING REMARKS

The central theme of this note has been the process of competitive, evolutionary selection in the presence of increasing returns: selection, that is, with positive feedback. Irrespective of the degree of returns to scale, evolution reflects Fisher's principle that the rate of change of the average competitive behaviour is a function of the degree of variety of behaviour. In this it reflects the central evolutionary concern with change at the population level.

It will be obvious to any reader that we have provided only the barest sketch of an evolutionary process. Such processes are inevitably open–ended and unpredictable for the very obvious reason that the mainspring of economic evolution is human novelty and creativity. It is because these complex behaviours have an institutional context that the policy maker may see a way to stimulate the process of regional development within the context of market selection environments.

IX. APPENDIX

The following proof of a maximum principle for a theory of competitive selection is based on Kimura (1958) [Crow and Kimura (chapter 5), Hofbauer and Sigmund (chapter 6 also)].

The theorem is as follows: for a short time interval δt, competitive selection leads to changes in market shares $\delta s_1, \delta s_2, \delta s_3 .... \delta s_n$ such that these changes produce the greatest possible reduction in average practice unit cost, $-\delta \bar{h}$, subject to the prevailing constraints.

Define $\quad \bar{h} = \sum_i s_i h_i \quad$ then $\quad -\delta \bar{h} = -\sum_i h_i (\delta s_i)$.

This is to be maximized subject to two constraints:–

a). $\quad \sum_i \delta s_i = 0$

b). $\quad$ From the definition of $V_s(g)$ we have

$$V_s(g) = \sum_i s_i (g_i - g_s)^2 = \sum_i \frac{1}{s_i} \left(\frac{ds_i}{dt}\right)^2 \text{ and } V_s(g) = f^2 V_s(h).$$

Thus $\quad \sum_i \frac{1}{s_i} (\delta s_i)^2 = f^2 V_s(h)(\delta t)^2$ defines the second relation between δs_i and δt which also constrains the movement of the market shares.

Define the Lagrangian L with undetermined multipliers λ_1, λ_2

$$L = -\delta \bar{h} + \lambda_1 \left[\sum_i \frac{1}{s_i}(\delta s_i)^2 - f^2 V_s(h)(\delta t)^2\right] + \lambda_2 \sum_i \delta s_i$$

Then $\quad \dfrac{\partial L}{\partial(\delta s_i)} = -h_i + \dfrac{2\lambda_1(\delta s_i)}{s_i} + \lambda_2 = 0, (i=1....n)$ for a stationary point.

Summing by s_i and taking account of constraint a) gives $\lambda_2 = \bar{h}$.

Therefore $\delta s_i = \dfrac{s_i(\bar{h} - h_i)}{2\lambda_1}$.

Squaring and substituting into constraint b) yields $\displaystyle\sum_i \frac{s_i(h_i - \bar{h})^2}{4\lambda_1^2} = f^2 V_s(h)(\delta t)^2$, that is, $\dfrac{1}{2\lambda_1} = \pm f(\delta t)$

To maximize the decline in $\bar{h}$ we require $\lambda_1 > 0$ hence $\dfrac{\delta s_1}{\delta t} = f s_1(\bar{h} - h_i)$

which is exactly the required result as the interval δt contracts to zero. The distance from mean, or replicator, principle maximizes the rate of decline in average practice unit cost. In this sense our competitive mechanism is optimal, and is so independently of whether the individual h_i reflect optimizing processes at the level of the individual firm.

REFERENCES

ANDERSEN, E.S. (1992), **Schumpeter and the Analysis of Economic Evolution**, mimeo, University of Aalborg.

AMENDOLA, M.; GAFFARD, J. (1990), **The Innovative Choice**, Basil Blackwell.

ARTHUR, W.B. (1989), "Competing Technologies and Lock–in by Historical Events", **Economic Journal**, Vol. 99, pp. 116–131.

ARTHUR, W.B. (1990), ""Silicon Valley" Locational Clusters: When do Increasing Returns Imply Monopoly", **Mathematical Social Sciences**, Vol. 19, pp. 235–251.

BANERJEE, A.; WEIHALL, J. (1992) "Evolution and Rationality: Some Recent Game–Theoretic Results", (Working Paper no. 345), mimeo, Industrial Institute of Economic and Social Research, Stockholm.

CRESSMAN, R. (1990), "Strong Stability and Density Dependent Evolutionary Stable Strategies", **Journal of Theoretical Biology**, Vol. 44, pp. 319–330.

CROW, J.F. (1990), "Fisher's Contributions to Genetics and Evolution", **Theoretical Population Biology**, Vol. 38, pp. 263–275.

CROW, J.F.; KIMURA, M. (1970), **An Introduction to Population Genetics Theory**, Harper and Row.

DAVID, P.A. (1986), "Narrow Windows, Blind Giants and Angry Orphans", mimeo CEPR, Stanford University.

DIXON, R.; THIRLWALL, A. (1975), "A Model of Regional Growth Rate Differences in Kaldorian Lines", **Oxford Economic Papers**, Vol. 27, pp. 201–214.

DOSI, G. (1984), **Technical Change and Industrial Transformation**, Macmillan, London.

DOWNIE, J. (1955), **The Competitive Process**, Duckworth, London.

EDWARDS, A.W.F. (1990), "Fisher, $\bar{W}$ and the Fundamental Theorem", **Theoretical Population Biology**, Vol. 38, pp. 276–284.

ELIASSON, G. (1991), "Deregulation, Innovation, Entry and Structural Diversity as a Source of Stable and Rapid Economic Growth", **Journal of Evolutionary Economics**, Vol. 1, pp. 49–63.

EWENS, W.J. (1989), "An Interpretation and Proof of the Fundamental Theorem of Natural Selection", **Theoretical Population Biology**, Vol. 36, pp. 167–186.

FINDLAY, S. (1990), "Fundamental Theorem of Natural Selection in Biocultural Populations", **Theoretical Population Biology**, Vol. 38, pp. 367–384.

FINDLAY, S. (1992), "Secondary Theorem of Natural Selection in Biocultural Populations", **Theoretical Population Biology**, Vol. 41, pp. 72–89.

FISHER, R.A. (1930), **The Genetical Theory of Natural Selection**, Oxford University Press, Oxford.

FRIEDMAN, D. (1991), "Evolutionary Games in Economics", **Econometrica**, Vol. 59, no. 3, pp. 637–666.

HANNAN, M.T.; FREEMAN, J. (1989), **Organizational Ecology**, Harvard University Press.

HANSEN, R.; SAMUELSON, L. (1988), "Evolution in Economic Games", **Journal of Economic Behaviour and Organization**, Vol. 10, pp. 315–338.

HOFBAUER, J.; SIGMUND, K. (1988), **The Theory of Evolution and Dynamical Systems**, Cambridge University Press, Cambridge.

KALDOR, N. (1966), **Causes of the Slow Rate of Economic Growth of the UK**, Cambridge University Press.

KALDOR, N. (1970), "The Case for Regional Policies", **Scottish Journal of Political Economy**, Vol. XVIII, no. 3, pp. 337–348.

KALDOR, N. (1985), **Economics without Equilibrium**, University College Cardiff Press, Cardiff.

KIMURA, M. (1958), "On The Change of Population Fitness by Natural Selection", **Heredity**, Vol. 12, pp. 145–167.

LUCAS, R.E. (1988), "On the Mechanics of Economic Development", **Journal of Monetary Economics**, pp. 3–42.

MANI, G.S. (1991), "Is there a General Theory of Biological Evolution" in P.

Saviotti and J.S. Metcalfe (eds), **Evolutionary Theories of Economic and Technological Change**, Harwood, London.

METCALFE, J.S.; GIBBONS, M. (1989), "Technology, Variety and Organization", **Research on Technological Innovation, Management and Policy**, Vol. 4, pp. 153–193.

METCALFE, J.S. (1989), "Evolution and Economic Change" in A. Silberston (ed.), **Technology and Economic Progress**, Macmillan, London.

METCALFE, J.S. (1992), "Variety, Structure and Change: An Evolutionary Perspective on the Competitive Process", **Revue D'Economie Industrielle**, no. 59.

MOKYR, J. (1992), "Technological Inertia in Economic History", **Journal of Economic History**, Vol. 52, pp. 325–338.

NAGYLAKI, T.; CROW, J.F. (1974), "Continuous Selective Models", **Theoretical Population Biology**, Vol. 5, pp. 257–283.

NELSON, R. AND WINTER, S. (1982), **An Evolutionary Theory of Economic Change**, Belknap, Harvard University Press.

PARRINELLO, S. (1993), "Non Pure Private Goods in the Economics of Production Processes", **Metroeconomica**, Vol. 44, no. 3, pp. 195–214.

PENROSE, E. (1961), **The Theory of the Growth of the Firm**, Blackwell.

PRICE, G.R. (1972), "Fisher's 'Fundamental Theorem' Made Clear", **Annals Human Genetics**, Vol. 36, pp. 129–140.

ROBERTSON, P.L.; LANGLOIS, R.N. (1992), "Institutions, Inertia and Changing Industrial Leadership", mimeo Department of Economics and Management, University College, University of New South Wales Australian Defence Force Academy.

ROBSON, A.J. (1990), "Efficiency in Evolutionary Games: Darwin, Nash and the Secret Handshake", **Journal of Theoretical Biology**, Vol. 144, pp. 379–396.

ROMER, P.M. (1986), "Increasing Returns and Long–Run Growth", **Journal of Political Economy**, pp. 1102–1037.

ROMER, P.M. (1990), "Endogenous Technical Change", **Journal of Political**

Economy, pp. S71–S102.

SILVERBERG, G. (1988), "Modelling Economic Dynamics and Technological Change, Mathematical Approaches to Self Organization and Evolution" in G. Dosi et al (eds.), **Technical Change and Economic Theory**, Pinter, London.

SILVERBERG, G.; DOSI, G.; ORSENIGO, G. (1988), "Innovation Diversity and Diffusion, A Self Organisation Model", **Economic Journal**, Vol. 98, pp. 1032-1054.

SOBER, E. (1985), **The Nature of Selection**, Harvard University Press.

SMITH, S. (1991), "A Computer Simulation of Economic Growth and Technical Progress in a Multi Sectoral Economy" in P. Saviotti and J.S. Metcalfe (eds.), **Evolutionary Theories of Economic and Technological Change**, Harwood, London.

STEINER, M (1990), "'Good' and 'Bad' Regions?", **Built Environment**, Vol. 16, pp. 52–68.

STEINER, M. (1991), "Technology Life Cycles and Regional Types: An Evolutionary Interpretation and Some Stylized Facts", **Technovation**, Vol. 11, pp. 483–498.

TURNER, J.R.G. (1969), "The Basic Theories of Natural Selection: a Naive Approach", **Heredity**, Vol. 24, pp. 75–84.

WITT, U. (1993a), "Path Dependence in Institutional Change" in P. David and C. Antonelli (eds.), **Path Dependence and its Implications for Economic Policy Analysis**, Basil Blackwell, New York.

WITT, U. (1993b), "The Theory of Societal Evolution – Hayek's Unfinished Legacy" in J. Birner (ed.), **In Memoriam of F A Hayek**, Rutledge, London.

4 UNEVEN TECHNOLOGICAL DEVELOPMENT[16]

Parimal Patel and Keith Pavitt
Science Policy Research Unit
University of Sussex

I. TECHNICAL CHANGE IS NEITHER AUTOMATIC NOR EASY

Until the early 1970s, it was commonly assumed that the open trading system would allow the rapid international diffusion of technology, both as easily transmissible information (e.g. blueprints and operating instructions), and embodied in machinery. As a consequence, the catching up of Western Europe and Japan to the levels of technology and efficiency of the world's leading country (the USA) would be relatively smooth. The same opportunities for technical change and growth were in principle open to the so–called developing countries, but many of these were constrained by traditions and cultures that did not value highly material acquisitiveness: Catholics and Confucians, for example, were disadvantaged in this respect, compared to Protestants.

These expectations have not been fulfilled. Technology–related events since then have turned out to be anything but smooth. There has been uneven development amongst sectors, in part as a consequence of the micro–electronics revolution. There has also been uneven development amongst countries. Some (e.g., the UK) have caught up only very partially, whilst others (e.g., FR Germany and Japan) have actually overtaken the world's technological leading country –the USA– in certain important sectors [Nelson (1990)]. These differences are reflected in the technological and competitive performance of companies: we shall see that, in some fields, the world pecking order in technology has hardly shifted at all over the past 20 years, whilst others have seen major changes amongst the leading companies. And amongst the developing countries, the (Southern European) Catholics and the (East Asian) Confucians have been relatively successful in closing the technological and economic gap with the world leaders.

This is because the international diffusion of technology is neither automatic

---

[16] This paper draws heavily on the results of research undertaken in the ESRC (Economic and Social Research Council)–funded Centre for Science, Technology, Energy and the Environment Policy (STEEP) at the Science Policy Research Unit (SPRU).

X. Vence-Deza and J. S. Metcalfe (eds.), Wealth from Diversity, 39–73.

nor easy [see Bell and Pavitt (1993)]. Both material artefacts and the knowledge to develop and operate them are complex, involving multiple dimensions and constraints that cannot be reduced entirely to codified knowledge, whether in the form of operating instructions, or a predictive model and theory. Tacit knowledge –underlying the ability to cope with complexity– is acquired essentially through experience, and trial and error. It is misleading to assume that such trial and error is either random, or a purely costless by–product of other activities like "learning by doing" or "learning by using". Tacit (and other forms of) knowledge are increasingly acquired within firms through deliberately planned and funded activities in the form of product design, production engineering, quality control, staff training, research, or the development and testing of prototypes and pilot plant. Differences amongst countries in the resources devoted to such deliberate learning –or "technological accumulation"– have led to international technological gaps which, in turn, have led to international differences in economic performance. The "neo-technology" theories of trade and growth pioneered by Posner (1961) and Vernon (1966) in the 1960s have been amply confirmed by the events of the 1970s and 1980s, by related econometric analyses of Soete (1981) and Fagerberg (1987; 1988), and by the company–based analysis of Franko (1989): hence the growing interest in implications of international technology gaps for policy [Ergas (1984)] and for theory [Dosi et al. (1990)].

But whilst international differences in technology are clearly important, commentary and analysis about them has left much to be desired. During the Cold War, descriptions and predictions about competitors' achievements in military technology often turned out to be inaccurate, because they were either ill–informed or self–serving, or both. The same often holds in debates about capitalist countries' civilian technology. J.–J. Servan–Screiber's "American Challenge" of the 1960s quickly evaporated. Japan's growing technology–based competitiveness in the 1970s was often ignored, or described as nothing more than a cunning mixture of copying, meek trade unions, and cheating. And the West German–based "Eurosclerosis" of the early 1980s has turned out to be only very partial, as subsequent events in Central and Eastern Europe testify.

One reason for these shortcomings is the very real difficulty of measuring and comparing technological accumulation across firms, sectors and countries. Given that the activities contributing to technological accumulation are complex and varied, all measures are bound to be imperfect. However, as a result of the growing demands from public and private policy–makers for better data, progress has been made in both measurement and conceptualisation. The advantages and drawbacks of the various measures have been extensively reviewed elsewhere [Freeman (1987), Grilliches (1990), Narin and Olivastro (1988), Pavitt and Patel (1988), Pavitt (1988a)]. In particular, we have shown in our earlier work that the combined use of data on R&D activities, and on patenting in the USA by country of origin, gives a plausible and consistent picture of technological activities at the world's technological frontier.

R&D is a better measure of rates of change in real resources over time, but it measures technological activities in small firms only very imperfectly. US patenting is a better measure of technological activities in small firms and can be broken down quite finely to specific firms and specific technical fields. Neither measure is satisfactory for software technology, but no alternative yet exists. And neither measure captures all the activities that lead to product and process innovations, such as design, management, production engineering, marketing and learning by doing.

II. THE EVIDENCE OF UNEVEN (AND DIVERGENT) TECHNOLOGICAL DEVELOPMENT

II.1 Amongst Countries in the Volume of Technological Activities

The data on R&D and US patenting activities show no evidence of convergence in national capacities for technological accumulation since the early 1970s, and some evidence of divergence in the 1980s[17].

Table 1 presents trends in the percentage of Gross Domestic Product spent by business on R&D activities in 16 OECD countries since 1967[18]. These show a certain stability in the rankings throughout the period at the two ends of the spectrum: Switzerland has remained with the highest share, and Ireland, Spain and Portugal with the three lowest shares. Otherwise there are countries who started near the top but have moved down the rankings: Canada, Netherlands and –above all– the UK; there are also countries that have improved their positions: FR Germany, Sweden, Japan and –above all– Finland. Overall, there are no statistical signs of convergence in the industry–funded shares over time, since the standard deviation of the distribution has not decreased over the period: on the contrary, it has increased markedly in the 1980s, suggesting technological divergence amongst countries. In this context, it is worth noting that the US share began slipping progressively below that of FR Germany, Japan, Sweden and Switzerland in the 1970s, and that the gap grew much larger in the 1980s.

---

[17] For a somewhat different interpretation, see Archibugi and Pianta (1992).

[18] Government funded R&D performed in industry is excluded. This is concentrated in defence –related activities, and in few countries: principally the USA, UK, France and FR Germany, where it has clearly stimulated accumulation in defence– related technologies (see Tables 5, 7 and 8 below). Its wider effects on technological accumulation are a matter of debate. Our own conclusion is that defence R&D has considerable opportunity costs, particularly in electronics, where leading–edge technologies and markets have shifted to civilian applications.

Table 1.– Trends in Industry Financed R&D as a percentage of GDP in 17 OECD Countries: 1967 to 1989

	1967	1969	1971	1975	1977	1979	1981	1983	1985	1987	1989
Belgium	0.66	0.64	0.71	0.84	0.91	0.95	1.05	1.09	1.20	1.23	1.23
Canada	0.40	0.39	0.38	0.33	0.32	0.39	0.49	0.46	0.56	0.55	0.53
Denmark	0.34	0.39	0.41	0.41	0.41	0.42	0.46	0.54	0.61	0.68	0.71
Finland	0.30	0.32	0.44	0.44	0.49	0.53	0.59	0.69	0.85	0.97	1.07
France	0.60	0.64	0.67	0.68	0.69	0.75	0.79	0.88	0.92	0.93	0.98
FR Germany	0.94	1.03	1.13	1.11	1.12	1.32	1.39	1.48	1.65	1.79	1.79
Ireland	0.19	0.23	0.30	0.23	0.22	0.23	0.26	0.27	0.35	0.41	0.45
Italy	0.33	0.38	0.44	0.43	0.37	0.40	0.43	0.42	0.49	0.49	0.56
Japan	0.83	1.00	1.09	1.12	1.11	1.19	1.38	1.60	1.84	1.85	2.05
Netherlands	1.12	1.04	1.02	0.97	0.87	0.86	0.85	0.91	0.99	1.13	1.09
Norway	0.35	0.39	0.41	0.49	0.49	0.50	0.50	0.61	0.80	0.88	0.81
Portugal	0.04	0.06	0.09	0.05	0.04	0.09	0.10	0.11	0.11	0.11	0.11
Spain	0.08	0.08	0.11	0.18	0.18	0.18	0.18	0.22	0.26	0.30	0.34
Sweden	0.71	0.69	0.80	0.96	1.07	1.11	1.24	1.45	1.71	1.74	1.57
Switzerland	1.78	1.78	1.67	1.67	1.71	1.74	1.67	1.67	2.16	2.13	2.07
United Kingdom	1.00	0.92	0.81	0.80	0.80	0.82	0.91	0.87	0.96	1.03	1.04
United States	0.99	1.03	0.97	0.98	0.98	1.05	1.18	1.33	1.44	1.38	1.37
Standard Deviation											
All Countries	0.46	0.46	0.43	0.43	0.45	0.47	0.48	0.52	0.61	0.61	0.60
Excluding US	0.47	0.46	0.43	0.44	0.45	0.47	0.48	0.52	0.62	0.62	0.61

Source: OECD.

Table 2 shows trends in per capita national patenting in the USA for the same 17 OECD countries. At first sight the evidence is more ambiguous. When the USA is included, the standard deviation of the population increases between the late 1960s and the early 1970s –thereby suggesting divergence– but then decreases until the mid–1980s, after which it increases again to its original level. However, there are well–known reasons for excluding the USA from such a comparison, since for firms in the USA, we are measuring domestic patenting, whereas for firms in other countries we are measuring foreign patenting. Given the propensity of firms to seek patent protection more intensely in their home country [Bertin and Wyatt (1988)], the rate of technological accumulation in the USA in overestimated. At the same time, given the tendency of firms to give increasing attention to patenting in foreign markets, trends over time will tend to overestimate any decline in US performance [Kitti and Schiffel (1978)], and thereby show a spurious degree of convergence

Table 2.– Trends in Per Capita Patenting in the US from 17 OECD Countries. US patents per million popoulation

	1963-68	1969-74	1975-80	1981-85	1986-90
Switzerland	138.03	197.09	207.12	179.23	193.63
United States	236.62	244.74	181.31	156.69	177.87
Japan	9.94	38.88	55.98	82.50	139.37
Germany	54.71	86.61	91.77	97.74	122.17
Sweden	64.16	94.69	100.92	87.14	99.56
Canada	41.33	55.81	48.95	45.75	63.07
Netherlands	35.93	48.81	46.82	47.01	60.40
Finland	5.14	14.84	22.30	30.82	50.43
France	26.11	41.02	39.71	38.80	49.64
United Kingdom	43.32	55.76	46.63	40.06	47.77
Denmark	18.39	31.87	29.74	27.87	35.67
Belgium	16.32	29.27	26.45	23.78	30.41
Norway	12.70	20.29	23.05	19.43	27.20
Italy	7.80	13.06	13.03	13.99	20.20
Ireland	1.97	6.47	5.06	6.74	12.93
Spain	1.23	2.06	2.27	1.59	3.07
Portugal	0.35	0.57	0.31	0.40	0.57
Standard Deviation					
All Countries	**60.56**	**67.56**	**59.44**	**52.05**	**59.26**
Excluding US	**35.05**	**48.91**	**51.44**	**46.13**	**53.58**

When the USA is excluded, the evidence in Table 2 on the whole confirms that in Table 1. Throughout the period, Switzerland stays at the top, and Ireland, Spain and Portugal at the bottom. Britain's relative position declines, whilst Finland, FR Germany and Japan improve. The standard deviation increases over the period, indicating international divergence. The one anomaly is the reduction in the standard deviation in 1981–85, but this may reflect the reduction in the overall number of patents granted, following a reduction in the number of patent examiners [See Grilliches (1990)].

Table 3 shows that, even within the European framework, there is no sign of technological convergence amongst countries. Here we show trends in the share of each country in total European patenting in the USA. Again we see relative decline of the UK, and the spectacular relative growth of Finland. We also see more clearly the absolute technological strength of Germany, which has increased its share of the total to more than 40%, and has diverged steadily from the other three countries of equivalent size: France, Italy and the UK.

Table 3.– Shares of West European Patenting in the US: 1963 to 1990

	1963-68	1969-74	1975-80	1981-85	1986-90
Germany	34.21	36.37	38.60	41.56	41.35
France	13.62	14.46	14.46	14.70	15.20
United Kingdom	25.15	21.36	17.93	15.66	14.94
Switzerland	8.78	8.58	9.00	8.05	7.08
Italy	4.34	4.85	4.98	5.50	6.36
Netherlands	4.74	4.43	4.45	4.68	4.88
Sweden	5.31	5.25	5.70	5.03	4.61
Belgium	1.65	1.94	1.78	1.62	1.65
Finland	0.25	0.47	0.72	1.04	1.37
Denmark	0.94	1.09	1.04	0.99	1.00
Spain	0.42	0.49	0.57	0.42	0.65
Norway	0.51	0.54	0.64	0.56	0.63
Ireland	0.06	0.13	0.11	0.16	0.25
Portugal	0.03	0.04	0.02	0.03	0.03
	100.00	100.00	100.00	100.00	100.00

Finally, Table 4 presents recent trends in patenting in the USA by a number of developing countries. We are very much aware of the inadequacies of US patenting as a measure of the largely imitative activities in technological accumulation that are performed in developing countries. Studies using other approaches have shown the superior performance of East Asian countries, compared to those of Latin America and to India [see, for example, Dahlman et al. (1987)]. Table 4 simply shows that, whilst most of the developing countries have continued with a very low level of US patenting, Taiwan and S. Korea have both seen massive increases in their US patenting, similar to that seen by Japan in the late 1950s. This indicates that technology in Taiwan and S. Korea is now attaining world best practice levels in an increasing number of fields –a striking example of technological divergence, compared to other developing countries.

II.2 In the Sectoral Composition of National Technological Activities

So far, we have compared countries' aggregate technological performance. Table 5 summarises the sectoral patterns of technological advantage of the USA, Western Europe and Japan. Based on the US patent classifications, it divides technologies into 11 fields. The content of most of them will be clear from their titles: technologies for extracting and processing raw materials are related mainly to food, oil and gas; defence–related technologies are defined as aerospace and munitions. For each country–region and technological field, we have calculated an index of "Revealed Technology Advantage" (RTA) in 1963–8 and 1984–8. RTA is defined as a country's or region's (or firm's) share of all US patenting in a technological field, divided by its share of all US patenting in all fields. An RTA of more than one therefore shows a country's or region's relative strength in a technology, and less than one its relative weakness. These measures correspond broadly to the measures of comparative advantage used in trade analyses.

Table 5 shows markedly different patterns and trends amongst the three main, technology–producing regions of the world in their fields of technological advantage and disadvantage. The USA has seen rapid decline in motor vehicles and consumer electronics; growing relative strength in technologies related to weapons, raw materials and telecommunications; and an improving position in chemicals. In Japan, almost the opposite has happened: growing relative strength in electronic consumer and capital goods and motor vehicles, together with rapid relative decline in chemicals, and continued weakness in raw materials and weapons. In Western Europe, the pattern is different again, and very close to that of its dominant country –FR Germany: continuing strength in chemicals, growing strength in weapons, continued though declining strength in motor vehicles, and weakness in electronics.

45

Table 4.– US Patenting Activities of Selected Developing Countries: 1969 to 1989.

Country	69	70	71	72	73	74	75	76	77	78	79
Taiwan	0	0	0	0	1	0	23	28	52	29	38
South Korea	0	3	2	7	5	7	11	7	5	12	4
China P. Rep.	5	6	15	8	10	22	1	6	1	0	2
Hong Kong	7	8	19	7	15	9	10	20	9	21	13
Mexico	67	43	63	43	42	51	66	78	42	24	36
Brazil	18	17	14	16	18	21	17	18	21	24	19
Venezuela	6	3	13	7	5	7	0	0	0	2	11
Argentina	17	23	22	29	27	24	24	24	20	21	24
Singapore	2	0	4	4	7	6	1	3	3	2	0
India	18	16	10	19	21	17	13	17	13	14	14

Country	80	81	82	83	84	85	86	87	88	89
Taiwan	65	79	88	65	97	174	208	343	457	589
South Korea	8	15	14	26	29	38	45	84	95	157
China P. Rep.	1	3	0	1	2	1	9	23	47	51
Hong Kong	27	33	18	14	24	25	30	34	41	47
Mexico	41	43	35	32	42	32	37	49	44	39
Brazil	24	23	27	19	20	30	27	34	29	36
Venezuela	11	12	10	5	11	15	21	24	20	23
Argentina	18	25	18	21	20	11	17	18	16	20
Singapore	3	4	3	5	4	9	3	11	6	18
India	4	6	4	14	12	10	18	12	14	14

Table 5.– Sectoral Patterns of Revealed Technological Advantage: 1963–68 to 1985–90.

	United States		Japan		Western Europe	
	1963-68	1985-90	1963-68	1985-90	1963-68	1985-90
Fine Chemicals	0.89	0.97	2.95	0.72	1.34	1.33
Industrial Chemicals	0.93	0.98	1.62	0.92	1.29	1.19
Materials	1.04	0.95	1.02	1.42	0.86	0.83
Non-Electrical Machinery	1.01	0.99	0.77	0.85	0.99	1.13
Motor Vehicles	0.89	0.55	0.83	2.21	1.48	1.02
Electrical Machinery	1.00	1.01	1.17	1.08	1.00	0.92
Electronic Capital Goods	1.02	0.97	1.47	1.65	0.92	0.61
Telecommunications	1.03	1.04	1.06	0.97	0.91	0.94
Electronic Consumer Goods	0.94	0.65	1.99	2.50	1.26	0.59
Raw Materials Based Technologies	1.08	1.28	0.44	0.37	0.61	0.83
Defence Related Technologies	0.99	1.15	0.36	0.09	1.14	1.40

Note: For the definition of the Revealed Technology Advantage Index see text.

47

Table 6.– Stability and Similarities Amongst Countries in their Sectoral Specialisations: Correlations of Revealed Technology Advantage Indicies across 34 Sectors.

Stability: Correlations Over Time : 1963-68 to 1983-90

Australia	Austria	Belgium	Canada	Denmark	Finland	France	Germany	Ireland	Italy	Japan	Netherland	Norway	Portugal	Spain	Sweden	Switzerland	UK	USA
0.28	0.76	0.54	0.67	0.47	0.59	0.82	0.35	0.05	0.32	0.45	0.66	0.35	0.25	0.53	0.73	0.83	0.23	0.55

Similarities: Correlations Amongst Countries: 1983-90

	Australia	Austria	Belgium	Canada	Denmark	Finland	France	Germany	Ireland	Italy	Japan	Netherland	Norway	Portugal	Span	Sweden	Switzerland	UK	USA
Austria	0.36																		
Belgium	-0.09	-0.14																	
Canada	0.52	0.47	0.05																
Denmark	0.18	-0.03	0.33	0.32															
Finland	0.47	0.45	0.20	0.54	0.45														
France	-0.27	-0.16	0.10	-0.14	0.10	-0.15													
Germany	0.27	0.05	0.22	-0.18	0.21	0.32	0.29												
Ireland	0.07	-0.10	0.09	0.21	0.14	0.03	-0.09	0.31											
Italy	0.28	0.28	0.06	0.34	0.30	0.53	-0.23	0.22	0.28										
Japan	0.43	-0.07	0.06	-0.44	-0.22	-0.26	-0.44	-0.20	-0.13	-0.13									
Netherlands	0.24	-0.18	-0.03	0.06	-0.04	0.07	-0.33	0.38	0.27	0.08	0.24								
Norway	0.36	0.36	-0.20	0.62	0.22	0.28	0.02	0.12	0.02	0.03	-0.50	0.23							
Portugal	0.32	0.48	0.17	0.31	0.11	0.43	-0.09	0.15	0.11	0.35	-0.20	-0.04	0.06						
Span	0.32	0.13	-0.11	0.34	0.68	0.38	0.00	0.28	-0.07	0.38	-0.23	-0.28	0.41	0.20					
Sweden	0.25	0.46	-0.05	0.38	0.40	0.53	0.36	0.26	0.07	0.19	-0.38	-0.35	0.30	0.30	0.45				
Switzerland	0.35	-0.12	0.11	0.21	0.01	0.07	0.06	0.72	-0.12	0.17	-0.19	-0.26	0.14	0.04	0.13	0.04			
UK	0.08	-0.15	-0.04	-0.10	0.40	0.03	0.23	0.30	-0.02	-0.03	-0.34	-0.20	0.17	0.00	0.32	0.10	0.20		
USA	0.22	-0.03	0.20	0.42	-0.02	-0.01	0.23	0.37	0.25	-0.09	-0.81	0.03	0.50	0.06	0.02	0.13	0.26	0.11	
West Europe	0.26	0.04	0.21	0.08	0.33	0.34	0.49	0.93	-0.19	0.27	0.41	-0.37	-0.02	0.17	0.35	0.39	0.73	0.45	0.19

Notes: For the definition of the Revealed Technology Advantage Index see text. For a definition of Western Europe Table 3.

* Denotes Correlation Coefficient significantly different from zero at the 5% level.

Table 6 examines the similarities and differences amongst countries' technological specialisations in greater and more systematic detail. It uses correlation analysis to measure both the stability over time of each country's sectoral strengths and weaknesses in technology (first row), and the degree to which they are similar to those of other countries (correlation matrix). The first row shows that, with five exceptions (Australia, Ireland, Italy, Portugal and the UK), most OECD countries have a statistically significant degree of stability in their technological strengths and weaknesses between the 1960s and the 1980s: 10 at the 1% level, and a further 4 at the 5% level, thereby confirming the path–dependant nature of national patterns of accumulation of technological knowledge.

The correlation matrix also confirms the differentiated nature of technological knowledge, with the very different strengths and weaknesses in Japan, the USA and Western Europe: each is negatively correlated with the other two, and significantly so in two cases out of three (the USA with the other two regions). More generally, it confirms that countries tend to differ markedly in their patterns of technological specialisation. Of the 171 correlations amongst pairs of countries in Table 6, only 31 (18%) are positively and significantly correlated at the 5% level, of which 8 at the 1% level. Amongst the latter, we find FR Germany similar to Switzerland (chemicals and machinery), and Canada similar to Australia, Finland and Norway (raw material–based technologies). Japan has a unique pattern of specialisation, with no significant positive correlations with other countries, but plenty of negative ones.

II.3 Amongst Large Firms

Table 7 shows the evidence of uneven technological development since the late 1960s amongst the world's largest firms. It is based on data that we have compiled on more than 600 of the world's largest, technologically active firms, as measured by their patent activity in the USA [see Patel and Pavitt (1991a)]. It shows, in the same 11 technological fields as in Table 5, trends in the total US patenting of the world's top 20 firms. Three major conclusions emerge from this Table.

First, the share of our large firms in the world's technological activities varies greatly amongst fields. For all combined it is about 49%, but the share increases to nearly 70% in electronic capital and consumer goods, and to more than 60% in motor vehicles, and more than 50% in fine and industrial chemicals, materials and telecommunications. The top 20 firms in motor vehicles and electronic consumer goods have more than 50% of the total. Mechanical engineering (including instruments) is the least concentrated field.

49

Table 7.– Shares of US Patenting of Top 20 Firms in 11 Technological Areas. Sorted According to the 1986–90 Share.

Technology Class 1: Fine Chemicals	69-74	75-80	81-85	86-90
Bayer (Germany)	2.84	4.06	3.39	3.72
Hoechst (Germany)	1.61	2.73	2.47	2.54
Merck (USA)	2.57	2.91	3.02	2.43
Ciba-Geigy (Switzerland)	4.33	2.89	2.82	2.23
Imperial Chemical Industries Plc (UK)	1.98	2.75	2.24	1.90
Du Pont (USA)	1.48	1.19	1.47	1.78
Warner-Lambert (USA)	0.71	0.68	0.68	1.66
Eli Lilly (USA)	1.67	1.78	2.04	1.34
Dow Chemical (USA)	1.21	1.39	1.25	1.24
BASF (Germany)	0.58	0.83	1.37	1.13
Boehringer (Germany)	1.00	1.60	1.22	1.05
Johnson & Johnson (USA)	0.36	0.83	1.07	1.05
American Cyanamid (USA)	2.43	1.69	1.24	1.03
Pfizer (USA)	1.17	1.32	1.34	1.01
Hoffmann-La Roche (Switzerland)	1.59	1.49	1.16	0.98
Smithkline Beckman (USA)	1.40	2.09	1.19	0.93
Monsanto (USA)	2.87	1.29	1.09	0.90
Takeda Chemical (Japan)	1.21	0.62	0.79	0.84
Squibb (USA)	1.03	1.27	0.98	0.82
American Home Products (USA)	0.95	1.88	1.06	0.80
Share of the top 20	**32.99**	**35.30**	**31.89**	**29.36**
Share of all large firms	**62.70**	**63.94**	**59.25**	**50.33**

Technology Class 2: Other Chemicals	69-74	75-80	81-85	86-90
Bayer (Germany)	2.57	3.14	3.16	2.78
Dow Chemical Company (USA)	2.40	1.79	2.09	2.72
Hoechst (Germany)	2.37	2.61	2.52	2.46
BASF (Germany)	1.40	1.88	2.16	2.36
Ciba-Geigy (Switzerland)	2.54	2.66	2.08	1.88
General Electric (USA)	1.39	1.61	2.37	1.73
Du Pont (USA)	3.29	2.23	1.71	1.70
Imperial Chemical Industries Plc (UK)	2.14	1.97	1.32	1.17
Shell (Netherlands)	0.85	0.61	0.79	1.14
Eastman Kodak (USA)	1.33	1.07	1.04	1.06
Henkel (Germany)	0.35	0.44	0.59	0.81
Exxon (USA)	0.79	0.74	0.93	0.78
Union Carbide (USA)	1.11	1.01	1.27	0.77
Allied-Signal (USA)	1.62	1.04	1.03	0.77
Rhone-Poulenc (France)	0.63	0.69	0.64	0.68
Sumitomo Chemical Company (Japan)	0.55	0.76	0.63	0.66
3M (USA)	0.52	0.43	0.50	0.61
Texaco (USA)	0.41	0.75	0.75	0.58
Air Products and Cheimcals (USA)	0.17	0.17	0.38	0.58
Philips Petroleum (USA)	1.46	1.21	1.13	0.57
Share of the top 20	**27.89**	**26.82**	**27.06**	**25.82**
Share of all large firms	**68.13**	**68.07**	**66.00**	**61.27**

Technology Class 3: Materials	69-74	75-80	81-85	86-90
3M (USA)	1.36	1.56	1.55	2.34
Fuji Photo Film (Japan)	0.31	0.69	1.60	2.28
PPG Industries (USA)	2.72	4.02	2.77	1.72
General Electric (USA)	2.32	2.38	2.54	1.65
Du Pont (USA)	3.48	2.03	1.54	1.63
Hitachi (Japan)	0.20	0.58	1.05	1.45
Dow Chemical Company (USA)	1.26	0.69	0.56	1.21
Hoechst (Germany)	1.01	0.74	0.98	1.03
Corning Glass Works (USA)	2.77	2.95	2.36	1.03
Saint-Gobain (France)	0.96	0.80	0.60	0.98
TDK (Japan)	0.10	0.19	1.04	0.94
Allied Signal (USA)	0.65	0.70	0.66	0.84
Emhart Corporation (USA)	0.46	0.33	0.72	0.84
Kimberly-Clark Corporation (USA)	0.61	0.17	0.23	0.82
Sumitomo Electric Industries (Japan)	0.06	0.24	0.43	0.80
Toshiba (Japan)	0.38	0.21	0.63	0.80
W. R. Grace (USA)	0.69	0.50	0.36	0.75
Owens-Corning Fiberglas (USA)	1.83	1.71	2.35	0.71
Canon (Japan)	0.02	0.08	0.10	0.70
Philips (Netherlands)	0.54	0.65	0.72	0.68
Share of the top 20	**21.74**	**21.22**	**22.76**	**23.20**
Share of all large firms	**63.30**	**60.64**	**59.08**	**57.00**

Technology Class 4: Non-Electrical Machinery	69-74	75-80	81-85	86-90
General Motors (USA)	1.23	0.88	0.72	0.95
Hitachi (Japan)	0.18	0.45	0.72	0.89
General Electric (USA)	1.22	1.18	1.04	0.78
Canon (Japan)	0.05	0.21	0.43	0.75
Toshiba (Japan)	0.08	0.15	0.48	0.69
Siemens (Germany)	0.29	0.59	0.67	0.66
Philips (Netherlands)	0.39	0.41	0.43	0.57
United Technologies (USA)	0.47	0.49	0.54	0.55
Westinghouse (USA)	0.56	0.48	0.37	0.53
Honda (Japan)	0.03	0.08	0.26	0.51
Nissan (Japan)	0.12	0.27	0.66	0.50
Toyota (Japan)	0.10	0.31	0.46	0.50
Allied Signal (USA)	0.78	0.68	0.78	0.47
Fuji Photo Film (Japan)	0.09	0.10	0.18	0.44
Mitsubishi Electric (Japan)	0.03	0.05	0.17	0.39
IBM (USA)	0.46	0.38	0.44	0.36
ITT (USA)	0.35	0.35	0.52	0.33
Aisin Seiki (Japan)	0.10	0.16	0.26	0.32
Robert Bosch (Germany)	0.24	0.31	0.43	0.32
Eastman Kodak (USA)	0.43	0.16	0.15	0.32
Share of the top 20	**7.22**	**7.68**	**9.70**	**10.85**
Share of all large firms	**32.33**	**33.75**	**36.49**	**34.44**

Technology Class 5: Motor Vehicles and Parts.	69-74	75-80	81-85	86-90
Honda (Japan)	0.91	2.33	5.63	9.10
Nissan (Japan)	1.74	5.84	7.10	5.55
Toyota (Japan)	0.76	6.60	6.01	4.60
Robert Bosch (Germany)	3.79	5.65	6.37	4.08
Mazda (Japan)	0.78	0.85	1.28	3.08
Mitsubishi Electric (Japan)	0.21	0.12	0.89	3.00
General Motors (USA)	5.21	4.17	3.21	2.84
Fuji Heavy Industries (Japan)	0.06	0.34	1.19	2.36
Nippondenso (Japan)	1.27	3.20	4.25	2.23
Yamaha Motor (Japan)	0.32	0.67	1.82	1.96
Hitachi (Japan)	0.38	1.11	1.53	1.90
Daimler-Benz (Germany)	2.71	1.63	1.28	1.50
Ford (USA)	2.41	1.47	1.57	1.46
Brunswick Corporation (USA)	0.42	0.42	0.69	1.23
Aisin Seiki (Japan)	0.34	0.45	0.79	1.08
Porsche (Germany)	0.42	0.20	0.37	0.88
Kawasaki Heavy (Japan)	0.08	0.28	0.40	0.82
Lucas Industries (UK)	0.95	1.68	1.48	0.82
Outboard Marine Corporation (USA)	0.70	0.61	0.89	0.84
Mitsubishi Heavy Industries (Japan)	0.19	0.55	0.29	0.77
Share of the top 20	**23.67**	**38.17**	**47.04**	**50.06**
Share of all large firms	**44.58**	**58.38**	**64.71**	**64.99**

Technology Class 6: Electrical Machinery	69-74	75-80	81-85	86-90
General Electric (USA)	5.77	5.44	4.93	2.85
Westinghouse (USA)	3.22	2.98	2.27	2.82
Philips (Netherlands)	1.44	1.81	1.60	2.16
AMP Corporation (USA)	1.10	1.34	1.43	2.11
Mitsubishi Electric (Japan)	0.20	0.38	1.32	1.96
Siemens (Germany)	1.47	2.11	2.04	1.92
Hitachi (Japan)	0.53	1.32	2.04	1.84
Toshiba (Japan)	0.41	0.52	1.71	1.44
General Motors (USA)	1.81	1.22	0.91	1.38
GTE Corporation (USA)	1.20	1.44	1.56	1.20
Motorola Inc. (USA)	0.51	0.56	0.80	0.99
Matsushita Electric Industrial (Japan)	0.85	0.96	0.91	0.88
ABB (Switzerland)	0.83	0.94	0.99	0.78
NEC (Japan)	0.22	0.26	0.34	0.67
United Technologies (USA)	0.93	0.75	0.87	0.58
Canon (Japan)	0.07	0.20	0.22	0.55
Honeywell (USA)	0.76	0.47	0.63	0.50
Sharp (Japan)	0.01	0.12	0.25	0.50
IBM (USA)	0.80	0.62	0.65	0.48
ITT (USA)	0.92	0.80	0.65	0.46
Share of the top 20	**23.05**	**24.25**	**26.13**	**26.10**
Share of all large firms	**48.72**	**50.38**	**52.71**	**47.70**

Technology Class 7: Electronic Capital Goods and Components.	69-74	75-80	81-85	86-90
Toshiba (Japan)	0.53	1.93	4.30	5.26
IBM (USA)	8.83	7.56	7.83	5.09
Hitachi (Japan)	1.71	3.23	3.62	4.88
Mitsubishi Electric (Japan)	0.13	0.50	0.68	2.93
NEC (Japan)	0.97	0.98	1.59	2.87
Motorola (USA)	2.15	2.62	2.77	2.86
Texas Instruments (USA)	1.97	2.74	2.70	2.83
Fujitsu Limited (Japan)	0.38	0.57	2.80	2.53
Philips (Netherlands)	2.81	3.15	2.54	2.45
General Electric (USA)	6.77	6.36	4.14	2.38
ATT (USA)	6.05	3.53	3.11	1.95
Siemens (Germany)	1.75	2.92	2.34	1.73
Sharp (Japan)	0.05	0.39	0.75	1.08
Canon (Japan)	0.04	0.37	0.50	1.04
Honeywell (USA)	2.23	2.68	2.21	0.99
General Motors (USA)	1.36	1.12	0.91	0.94
Tektronix Inc. (USA)	0.26	0.21	0.41	0.92
Unisys Corporation (USA)	3.47	2.67	2.59	0.85
Hewlett-Packard (USA)	0.60	0.96	0.58	0.78
Sony Corporation (Japan)	0.43	1.22	0.81	0.78
Share of the top 20	**42.47**	**45.71**	**47.16**	**45.15**
Share of all large firms	**68.64**	**71.39**	**73.94**	**69.23**

Technology Class 8: Telecommunications	69-74	75-80	81-85	86-90
ATT (USA)	5.97	4.05	4.38	4.13
Siemens (Germany)	2.22	3.63	4.00	3.27
General Electric (USA)	4.29	4.31	4.18	2.83
Motorola Inc.(USA)	1.02	1.59	2.09	2.62
NEC (Japan)	0.61	0.67	1.17	2.61
Philips (Netherlands)	1.54	1.82	2.47	2.44
Westinghouse (USA)	3.30	3.10	2.31	1.75
Toshiba (Japan)	0.26	0.60	1.17	1.72
Mitsubishi Electric (Japan)	0.21	0.36	1.08	1.59
General Motors (USA)	1.57	1.62	0.85	1.41
Hitachi (Japan)	0.43	1.13	1.40	1.40
ITT (USA)	3.55	3.00	2.13	1.26
Thomson-CSF (France)	0.82	1.19	1.88	1.11
IBM (USA)	1.19	0.84	0.90	1.10
GTE Corporation (USA)	1.49	1.87	1.92	1.03
Northern Telecom (Canada)	0.54	0.69	1.08	0.94
Fujitsu Limited (Japan)	0.20	0.26	0.53	0.90
Alps Electric (Japan)	0.13	0.25	0.54	0.79
Compagnie Generale D'Electricite (France)	0.58	0.72	0.93	0.75
Rockwell International Corporation (USA)	0.63	0.99	1.43	0.70
Share of the top 20	**30.57**	**32.60**	**36.42**	**34.33**
Share of all large firms	**53.16**	**55.18**	**58.92**	**56.47**

Technology Class 9: Electronic Consumer Goods and Photography	69-74	75-80	81-85	86-90
Canon (Japan)	0.95	2.51	5.26	6.65
Fuji Photo Film (Japan)	2.12	3.68	3.93	6.36
Eastman Kodak (USA)	6.24	3.28	3.26	3.45
Toshiba (Japan)	0.42	0.52	1.86	3.29
Philips (Netherlands)	2.38	2.72	2.77	3.02
Sony Corporation (Japan)	1.02	2.23	2.51	2.94
Hitachi (Japan)	0.62	1.53	2.30	2.92
General Electric (USA)	3.97	4.41	5.18	2.71
Minolta Camera (Japan)	0.88	1.46	1.44	2.67
Xerox Corporation (USA)	3.79	5.65	2.95	2.18
Ricoh (Japan)	1.00	1.22	1.60	1.95
Konica Corporation (Japan)	0.52	0.91	1.95	1.88
Sharp (Japan)	0.00	0.18	0.37	1.78
Matsushita Electric Industrial (Japan)	1.20	1.67	1.45	1.67
Pioneer (Japan)	0.18	0.37	0.98	1.45
IBM (USA)	2.92	2.17	1.86	1.42
Mitsubishi Electric (Japan)	0.07	0.12	0.35	1.21
NEC (Japan)	0.36	0.53	0.55	1.12
Olympus Optical (Japan)	0.14	0.81	2.20	1.12
Siemens (Germany)	0.59	0.80	1.05	1.00
Share of the top 20	**29.38**	**36.78**	**43.83**	**50.81**
Share of all large firms	**61.00**	**62.32**	**66.46**	**69.91**

Technology Class 10: Raw Materials Related Technologies	69-74	75-80	81-85	86-90
Mobil Oil (USA)	2.17	3.12	4.46	4.75
Exxon (USA)	3.00	2.66	2.74	2.18
Halliburton (USA)	0.60	0.75	1.54	1.68
Chevron (USA)	2.66	2.31	2.26	1.42
Nabisco Brands (USA)	0.32	0.30	0.57	1.40
Baker Hughes (USA)	0.41	0.48	1.45	1.39
Amoco Corporation (USA)	0.18	0.76	1.24	1.35
Shell (Netherlands)	2.13	1.05	0.99	1.30
Texaco (USA)	2.49	2.82	2.24	1.25
Philip Morris (USA)	1.32	1.45	1.99	1.20
Allied Signal (USA)	3.00	3.08	1.73	1.16
Atlantic Richfield Company (USA)	0.89	0.83	1.09	1.05
Deere (USA)	1.04	0.63	0.85	0.89
Union Oil Of California (USA)	0.57	0.53	0.57	0.83
Du Pont (USA)	0.87	0.98	0.73	0.67
Nissan (Japan)	0.03	0.12	0.31	0.61
Schlumberger (USA)	0.85	0.34	0.33	0.56
Nestle (Switzerland)	0.15	0.40	0.56	0.56
United Technologies (USA)	0.26	0.41	0.69	0.53
BAT (UK)	0.19	0.29	0.51	0.52
Share of the top 20	**23.13**	**23.29**	**26.84**	**25.28**
Share of all large firms	**47.96**	**46.79**	**49.06**	**43.21**

Technology Class 11: Defence-Related Technologies	69-74	75-80	81-85	86-90
Boeing Company (USA)	1.06	3.66	4.61	4.34
MBB (Germany)	1.18	1.54	1.56	2.74
General Electric (USA)	1.58	1.97	1.94	1.45
Oerlikon-Buhrle (Switzerland)	0.89	1.00	1.04	1.37
British Aerospace (UK)	0.66	0.20	0.69	1.35
Feldmuhle Nobel (Germany)	1.56	1.12	0.93	1.09
Morton Thiokol (USA)	0.93	1.00	0.97	1.07
Honeywell (USA)	0.31	0.05	0.52	1.04
General Dynamics (USA)	0.29	0.87	1.59	0.97
Imperial Chemical Industries Plc (UK)	1.33	1.15	1.63	0.91
General Motors (USA)	0.54	0.27	0.66	0.86
Olin Corporation (USA)	1.16	0.47	0.42	0.79
Westinghouse (USA)	0.15	0.25	0.90	0.79
United Technologies (USA)	1.04	0.85	0.83	0.71
Aerospatiale (France)	0.35	0.35	0.97	0.69
Grumman Corporation (USA)	0.02	0.22	0.35	0.66
Lockheed Corporation (USA)	0.79	0.55	0.83	0.53
Sundstrand Corporation (USA)	0.02	0.02	0.07	0.53
Halliburton Co. (USA)	0.12	0.20	0.21	0.51
Rockwell International (USA)	0.83	0.35	0.52	0.48
Share of the top 20	**14.79**	**16.10**	**21.23**	**22.89**
Share of all large firms	**31.07**	**27.91**	**33.73**	**33.12**

Second, the technological strengths and weaknesses of each region, shown in Table 5, are in general reflected in the number of nationally based large firms appearing in the top 20 in Table 7. Thus, as summarised in Table 8, Japanese firms make up 12 of the top 20 firms in motor vehicles and 14 in consumer electronics and photography, US firms make up 17 of the top 20 in raw materials and 15 in defence, whilst European firms have their largest numbers in chemicals.

Table 8.– Nationalities of the Top 20 Firms in US Patenting: 1986–90.

	Japan	United States	West Europe	Increase in % Share of Top 5 Firms since 1969-74
Defence Related Technologies	0	15	5	5.9
Raw Materials Based Technologies	1	17	3	5.7
Fine Chemicals	1	12	7	- 0.2
Industrial Chemicals	1	11	8	0.9
Materials	6	11	3	- 0.6
Telecommunications	6	9	4	1.4
Electrical Machinery	7	10	3	0.2
Electronic Capital Goods	7	10	3	8.9
Non-Electrical Machinery	9	8	3	1.3
Motor Vehicles	12	4	5	20.2
Electronic Consumer Goods	14	4	2	10.6

Third, in addition to uneven development of large firms in each technological field according to their nationality, there has also been an uneven degree of stability (or instability) in the firms' shares and rankings within each technological field. A casual reading of Table 7 shows that in some fields, the leaders of the early 1970s continued to be so into the late 1980s, whilst in others new leaders emerged during the period. This is shown in the final column of Table 8, which presents the increase in the percentage share of total patenting held by the top five firms in 1986–90, compared to their share in 1969–74: a small increase (or decrease) in share reflects stability in the ranking and shares of the top five, whilst a large increase reflects the emergence of new leaders or the reinforcement of the dominance of existing ones.

Thus, the massive increase in the shares of the top 5 in motor vehicles, and in electronic capital and consumer goods, mainly reflects the emergence of Japanese firms as technological leaders in these fields, whist in raw materials and defence technologies, they reflect mainly a re–enforcement of the dominance of established US and some European firms. The more stable shares of the top 5 in the two chemical sectors reflect the continuing strength of mainly European firms.

III. THE CAUSES OF UNEVEN DEVELOPMENT

III.1 The Starting Point: "Localised Learning" rather than "Information Asymmetries" or "Market Failure"

So far, we have presented evidence of uneven development, across both countries and large firms, in the volume and the composition of the technological activities that are an essential basis for their dynamic efficiency. We now turn to explanations, and in doing so we must remind the reader that the essential characteristics of these technological activities are localised learning, involving trial and error, and the accumulation of often tacit knowledge, not only in individuals, but also and especially in institutions like business firms.

For this reason, we reject explanations based on "information asymmetries", since "information' is easily transferred, and are therefore unlikely to be the basis for persistent differences amongst companies and countries (and if the definition of information is broadened to cover all knowledge —including tacit knowledge— then the concept becomes an empty tautology). We are also sceptical that the "market failure" approach can explain by itself such persistent differences amongst market economies, especially when a good part of the tacit (and specific and differentiated) knowledge can be appropriated by the business firms that finance their acquisition. Instead, we shall argue that the main reasons for uneven development must be sought in the nature of the localised learning activities that underlie technical change.

.

III.2 "Institutional Failure" in the Competence to Evaluate and Benefit from Technological Activities

We begin by proposing that the growing international differences in the volume of technological activities reflect institutional failures in the competence to evaluate and benefit from investments that are increasingly specialised and professio- nalised in nature (e.g. industrial R&D laboratories employing highly qualified specialists in a variety of fields of science and engineering), and are long –term and complex in their economic impact (e.g. from research on photons, through the laser, to the compact disc, over a period of 25 years). For purposes of exposition, we have found it useful to distinguish between national systems of innovation that we define as *myopic*, and those that we define as *dynamic* [Pavitt and Patel (1988)].

Briefly stated, *myopic* systems treat investments in technological activities just like any conventional investment: they are undertaken in response to a well- defined market demand, and include a strong discount for risk and time. As a consequence, technological activities often do not compare favourably with conventional investments. *Dynamic* national systems of innovation, on the other hand, recognise that technological activities are not the same as any other investment. In addition to tangible outcomes in the form of products, processes and profits, they also entail the accumulation of important but intangible assets, in the form of irreversible processes of technological, organisational and market learning, that enable them to undertake subsequent investments, that they otherwise could not have done.

The archetypal dynamic national systems of innovation are those of FR Germany and Japan, whilst the myopic systems are the UK and the USA. The essential differences can be found in three sets of institutions:

First, in the financial system underlying business activity: in Germany and Japan, these give greater weight to longer–term performance, when the benefits of investments in learning begin to accrue. And they generate both the information and the competence to enable firm–specific intangible assets to be evaluated by the providers of finance [Corbett and Mayer (1991)].

Second, there are the methods of management, especially those employed in large firms in R&D–intensive sectors: in the UK and USA, the relatively greater power and prestige given to financial (as opposed to technical) competence is more likely to lead to incentive and control mechanisms based on short–term financial performance, and to decentralised divisional structures insensitive to new and longer term technological opportunities that top management is not competent to evaluate [Abernathy and Hayes (1980), Lawrence (1980)].

Third, there is the system of education and training: the German and Japanese systems of widespread yet rigorous general and vocational education provide a strong basis for cumulative learning. The British and US systems of higher education have performed relatively well, but the other two–thirds of the labour force are less well trained and educated than their counterparts in continental Europe and East Asia [Prais (1981), The Independent (1992)].

III.3 Local Learning and Local Inducement Mechanisms

We further propose that the observed international differences in the sectoral patterns of technological accumulation emerge from the localised nature of technological accumulation, and the consequent importance of the local inducement mechanisms that guide and constrain firms along cumulative technological trajectories. We know from earlier debates about the relative importance of "technology push" and "demand pull" that these inducement mechanisms are numerous, and that their relative importance varies amongst sectors. It is nonetheless possible to distinguish three mechanisms: (i) factor endowments; (ii) directions of persistent investment, especially those with strong intersectoral linkages; (iii) the cumulative mastery of core technologies and their underlying science bases. The relative significance of these mechanisms change over time. In the early stages, the directions of technical change in a country or region are strongly influenced by local market inducement mechanisms related to scarce (or abundant) factors of production and local investment opportunities. At higher levels of development, the local accumulation of specific technological skills itself becomes a focusing device for technical change.

Factor Endowments. The most obvious of the local inducement mechanisms has been the search to alleviate a relative factor scarcity: in this context technical change can be seen as the search to lengthen an existing production function, in order to save further on an expensive factor of production. Its historical importance in the development of labour–saving techniques in the USA (and elsewhere in the developed world) is well documented, as is its importance in generating technical responses to differing natural resource endowments: see for example, the effects of different fuel prices on the development paths of automobile and related mechanical technologies in the United States, on the one hand, and in Europe and East Asia, on the other.

Investment–led inducements and inter–sectoral linkages. A second set of inducement mechanisms are reflected in another analytical tradition that stresses the importance of the investment–induced nature of technical change [Schmookler (1966)], and of technical linkages and imbalances among firms and sectors [see Carlsson and Henriksson (1991), Justman and Teubal (1991)]. The exploitation of abundant natural resources is one variant, creating opportunities for local technical change, technological accumulation and competitiveness in upstream extraction and downstream processing. Witness the effects of abundant natural resources in Canada, the USA and Scandinavia, where the development of natural resource–based sectors has contributed to competitiveness in the capital goods used in these sectors [Patel and Pavitt (1991b)]. Other important linkages have included those from investment in mass–produced automobiles (and some other consumer durables) to technological accumulation in associated capital goods producing–sectors [Lee (1993)]; and those related to government investment programmes –see, for example, their effects on shipbuilding, and railways and communications equipment in the early modernisation of Japan [Nakaoka (1987)].

Mastery of Core Technologies. Cost and investment–led inducement mechanisms cannot explain the emergence of all areas of technology–based competitiveness. For example, Swiss competitiveness in marine diesels has little obvious link with any national endowment in maritime resources, but has a more

64

obvious connection with the engineering competence developed initially in the production of textile machinery. At the higher levels of technological accumulation in today's developed countries, the central inducement mechanism has often become the cumulative mastery and exploitation on world markets of core technologies with multiple potential applications. When the core technologies are science–based, and the linkage horizontal rather than vertical: in other words, trajectories are not traced through inducements from users to producers of capital goods, but through diversification into new product markets from an R&D base. Thus, Swiss strength in pharmaceuticals began in dyestuffs, just as Germany has retained its strength through the successive waves of major innovations in chemicals. The Swedish pattern is more complex, having begun in mining technology and ended up in (among other things) robots –the common core being machinery and metals. Contemporary patterns of technological advantage in the industrially advanced countries therefore tend to reflect relative endowments in technological (human) capital, in addition to other factors of production. Thus the relative advantage of the UK in fine chemicals, and the USA in software technology, reflect their strong endowments in technologies dependant on good university research and graduates, whilst their weakness in automobiles and other engineering industries reflects their weakness in activities requiring a work force with a high level of skills.

Competitive Rivalry. Finally, the statistical evidence suggests that the above national inducement mechanisms are more likely to result in accumulated technological competence under conditions of competitive rivalry amongst a number of large firms, rather than of the concentration of technological activities in fewer but larger firms. Amongst the world's largest companies, R&D intensity and patenting intensity do not increase significantly with firm size; and although there is slight increase in technological diversity with firm size, this is least pronounced in the R&D–intensive sectors. In addition, sectoral technological advantage in Japan, the USA and W. Europe is significantly and positively associated with the *number* of large firms in all three regions, but with their *size* only in Western Europe.

These results support analysts stressing the importance of competitive rivalry in stimulating innovative performance, and advocating a strong competition policy discouraging mega–mergers [Sharp (1989), Porter (1990)]. However, deeper empirical understanding is still required of the dynamic interactions amongst technology–intensity, size and rivalry. In particular, does lack of rivalry reflect the absence of a tough competition policy imposed by public authorities?. Or does it reflect the prevalence of financially based management that enables ever larger and more diversified firms to be effectively controlled, but that at the same time either overloads or stifles their entrepreneurial judgements and results in stagnation?.

65

III.4 The Localised Production of Large Firms' Global Technology

The similarities between Tables 5 and 8 have already been remarked upon: country' technological specialisations are reflected in the specialisations of home-based large firms. More generally, we have shown elsewhere that, in addition to the fields of specialisation of large firms, both the level and the rate of growth of national technological activities are reflected in those of large firms [Patel and Pavitt (1991a)]. This is because the overwhelming proportion of the technological activities of the world's largest firms continues to be performed in their home country. A recent update by one of us confirms that this continues to be the case, whatever the rhetoric about "techno–globalism" [Patel (1993)]. The basis for our analysis is the address (i.e. country of work) of the inventor shown in each of the US patents granted to the large firms in our database. The results are summarised in Tables 9 and 10. They are entirely consistent with the less systematic and detailed information that has been published on the international R&D activities of large firms.

Table 9 shows that, in the second half of the 1980s, 89% of the technological activities of the world's largest firms continued to be performed in their home country –a 1% increase over the previous five–year period. Not unsurprisingly, the share performed in foreign countries by large firms based in smaller countries tends to be higher, although the proportion for Finnish firms (18.3%) compared to that for British firms (45.1%) shows that other factors are at work. In particular, Cantwell (1992) has shown that the share of foreign in total production is the most important factor explaining differences amongst firms in the location of their technological activities, but that foreign patenting shares are smaller than foreign production shares: in other words, the technology intensity of foreign production is consistently and significantly less than that of home production.

Table 9 also shows that the foreign technological activities of large firms are not globalised, but concentrated almost exclusively in the "triad" countries –especially the USA and Europe (and, more specifically, Germany). More detailed data shows that the largest proportionate increases in foreign technological activities in the 1980s were in British and Swedish firms –especially in the USA, and in a number of smaller European countries outside the EEC –Switzerland, Finland, Norway– all increasing their share within the EEC. The firms based in countries with the largest technological activities –the USA, Japan and FR Germany– had amongst the lowest proportionate increases in foreign technological activities. In firms from most countries, most of any increase in foreign technological activities came as a by–product of take–overs and divestitures, rather than an explicit re-location of technological activities.

Table 9.– Geographic Location of Large Firms' US Patenting Activities, According to Nationality: 1985–90.

Percentage Shares

Firms' Nationality	Home	Abroad	Of Which			
			USA	Europe	Japan	Other
Japan (143)	98.9	1.1	0.8	0.3	-	0.0
USA (249)	92.2	7.8	-	6.0	0.5	1.3
Italy (7)	88.1	11.9	5.4	6.2	0.0	0.3
France (26)	86.6	13.4	5.1	7.5	0.3	0.5
Germany (43)	84.7	15.3	10.3	3.8	0.4	0.7
Finland (7)	81.7	18.3	1.9	11.4	0.0	4.9
Norway (3)	68.1	31.9	12.6	19.3	0.0	0.0
Canada (17)	66.8	33.2	25.2	7.3	0.3	0.5
Sweden (13)	60.7	39.3	12.5	25.8	0.2	0.8
UK (56)	54.9	45.1	35.4	6.7	0.2	2.7
Switzerland (10)	53.0	47.0	19.7	26.1	0.6	0.5
Netherlands (9)	42.1	57.9	26.2	30.5	0.5	0.6
Belgium (4)	36.4	63.6	23.8	39.3	0.0	0.6
All Firms (587)	**89.0**	**11.0**	**4.1**	**5.6**	**0.3**	**0.9**

Note: The parenthesis contains the number of firms based in each country.

This surprising (and unfashionable) conclusion suggests that, globalisation of markets and (increasingly) production notwithstanding, there remains at least one compelling reason for companies to concentrate a high proportion of their technological activities in one location. The development and commercialisation of major innovations requires the mobilisation of a variety of often tacit (person-embodied) skills, and involves high uncertainties. Both are best handled through intense and frequent personal communications and rapid decision–making –in other words, through geographical concentration. In this context, it is worth noting that the rapid product development times in Japanese firms [Clark et al. (1987)] have been achieved from an almost exclusively Japanese base, whilst the strongly globalised R&D activities of the Dutch Philips company are said to have slowed down product development.

Table 10.– Geographic Location of Large Firms' US Patenting Activities, According to Product Group: 1985–90.

Percentage Shares.

Product Group	Abroad	Of Which			
		USA	Europe	Japan	Other
Drink & Tobacco (18)	30.8	17.5	11.1	0.4	1.8
Food (48)	25.0	14.8	8.5	0.1	1.7
Building Materials (28)	20.6	9.1	9.8	0.1	1.6
Other Transport (5)	19.7	2.0	6.8	0.0	10.9
Pharmaceuticals (25)	16.7	5.5	8.3	1.1	1.7
Mining & Petroleum (47)	15.0	9.7	3.5	0.1	1.6
Chemicals (72)	14.4	8.0	5.1	0.3	1.0
Machinery (68)	13.7	3.5	9.1	0.1	1.1
Metals (57)	12.8	5.4	5.7	0.1	1.6
Electrical (58)	10.2	2.6	6.8	0.3	0.4
Computers (17)	8.9	0.1	6.6	1.1	1.1
Paper & Wood (34)	8.1	2.4	4.9	0.1	0.7
Rubber & Plastics (10)	6.1	0.9	2.4	0.4	2.4
Textiles etc. (18)	4.7	1.4	1.8	0.8	0.6
Motor Vehicles (43)	4.4	0.9	3.2	0.1	0.2
Instruments (20)	4.4	0.4	2.8	0.5	0.8
Aircraft (19)	2.9	0.3	1.8	0.1	0.7
All Firms (587)	**11.0**	**4.1**	**5.6**	**0.3**	**0.9**

Note: The parenthesis contain the number of firms in each product group.

Furthermore Table 10 shows that, again contrary to current fashion, firms making products with the highest technology intensities are amongst those with the lowest degrees of internationalisation of their underlying technological activities: those producing aircraft, instruments, motor vehicles, computers and other electrical products are all below the average for the population of firms as a whole. In all these products, links between R&D and design, on the one hand, and production, on the other, are particularly important in the launching of major new products, and benefit from geographical proximity. The one technology intensive exception is pharmaceutical products, where the share of foreign R&D is high, but where the from R&D and production are unimportant compared to the links with high quality basic research, and with nationally based agencies for testing and validation.

By contrast, we see in Table 10 a high proportion of foreign R&D in industries, where some localised technological activities are required, either to adapt products to differentiated local tastes, or to exploit local natural resources: food, drink and tobacco, building materials, mining and petroleum. All in all, these differences between product groups suggest a general theorem: *localised products require global R&D; global products do not.*

IV. CONCLUSIONS: THE IMPORTANCE OF LOCALISED (AND COMPLEX) LEARNING

The reasons for uneven technological development are to be found in the nature of technology itself. Technology is knowledge of how to make complex artefacts work. Its acquisition depends largely on *learning*: i.e. experience and trial and error. It is tacit (institution or person–embodied) rather than codified (easily transmitted information), and is cumulative and localised in its nature and effects. However, as in other economic activities, learning has become increasingly deliberate, differentiated, specialised and roundabout. This has at least three major implications for analysis and policy.

First, differences between firms and countries in the ***direction*** of learning should be considered the rule, rather than the exception, given the localised and cumulative nature of learning, and the variety of possible inducement mechanisms. Those in the home market are particularly important, given the continuing tendency for even very large and multinational firms to locate there a high proportion of their most strategic learning activities (R&D).

Second, differences between firms and countries in the ***rate*** of learning should also be considered the rule, rather than the exception, unless the laggards have a capacity for "learning to learn" [Stiglitz (1987)]. In operational terms, this means establishing the competencies, institutions and incentives that enable firms and countries to catch up, or keep up, with world best practice technology.

For firms, effective learning depends in part on spillovers –especially the mobility of experienced managers, engineers and workers amongst firms. But it also requires the allocation of resources to local (i.e. in–house) learning activities that are increasingly complex and difficult to understand, evaluate and manage.

For countries, the possibilities of positive spillovers are more limited. Multinational firms can and do make a contribution but –as we have seen– not very much through foreign based R&D activities. And the possibilities of learning through (foreign) example or experience of how –in financial institutions and management– to evaluate and manage complex and difficult learning activities are

69

more limited than within countries

     Third, theories and models of endogenous technical change should face up to the challenge of explaining the uneven rate of development amongst countries and associated firms that we observe in practice. In doing so, they should incorporate not just the incentives and externalities associated with localised learning, but also –and above all– the differing *institutional competencies* across countries to evaluate realistically and implement effectively the economic benefits of learning activities that are increasingly complex, long–term and roundabout.

REFERENCES

ABERNATHY, W.; HAYES, R. (1980), "Managing our Way to Economic Decline", **Harvard Business Review**, July/August.

ARCHIBUGI, D.; PIANTA, M. (1992), **The Technological Specialisation of Advanced Countries**, Dordrecht, Kluwer Academic Publishers.

BELL, M.; PAVITT, K. (1993), "Accumulating Technological Capability in Developing Countries", Proceedings of the World Bank Annual Conference on Development Economics, 1992. Washington, (forthcoming).

BERTIN, G.; WYATT, S. (1988), **Multinationals and Industrial Property**, Hemel Hempstead, Harvester–Wheatsheaf.

CANTWELL, J. (1989), **Technological Innovation and Multinational Corporations**, Oxford, Blackwell.

CARLSSON, B.; HENRIKSSON, R. (1991), **Development Blocks and Industrial Transformation: the Dahmenian Approach to Economic Development**, Stockholm, The Industrial Institute for Economic and Social Research (IUI).

CLARK, K.; FUJIMOTO, T.; CHEW, W. (1987), "Product Development in the World Auto Industry", **Brookings Papers on Economic Activity**, no. 3.

CORBETT, J.; MAYER, C. (1991), **Financial Reform in Eastern Europe**. Discussion Paper no. 603. Centre for Economic policy Research (CEPR), London.

DAHLMAN, C.; ROSS–LARSEN, B.; WESTPHAL, L. (1987), "Managing Technological Development: Lessons from Newly Industrialising Countries", **World Development**, 15, pp. 759–775.

DOSI, G.; PAVITT, K.; SOETE, L. (1990), **The Economics of Technical Change and International Trade**, Wheatsheaf.

ERGAS, H. (1984), **Why do some countries innovate more than others**, Centre for European Policy Studies, Paper no. 5, Brussels.

FAGERBERG, J. (1987), "A Technology Gap Approach to Why Growth Rates Differ" in C. Freeman (ed.), **Output Measurement in Science and Technology: Essays in Honour of Y. Fabian**, North–Holland, Amsterdam.

FAGERBERG, J. (1988), "International Competitiveness", **Economic Journal**, 98, pp. 355–374.

FRANKO, L. (1989), "Global Corporate Competition: who's winning, who's losing, and the R&D factor as one reason why", **Strategic Management Journal**, 10, pp. 449-474.

FREEMAN, C. (ed.) (1987), **Output Measurement in Science and Technology: Essays in Honour of Y. Fabian**, North–Holland, Amsterdam.

GRILLICHES, Z. (1990), "Patent Statistics as Economic Indicators: a Survey", **Journal of Economic Literature**, 28, pp. 1661–1707.

JUSTMAN, M.; TEUBAL, M. (1991), "A Structuralist Perspective on the Role of Technology in Economic Growth and Development", **World Development**, 19, pp. 1167–1183.

LEE, K. (1993), D. Phil. Thesis (in preparation), Science Policy Research Unit, University of Sussex.

KITTI, C.; SCHIFFEL, D. (1978), "Rates of Invention: International Patent Comparisons", **Research Policy**, 7, pp. 323–340.

LAWRENCE. P. (1980), **Managers and Management in W. Germany**, London, Croom Helm.

NELSON, R. (1990), "US Technological Leadership: where did it come from, and where did it go?, **Research Policy**, 19, pp. 117–132.

NAKAOKA, T. (1987), "On Technological Leaps of Japan as a Developing Country", **Osaka City University Economic Review**, 22, pp. 1–25.

NARIN, F.; OLIVASTRO, D. (1988), "Technology Indicators based on Patents and Patent Citations" in A. Van Raan (ed.), **Handbook of Quantitative Studies of Science and Technology**, Amsterdam, North Holland.

PATEL, P. (1993), "Localised Production of Technology for Global Markets", (mimeo), Sussex University, Science Policy Research Unit.

PATEL, P.; PAVITT, K., (1991a), "Large Firms in the Production of the World's Technology: an Important Case of 'Non–Globalisation'", **Journal of International Business Studies**, 22: 1–21.

PATEL, P.; PAVITT, K., (1991b), "The Limited Importance of Large Firms in Canadian Technological Activities" in D. McFetridge (ed.), **Foreign Investment, Technology and Economic Growth**, Calgary, University of Calgary Press.

PATEL, P.; PAVITT, K. (1992), "The Innovative Performance of the World's

Largest Firms: Some New Evidence", **The Economics of Innovation and New Technology**, 2: 91–102.

PAVITT, K. (1988a), "Uses and Abuses of Patent Statistics" in A. Van Raan (ed.), **Handbook of Quantitative Studies of Science and Technology**, North–Holland, Amsterdam.

PAVITT, K. (1988b), "International Patterns of Technological Accumulation" in N. Hood and J.–E. Vahlne (eds.), **Strategies in Global Competition**, Croom Helm, Beckenham.

PAVITT, K.; PATEL, P. (1988), "The International Distribution and Determinants of Technological Activities", **Oxford Review of Economic Policy**, 4, pp. 35–55.

PORTER, M. (1990), **The Competitive Advantage of Nations**, London, Macmillan.

POSNER, M. (1961), "International Trade and Technical Change", **Oxford Economic Papers**, 13, pp. 323–341.

PRAIS, S. (1981), "Vocational Qualifications of the Labour Force in Britain and Germany", **National Institute Economic Review**, 98, pp. 47–59.

ROMER, P. (1990), "Endogenous Technological Change", **Journal of Political Economy**, 98, pp. 71–102.

SCHMOOKLER, J. (1966), **Invention and Economic Growth**, Boston, Mass., Harvard UP.

SHARP, M., "Europe: a Renaissance?", **Science and Public Policy**, 18, pp 393-400.

SOETE, L. (1981), "A General Test of Technological Gap Trade Theory", **Weltwirtschaftliches Archiv.**, 117, pp. 638–666.

STIGLITZ, J. (1987), "Learning to learn, localised learning and technological progress" in P. Dasgupta and P. Stoneman (eds.), **Economic Policy and Technological Performance**, Cambridge, Cambridge University Press.

VERNON, R. (1966), "International Investment and International Trade in the Product Cycle", **Quarterly Journal of Economics**, 80, pp. 190–207.

5 THE CHANGING ECONOMICS OF TECHNOLOGICAL LEARNING: IMPLICATIONS FOR THE DISTRIBUTION OF INNOVATIVE CAPABILITIES IN EUROPE

R. Cowan (University of Western Ontario)
and Dominique Foray (CNRS, IRIS TS and OECD)

When a new technology first appears as a possible solution to economic or technical problems, it is typically not well–understood. The degree to which any particular technology is subject to uncertainty about its characteristics and costs varies with the extent to which the technology is new and novel, but for some technologies it can be extreme. When this is the case, developing the technology will involve a considerable amount of learning about its functions, performance, and operational characteristics.

We can characterize the process of learning about a technology as involving two essential dichotomies. The first concerns whether the knowledge is created through deliberate "off–line" experimentation or is generated "on line" as a by–product of economic activities through learning–by–doing and learning–by–using. The second dichotomy concerns the life cycle of the technology. Learning through diversity and learning from standardization are the two sources of learning, which can be thought of as operating at different phases in the development of a particular area of technological (or scientific) practice.

This characterization relates to traditional analyses of technical change through the concerns expressed in that research. The concerns with the under–supply of learning–by–doing, or the separation between production and learning, all depend on the first dichotomy. On the other hand, the concern with the effect of standardization and the loss of variety on technological change depends on the existence of the second.

While these two dichotomies have not generally been explicitly enunciated, until recently they have applied well to the generation of economically valuable knowledge. In this paper we argue that the evolution of learning technologies in recent years has undermined these classifications and is causing the collapse of both dichotomies. Changes in the technologies of learning have increased the variety and the complexity of the situations in which learning can occur. In particular, new opportunities of "learning continually", through the methods of "on–line" experiments, and of maintaining technological diversity, through the methods of "options generation" are the base of a new learning paradigm.

75

X. Vence-Deza and J. S. Metcalfe (eds.), Wealth from Diversity, 75–101.

These tendencies have, in turn, important implications for the international distribution of innovation capabilities within Europe. As it affected the periphery, E.C. industrial policy can be seen as having two effects [Foray, Rutsuaert and Soete (1994)]: providing endogeneous resources for manufacturing activities, as distinct from activities of knowledge production: and contributing to the maintenance of technological diversity within Europe, at the cost of keeping the periphery off the technological frontier. Clearly, these combine to effectively exclude the periphery from activities of knowledge production. The collapse of the old learning paradigm changes the rules of the game, however. In particular, the increasing value of "on-line" learning (learning occuring during the manufacturing process) may induce an increase in the relative weight of "countries– with little R&D" in the European distribution of innovative capabilities. If this is the case, then the emergence of the new learning paradigm may provide an opportunity for the European periphery to catch up with the centre.

We begin the paper with an analysis of the two dichotomies and relate them to the literature on the economics of technical change, and the concerns expressed therein. We then address the new technologies of learning and argue that changes from the old to the new learning paradigms will reduce the relevance of the two dichotomies. Finally, we suggest that while the collapse of the dichotomies may facilitate catching up for the European periphery, it creates new problems as well.

I. LEARNING FROM DELIBERATE EXPERIMENTATION AND ANALYSIS VERSUS LEARNING AS A BY–PRODUCT OF ECONOMIC ACTIVITIES

The first dichotomy can be seen as a description of "where" learning occurs. Learning can either be deliberate, stemming from activities pursued for the sake of gathering information, or it can be a by-product of activities pursued for other reasons. This distinction is often expressed in the physical location of the learning –either in some form of an R&D establishment, or at the place of production or consumption that uses the technology in question.

I.1 Learning from Deliberate Experimentation and Analysis

Any attempt to create a new, economically valuable product or process, which inevitably involves generating new knowledge, will take place in the context of some background knowledge[19]. This background knowledge can be characterized as a probability distribution summarizing information regarding the probabilities of success of all of the potential ways of generating the new knowledge. Deliberate learning can be seen as aimed at one of three things: improving the state of

---

[19] Consider, for example, an attempt to create a new alloy having particular properties. There will be background knowledge about things like the make–up of alloys having similar properties, and the reactions of compounds to things like heat and catalysts. The new knowledge needed is how exactly to produce a metal with the desired qualities.

background knowledge; improving the tools with which to develop or exploit that background knowledge; and attempting to exploit it.

In the pursuit of these aims, three types of technical knowledge are active: generic technologie[20]; infratechnolgies[21]; and applied knowledge[22]. While these types of knowledge depend on each other, the dependence in not linear. For instance, in order to produce generic knowledge, which appears to be the most basic of the three, and upon which the others depend, it is frequently necessary to overcome a critical lack of infratechnologies and research instrumentats. Thus a phase of acquisition of certain technical, and in some sense applied, knowledge, may precede basic, and applied, research[23].

Deliberate experimentation consists in developing prototypes and demonstrators and carrying out simulated and real experiments to collect and record the performance characteristics of technologies under examination. The resulting scientific and engineering knowledge forms a basis for systematic technological development. The knowledge produced by experiments and simulation is supposed to ensure the formation of generalizing rules and hypotheses and ultimately to support the construction of predictive models about the performance of a new technology. Thus experimentation plays an important role in deepening scientific understanding of technological operations and processes.

A central feature of this side of the dichotomy is the deliberate and controlled search for knowledge. But the discovery of new knowledge or information can have two sources: systematic, rational enquiry and observation; and what Schelling refers to as accidental discovery [Schelling (1994)]. Accidental learning arises from attempts to acquire directly the knowledge needed to produce an economically valuable process or product rather than by systematically exploring the probability distribution that would tell us the most likely place to look for it. Any such attempt may or may not produce the product or process, but it will produce information. Though our perception is of a continuum between the two extremes of exploring the probability distribution and attempting to produce the knowledge-output directly, it is useful to construct a conceptual distinction between these two modes of learning.

---

[20] The elaboration of generic knowledge is the first phase in technology research; the objective is to show that the concept for an eventual market application "works" in a laboratory environment and thus reduce the typically large technical risks before moving on to the more applied phases of R&D [See Tassey (1992)].

[21] Infratechnologies include practices and techniques, basic data, measurement methods, test methods, and measurement–related concepts which increase the productivity or efficiency of each phase of the R&D, production, and the market development stages of economic activity [Tassey (1992)].

[22] This knowledge supports the conversion of generic technology into specific prototype products and processes with fairly well–defined performance parameters. It involves production and cost considerations. The research output is a commercial prototype in the sense that proof of a commercial concept has been achieved [Tassey (1992)].

[23] According to Rosenberg (1992), "the conduct of scientific research generally requires some antecedent investment in specific equipment for purposes of enhancing the ability to observe and measure specific categories of natural phenomena".

We can consider the process of knowledge generation as a compound event (A) which consists of the joint events [Hirshleifer (1971)]:

– state (a) is true, which means, for example, that it is possible to create a new alloy (basic research allows agents to assign probabilities (Pr(a), 1–Pr(a)) to the underlying states of the world);

– and, A, this fact is successfully exploited (the alloy is actually created). Thus, Pr(A)<Pr(a).

Research by accident is considered here as a deliberate experiment, in which, however, the agent ignores the potential to generate information about Pr(a) during this attempt to make the alloy (there is perhaps little information about the probability Pr(a)). There is, we should point out, a difference between the process of learning from disaster[24] and the process of research by accident because the latter occurs within the framework of deliberate experimentation[25].

By contrast, a discovery based on a rational exploration corresponds to the compound event described above (discovery occurs after having gained information about Pr(a)). In this schema, specific functions are assumed by basic research (assigning probabilities to the states of the world) and by applied research, and development (exploiting successfully the basic information about the states of the world). Thus, the R&D manager must continually trade off between allocation of resources to basic research for improving the state of background knowledge, and thereby increasing the probability of success of applied research and the immediate ("in the dark") allocation of resources to a given domain of applied research. He must also trade off between the production of research tools (infratechnologies) and the undertaking of research, bounded by whatever limitations there may be in the availability of those tools.

I.2 Experiential Learning as By–Product of Economic Activities

The second side of the first dichotomy refers to by–product learning as the second main mechanism for reducing uncertainty about the characteristics of a new technology. On this side of the dichotomy, the motivation for acting is not to acquire new knowledge but rather some other activity: to produce or to consume. As a by-product of economic activities, this mode of learning has two main aspects: (a) learning–by–doing, which is a form of learning that takes place at the manufacturing

---

[24] "Learning from disaster" refers to the redirection of attention as a result of experience with unexpected events such as airplane crashes, nuclear power plants accidents, etc. resulting in the discovery of unanticipated phenomena. [See P. David, G. Rothwell and R. Maude–Griffin (1991)].

[25] "Accident" is often, colloquially at least, associated with something bad (as in "car accident"). This not part of the strict definition but the association is strong. We are not using it in that sense but rather in the sense of "something (good or bad) haphazard".

78

stage; and (b) learning–by–using, which is linked to the irreplaceable role of users and adopters in the process of knowledge creation. The user has specific, sometimes idyosyncratic knowledge and masters those situations requiring the local implementation of the technological processes and objects. By interacting with the producer, he will engender learning–by–using mechanisms, of which Rosenberg, Lundvall and Von Hippel have demonstrated the significance. Users and adopters are a critical link in the chain of positive feedbacks which is at the root of the dynamic evolution of technology.

Two reasons seem to explain the importance of this sort of learning in the process of technological change.

The first one deals with the fact that new technologies are typically very primitive at the time of their birth. Thus, the main part of knowledge about the potential functions and performances of a new technology is generated during the process of use and diffusion which will facilitate the elimination of "bugs", as well as generate a flow of improvements in technical and service characteristics. Rosenberg explores the magnitude and the effects of learning by using in the particular case of aeronautics: "The confidence of designers in the structural integrity of a new aircraft is an increasing function of elapsed time and use. Prolonged experience with a new design reduces uncertainties concerning perfor-mance and potential, and generates increasing confidence concerning the feasibility of design changes that improve the plane's capacity. The stretching of aircraft, so critical to the economics of the industry, has been closely tied to the growing confidence in performance generated by learning by using" [Rosenberg (1982)].

The second reason is clearly demonstrated by T. Schelling (1994). In the general case of an existing technology that causes problem and needs to be replaced with an alternative, engineers are preoccupied with simply accomplishing with the new technology what the old did, and, consequently, can miss opportunities inherent in the new technology. Only a period of long use can reveal all the effective functions with respect to which the new technology has the potential to be superior to the old one.

I.3 Functional Assignment

Having made a distinction between deliberate experimentation (whether controlled or "by accident"), referring to an organized, "off–line" research process, and experiential learning as a by–product of economic activities, occuring "on–line" in the course of the production and diffusion of new methods and products, we can define a sort of functional assignment, a division of labour among the various processes of learning, which is based on the trade–off between productivity and knowledge–production. According to Arrow (1969), deliberate experiments are situations in which the actual output (e.g. nylon) is of negligible importance (in motivating the actors) relative to the information. In the case of experiential learning as byproduct, the opposite holds; the motivation for engaging in the activity is the physical output, but there is an additional gain, which may be relatively small, in information which reduces the cost of further production.

Intermediate cases are possible, the most obvious example being the pilot plant, in which, while the main goal is to generate knowledge about production processes and to learn something about the probability distribution of outcomes for future repetitions of the activity, the output of the plant is known (or at least expected) to be economically valuable.

Figure 1: Functional Assignment in the Economics of Technological Learning

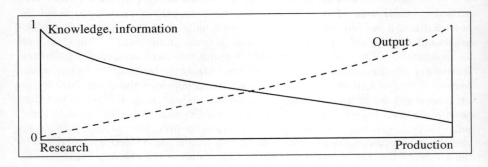

This figure describes a sort of world where the goals of improving static efficiency in production and consumption and generating new ideas and findings to put them in the economy are strictly separate. Here, the trade–off between production and knowledge acquisition is particularly stark–resources are allocated exclusively to one activity or the other. Within the management of R&D on the one hand, and of the manufacturing process on the other, no such trade–off problem exists. It is important to note that this "division of labour" was not only an abstract representation of the world; it was also a basic principle of the functioning of the real economy of mass production and consumption.

I.4 Economic Implications: Feedback Failures

The description of learning as having these two sources creates the vision of a system having several parts which are interlinked. Learning in one part of the system affects the benefits to learning activities pursued elsewhere. It is clear that to exploit such a system fully these links must be active. This is the simple fact that underlies concerns in the literature regarding knowledge feedbacks. The difficulties activating these links are worst when the first dichotomy is strong.

Complementarities between activities in the two spheres of learning have the potential to form many "virtuous circles", which can play a role in raising the performance of innovation systems. As David, Mowery and Steinmueller (1992) have demonstrated, one can impute to basic research an informational payoff that increases the efficacy of resources allocated to applied research: basic research produces knowledge like surveying; it generates maps that raise the return to further investment in exploration and exploitation. Applied research, in turn, produces instruments, prototypes and data (infratechnology) –as well as new observational phenomena– that improve the marginal social efficiency of investments in basic research. The production of scientific instrumentation lies at the heart of these feedback complexes between basic research and industrial development, and has contributed to accelerating the pace of productivity improvement in research activity itself, as both David (1993) and Brooks (1994) have pointed out. For instance, robotics or laser technology are potentially effective "bridging agents", which support the connection between different communities of researchers and engineers and may generate virtuous circles of learning.

There is another virtuous circle that can operate between users and producers. As a technology is used, users' experiences indicate potential improvements that can be made to it. If they are made, this will increase the value of the technology to other potential users, and will therby increase incentives to adopt it. Increases in adoption, of course, imply increases in use and further learning by using. This is, potentially at least, a second virtuous circle whereby learning increases use, and use increases learning.

The potential benefits of these feedback loops will only be realized, though, if there is a smooth flow in the intensive distribution of knowledge between the entities involved in the different forms of research, and between users and producers. With reference to the first communication, in an extremely dichotomous situation, we can observe a failure of this feedback stemming from the nature of the knowledge produced in the two locations. It is relatively tacit knowledge that is generated on–line, and this sort of knowledge is difficult to incorporate into the process of deliberate search[26]. Deliberate search tends to place more emphasis on

---

[26] Tacit knowledge cannot be dissociated from the work practices of research and production units. These forms of knowledge are acquired experientially, and transferred by demonstration, rather than being reduced immediately or even eventually to conscious and codified methods and procedures. Tacit knowledge therefore cannot be expressed outside

codified, (and therefore easily transmittable) knowledge[27].

The failure can occur even under a more subtle mechanism of assigning resources to different learning activities. The nature of pecuniary and non–pecuniary rewards will determine the extent to which and the way in which knowledge is codified [Dasgupta and David (1994)]. As a rough generalization, in deliberate search resources are allocated to the costly process of codification, since the rewards accrue from publication and dissemination, which is only possible if knowledge is codified. On the other hand, few resources are devoted to codifying the tacit knowledge acquired during production and consumption activities, since the rewards from dissemination (at least of knowledge learned by producers) is often negative.

Thus, especially from the point of view of those doing deliberate search, the activities of production and consumption are not thought of as generating knowledge that can be easily used elsewhere. There can be a sort of vicious circle. The part of the knowledge stock that is produced "outside" (not in places dedicated to knowledge production) remains invisible or tacit. Its invisibility implies that when resources are being allocated among learning activities, this location is overlooked. Thus it is "underfunded" and produces less than the optimal amount. This can generate a further decrease in the amount of learning that takes place "outside", and the cycle continues until "outside" locations are perceived as contributing nothing to technological advances. A system of by–product learning that is not explicitly financed means that market processes do not as a rule lead to socially optimal rates of learning: (a) private producers may be financially constrained to minimize short-run production costs, which makes them myopic in their evaluation of the learning component of production; (b) externalities and spillovers mean that (even with foresight and perfect capital markets) private strategic behavior does not yield socially optimal learning rates [Arrow (1962)].

A strict functional assignment implies a strong separation of functions, personnel, and location of learning. This, in its most extreme form, constitutes a feedback failure and an inability (or more generally a difficulty) in generating the virtuous circles described above.

---

the production context in which it is generated. [See P.A. David (1993), P.A. David and D. Foray (1994)].

[27] Codification of knowledge is a step in the process of reduction and conversion which renders the transmission, verification, storage and reproduction of information especially easy. Codified information typically has been organised and expressed in a format that is compact and standardized to facilitate and reduce the cost of such operations. The presentation of knowledge in codified form will depend on the cost and benefits of doing so of course. The standardization of language and expressions under different forms strongly reduces the costs of codification. This in an area in which the revolution in information technology, including such advances as the substitution of graphic representation for natural language, the development of expert systems, etc. has made itself very strongly felt. [See P.A. David (1993), P.A. David and D. Foray (1994)].

II. LEARNING FROM DIVERSITY VERSUS LEARNING FROM STANDARDIZATION

The second of the two dichotomies characterising technological learning concerns the life–cycle of the product or technology, and the different types of learning that tend to occur during different phases. When there is a large amount of uncertainty about the technical functions and economic merits of a new technology, its first introduction typically ushers in a period during which many variants are formulated, tried out and even tested with potential consumers or users. After this period of experimentation, one, or perhaps a small number of variants will emerge as "standard practice" or "dominant design". The selection can be passive, through the competitive market mechanism, for example; or active, as in the case when a dominant economic or political actor decides that a particular variant should become the standard. When the development of a technology conforms to this pattern, two types of learning are distinguishable.

II.1 Diversity

The first type of learning is extensive learning or "learning from diversity"; which involves experimentation with a variety of options, and through the results of the experimentation, leads to the elimination of certain avenues of development. In this phase, the objective is to gain broad knowledge of many possible avenues by which the problem at hand can be attacked. The point is that any problem has many potential solutions, and it is necessary to learn something about many of them before it is possible to make a sensible choice regarding which is likely to be the most effective. Thus at this stage, the goals of a decision–maker tend to be very broad and involve exploring many variants of the technology. The question being addressed, to speak as if there is a central authority steering the process (this is not always the case, of course), is which of the many possible paths to follow.

II.2 Standardization

The second type of learning can be called intensive learning or "learning from standardization", in which attention is concentrated on one technological variant, making it easier to identify empirical irregularities, anomalies and problem areas deserving further investigation, correction and elaboration [Cowan (1991), David and Rothwell (1993)]. Here, a choice has been made, either actively or passively, and one solution has been selected. The learning here involves discovery, no longer of broad implications regarding exploration of different solutions, but rather of the details about how to make this particular solution most effective. Here a path has been chosen, and the effect of learning is to make the most of this path, through careful exploitation of all of its potential.

II.3 From Diversity to Standardization

The optimal timing of a changeover from a diversity of technical solutions to standardisation on one technology depends on the amount of experimentation that has been done with different variants of the technology. Introducing a standard too early could prematurely end the period of experimentation and lead to the diffusion of an inferior technology whereas late introduction may result in excess of novelty, and the formation of wrong expectations about the chances of competing technologies –there will be users who adopt technologies which will not be selected as the future standard.

A shift similar to the one described here can be seen in the history of nuclear reactor technology. In 1955, at the first international conference on nuclear power in Geneva, about 100 types of reactors were discussed. Three years later, the number was down to about 12. When the U.S. Navy decided to produce a nuclear powered submarine, after initial experiments in the AEC with six technological variants, two vareties were considered, and after a single experience with each, only one was intensively explored and developed over the following decade [Cowan (1990)].

II.4 Economic Implications

As described above, technical change and evolution can be viewed as a process of exploring a wide range of technical options. Over time, the cumulative nature of the learning processes about the merits of competing technologies leads on the one hand to ε reduction of diversity and to a subsequent loss of development power of the system, while it leads on the other hand to an increase of the efficiency of the technology selected as the standard. Thus, standardization has costs and benefits. It allows the industry to decrease production costs (increasing returns to scale), and to facilitate the diffusion of innovation (due to the existence of technical standards). Standardization, however, entails a loss of diversity. Technological variants having unique properties may be lost and never properly explored. The long run effect is that the scope for future developments will be narrowed. Technological advances that depend on the prior development of these unique, and now lost, properties are put in jeopardy [Cowan (1991)]. This suggests a need for policies to maintain diversity, but with the foresight that at some point in time diversity can decrease benefits by preventing economies of scale, a reduction in costs through intensive learning about a technical option, and the potential for network externalities.

Given that a policy of encouraging diversity within a technological family will eventually have to be abandonned, one problem is how to select which technical options to support. A solution which alleviates some of the problems for future developments is to determine the technological distance between several technologies and to support those that are relatively more distant than others on the assumption that the loss of diversity among technologies that are very similar is acceptable. This calls for some measures of collective dissimilarity, as argued and demonstrated by Weitzman (1993):

Figure 2: Modes of Generation of Knowlwedge

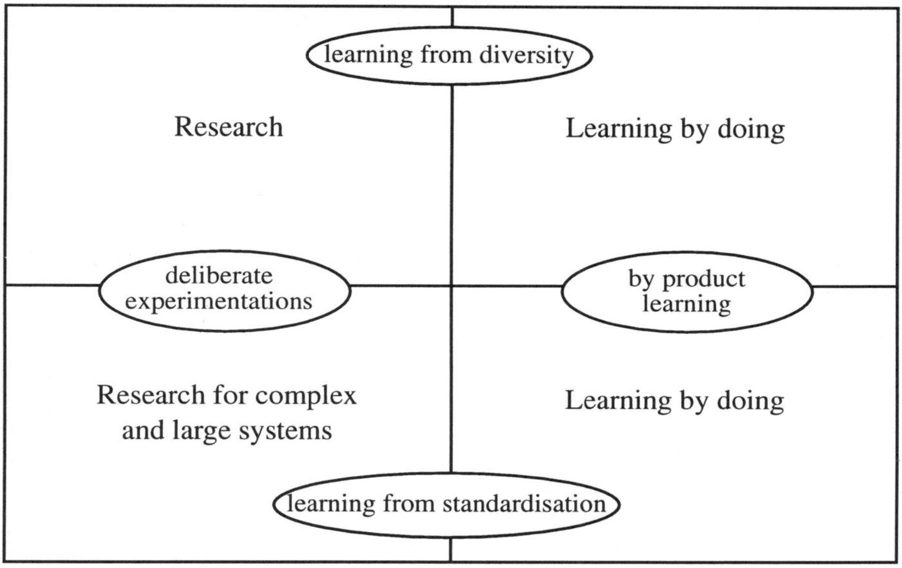

III. TYPES OF LEARNING: A TENTATIVE SCHEMA

Figure 2 represents a tentative schema within which we may locate different modes of generating knowledge. The vertical axis deals with the evolution from diversity to standardization, while the horizontal axis concerns the progression from experimentation to by–product (off–line to on–line) learning. There are two principal modes, combining diversity and experimentation (which corresponds to the basic definition of research, as producing a great deal of information) on the one hand, and standardization and by–product learning (which corresponds to the basic definition of learning–by–doing, as producing mainly actual outputs), on the other hand. It is possible to imagine two other modes, however, involving a mix of information and actual output: combining by–product learning and diversity means that some learning–by–using processes can lead to product differentiation for example [Von Hippel (1988)]; combining standardization and experimentation reveals a specific process of knowledge generation in the case of large complex system, where each new plant, or system or program –although produced in a recurring manner– in fact has many features of an experiment (each artefact differs in identifiable, predictable, and sometimes planned ways; and each one generates information that is associated with these differences).

The traditional economics of technological learning –concerned with under-investment in on–line learning, feedbacks failures between research and development, and the loss of variety– is developed on the basis of this double-dichotomy. It is our contention that today considerable changes in learning technologies are affecting the way learning takes place, and so are removing some of the old concerns, and replacing them with new ones.

IV. RECENT CHANGES IN THE TECHNOLOGY OF LEARNING

Recent changes in the technology of learning have led to the collapse, or partial collapse, of the two dichotomies discussed above. There are three main factors that contribute to this collapse: the convergence of learning and production technologies; our increased ability to codify knowledge; and the progress in information technology that is extending the power of electronic networks as research tools.

IV.1 Convergence of Learning and Production Technologies

In the last two decades the use of computers and computer technology has grown dramatically. This growth has occurred both in production technology and in R&D technologies. Computers are used to control and monitor experiments; they are also used in precisely the same way to control and monitor production processes (numerically controlled machine tools provide the most direct analogue). The digital telephone switch is simply a computer designed to process and transmit particular

86

types of information, but using relatively standard techniques. There has, thus, been a growing convergence between the technologies of learning and knowledge generation, and technologies of production. Initially, and most obviously, as computers spread throughout the production process, and are used widely in research, we observe convergence in the technologies used to process information. This, of course, implies the need for convergence in the way information is recorded and stored. In turn, this creates a situation in which the same types of information (since only certain types of information can be efficiently stored in machine readable format) are being used, generated, processed and recorded in the two places. This suggests that the transfer of information between production and consumption on the one hand and R&D activities on the other should be easier. As an illustration, consider the software industry where strong effects of learning by using occur, thanks to the flow of information stemming from the customers: many computer companies routinely provide extensive software support that involves software modification when bugs are discovered by customers [Rosenberg (1982)].

IV.2 Codification, Algorithmic Successes, etc.

The discussion in the previous sections draws attention to the increased codification of knowledge. In the move from crafts industries to automated fordist manufacturing, what was once tacit knowledge came to be embedded in machines. The digital revolution has continued and intensified this move towards codification. In this regard, we observe a self–reinforcing cycle.

Advances in information science and linguistics have increased our ability to codify and formalize knowledge and information. We have learned to describe more, and more complex, things digitally. This has improved our ability to test models and hypotheses through simulations (using virtual prototypes) rather than through practices, and to do so on the basis of an arbitrarily large range of assumed conditions. To understand these improvements, it is important to realize that many problem–solving tasks involve a degree of complexity far in excess of current computational capacity. It is a common feature of optimisation problems that the number of possible solutions and solution paths increases as an exponential function of the number of independent variables –so that the extrema rapidly become too numerous to handle. Therefore, the ability to solve these problems depends on the availability of algorithms which provide reasonable approximations to the analytical optimum. The development of these algorithms is one of the greatest achievements of modern mathematics; a characteristic they share is that while the analytical complexity of the problems being solved increases exponentially in the number of variables, the computational requirements of the algorithm increase by some linear function of the problem's size [Ergas (1994)]. These new research tools permit more quickly focussed and hence more productive search, thereby cutting the delay involved in going from the initial specification to the agreed–upon prototype. Moreover by reducing the time and cost required for new product and process design, these tools encourage producers to experiment across a broad front –to

87

develop and try many variants of a new design rather than only one or a few [Arora and Gambardella (1983)].

The ability to perform many simulations is of no value, however, unless the simulations are good models of reality. Thus the development of good predictive models of the world is crucial to making the previous two advances of value. But of course, these advances themselves make the development of good models easier and more feasible. Codified models can be simulated to see whether they give unreasonable results under reasonable parameter values. If they do, the source of the results can be found, and then changed before doing physical experiments.

Thus, advances in our ability to describe and codify knowledge, combined with advances in algorithms have increased our ability to generate good models of physical processes. Improvements in those models make advances in codification and algorithms more useful, as they can be implemented in a wider variety of places. A virtuous circle has formed which can be deployed in decreasing the costs of developing new products and processes.

Clearly, this explosion of the use of simulation methods does not eliminate the necessity of real experiments. In fact, the performances calculated by simulation can vary greatly according to the selected hypotheses and parameters. Therefore, an important task is the validation of those parameters via the correlation between simulation and real experimentation. The new research possibilities provided by the extension of simulation, however, can dramatically decrease the costs of basic research and increase the relative efficiency of systematic and controlled experiments relative to research by accident.

IV.3 Electronic Networks

The perception of an emerging new paradigm for technological learning is reinforced by considering ongoing developments in information and telecommunications technologies that are extending the power of electronic networks as research tools [David and Foray (1994)]. The network connects some information sources that are a mixture of publicly available (with and without access charge) information and private information shared by collaborators, including digitized reference volumes, books, scientific journals, libraries of working papers, images, video clips, sounds and voice recordings, raw data streams from scientific instruments and processed information for graphical displays, as well as electronic mail, and much else besides. These information sources, connected electronically as they are through the Internet (or the World Wide Web) represent components of an emerging, universally accessible digital library. What appears to lie ahead is the fusion of those research tools with enormously augmented capabilities for information acquisition and distribution beyond the spatial limits of the laboratory or research facility, and consequently a great acceleration of the potential rate of growth of the stocks of accessible of knowledge. It would seem to follow that cooperative research organizations will be best positioned to benefit from the information technology–intensive conduct of science and technology result.

These tendencies to technological convergence, knowledge codification and electronic networking induce the emergence of a new paradigm of knowledge acquisition which, in turn, greatly influence the economics of learning.

V. A NEW LEARNING PARADIGM

The changes in the technologies of learning described here imply changes in the economics of learning and industrial research. In general, the new tools of learning ease some of the problems in the economics of learning, through reducing costs of storing and transmitting knowledge. The schema (figure 2) presented above is particularly useful to assess the importance of these changes.

V.1 A More Complex "Socially Distributed" Knowledge Production System

It is our contention that the changes in learning technologies lead to the increasing importance of external capabilities (outside from the firm) in the innovation process by extending the "mobility" and the institutional diversity of knowledge. These changes conduce to a proliferation of new places having the explicit goal of producing knowledge and undertaking deliberate research activities. They conduce, thus, to a more complex socially distributed knowledge production system. This evolution can be seen as following two paths: on the one hand, the increasing value of on–line learning (by doing and by using); and on the other hand, the emergence of new forms of learning, which share features of both deliberate experimentations and by–product learning.

* The increasing value of on–line learning

As the technologies of research and production are converging, the value of on–line learning increases. First, if "running experiments" has become less costly, then the value or information in general increases. Information acquires value through being explored, refined, and integrated into the knowledge stock. If the cost of doing so decreases (due to an ability to replace some physical experimentation with simulation for example) then information, and activities that generate it, become more valuable. This implies, of course, that information generated on–line is of more value. Second, on–line learning is easier to codify in a way that is useful to knowledge generation that takes place off–line. Ease of integration of the two types of information arises for several reasons. As production controls become digitized they become easier to monitor. If output is similarly, monitored digitally (perhaps through automation of quality control procedures) it becomes relatively simple to generate considerable quantities of performance data. These data are in a form easily used in the off–line learning processes.

89

The convergence of research and production technologies implies an improvement in the feedback mechanisms between the two locations. As the information produced is more similar in form, it is easier to transmit from one location to the other, and it is more useful when it arrives.

We would expect, then, that since the value of on–line learning increases, and becomes more visible, there will be more of it. We should expect to see more learning in places where, formerly, learning appeared to be of little value in generating further knowledge. Further, if the cost of integrating the knowledge generated on–line into the off–line knowledge generation process decreases, they should become more and more integrated, and the first dichtomy should begin to collapse. With the collapse of the dichotomy, of course, concerns about feedback failures become less pressing.

* *"Off–line" by–product learning or "on–line" deliberate experimentation*

New forms of learning are emerging which are neither pure "off–line" experimentations nor pure "on–line" by–product learning. A major evolution in research and technological learning requires, however, a change in our representation. New forms of learning and experimentation are emerging with an ambiguous position. A characteristic they share is that they do not fit in with the functional assignment defined by Arrow. In certain circumstances (e.g. research in a large/complex technical system) an experiment whose only output is new knowledge may be too expensive to justify itself. This means that some kinds of experiments are feasible only with systems that are simultaneously producing salable products or services[28]. The operation of putting samples of new materials into a nuclear reactor, which is in operation, is a good case in point. Materials are exposed to various temperatures and conditions and then taken to be tested in the laboratory.

Maximizing the knowledge externality benefit, thus, requires the addition of instrumentation in order to take advantage of observational opportunities on the production line, or the slowing down of the production stream for the purpose of eliciting new knowledge that could not be obtained otherwise: either "off–line" by-product learning or "on–line" deliberate experiments or collateral experiments, those new forms of learning cannot be easily classified with respect to Arrow's dichotomy, since they produce both effective outputs and knowledge while tolerating a certain degree of deterioration of productivity. As the Japanese manufacturing experiences suggest, however, a great deal of added value in terms of knowledge may be obtained at very low cost with little sacrifice of product output by adding a certain amount of instrumentation and extra observing and reporting personnel to an otherwise routine production operation. In sum, these hybrid forms of learning allow one to generate knowledge about the performance characteristics of the technology

---

[28] This raises major questions about the economic viability of the French nuclear power plant "Super Phenix", which has been converted into a pure research instrument.

continuously, not only at the research stage but also during the total life cycle of the technology. Of course, the efficiency of those learning processes will be affected by the speed and integrity with which the information and data recorded are transmitted within the technological community involved and by the time lag between the perception (in real time) of a problem and the implementation of the relevant solution.

V.2 Option Generation and Recombination, and the Exploitation of Some Virtues of Both Diversity and Standardization

In the vertical dimension of the schema, the new R&D methods allows one to maintain the learning virtues of both diversity and standardization; i.e. to explore continually a large spectrum of technological variants without sacrificing the benefits derived from economies of scale, intensive learning about a technical option and network externalities. The new simulation methods improve the ability to generate a large range of alternatives (virtual diversity) to be explored and increase the efficiency of the process by which alternative design approaches are developed, tested and selected [Ergas (1994b)]. This implies the ability to do a more thorough exploration of diverse options before learning moves to the standardization phase. One finds in the literature on technical choice the suggestion that there is a tendency for standardization to occur too early in the life of a technology [Cowan (1991)]. The ability to do simulated experiments mitigates this tendency as it lowers the cost of learning through diversity. If some experiments can be simulated, and the information generated can be used to better focus resources devoted to physical experiments, we effectively lower the costs of experimenting, and effectively increase the total number of experiments that can be performed per unit calendar time. This has the effect of permitting a wider and more thorough exploration of available diversity, and of delaying the switch (in event time, which is what is important here) from diversity to standardization.

Here again, the new possibilities of simulation undermine the sharpness of this classification as well as the vision of a continuous and irreversible loss of diversity. On the basis of the virtual diversity which is created and explored by simulation, one can select which technical options to support, in order to deliberately narrow down to a small number of competing standards and carry these along in parallel on a provisional basis while continually carrying on learning–by–doing with both standards. Such parallel standards properly instrumented and monitored, can preserve some of the learning virtues of both diversity and standardization.

As our ability to describe the world and codify those descriptions increases, so does our ability to preserve diversity. This depends of course, on the quality of the models in which the description is located, but if these are good, we can keep variety alive in the form of simulation. The irrevocable loss of variety referred to above need no longer be so severe. It is less expensive to keep a simulation of a process running than it is to keep the process itself running. The more complex the technology the more this is pertinent. Secondly, if the simulation is well–written and documented, it is typically easier to re–start a simulation of a process, or simulated experiments, than it is to re–start the process itself. In the most extreme case, which, we grant, is never likely to be realized, variety need never be diminished, as preserving a simulation, the costs of which is small, will, because the knowledge is codified, effectively preserve the variety implicit therein.

Thus, new opportunities of "learning continually" (through the methods of "on–line" experimentations) and of maintaining technological diversity (through the methods of "options generation") are the basis of the new learning paradigm.

VI. IMPLICATIONS FOR THE DISTRIBUTION OF INNOVATIVE CAPABILITIES IN EUROPE

It is our contention that the collapse of both dichotomies has serious implications on the international distribution of innovative capabilities within Europe: the new learning paradigm may provide the European periphery with an opportunity to "catch up".

VI.1 The Catching up Process and the Conditions of Social Capabilities

According to M. Abramowitz (1986), the catching up hypothesis asserts that being backward in terms of level of productivity carries a potential for rapid advance. In a comparison across several countries, growth rates of productivity over a long period tend to be inversely related to initial levels of productivity.

This simple hypothesis, however, needs qualification. As observed by Abramowitz (1991), "*technological backwardness is not usually a mere accident. Tenacious societal characteristics normally account for a portion, perhaps a substantial portion of a country's past failure to achieve as high a level of productivity as economically more advanced countries have done. The same deficiencies, perhaps in attenuated form, normally remain to limit a backward country from making the full technological leap envisaged by the simple hypothesis. I have the name for these characteristics. I call them 'social capability'*".

With the expression "social capability", Abramowitz means the institutional and social requirements to take advantage of the opportunity provided by technological backwardness. According to Dosi and Freeman (1992), who carefully analysed the experience of East Asian countries, five channels of technological learning are particularly significant as necessary conditions for catching up: i) the education of large numbers of qualified engineers; ii) the efficient deployment of technical and scientific activities in productive sectors, i.e. the development of competence for technology acquisition and technical innovation within the enterprises themselves; iii) the private, public and hybrid infrastructure supporting

92

the coordination among firms in technological activities: high–tech consortia and technological clubs; a system of intellectual property right; institutions for establishing formal standards; as well as for producing infratechnologies, instrumentation, measurement and testing procedures; iv) practices of management combining R&D management with design, production and marketing operations; v) substantial investment in physical equipment (telecommunication infrastructure, computer hardware and software).

Summing up this discussion on catching up in new technology, the basis of success lies in a combination of institutional and social change, allowing a country to meet the requirements for organizational learning.

The Abramowitz argument is that a country, which is technologically backward, has potential for generating growth, provided its social capabilities are sufficiently great to permit successful exploitation of technology already in operation in richer countries.

Technological borrowing by followers is, of course, the essence of the catch–up hypothesis. According to Abramowitz, there are, however, other interactions between followers and leaders by way of the rivalry in trade, and by way of capital flows and population movements. Moreover the transfer of knowledge is not solely from leader to followers, and the net flows tend to shift as average productivity gaps become narrower: "*As technological gaps narrow, the direction of flows begins to change. Countries still a distance behind in average productivity move into the lead in particular branches and become sources of knowledge for older leaders*" [Abramowitz (1991)].

Thus, in a further step of the argument, the question which arises is no longer about the potential of the follower to exploit the technology already in operation elsewhere. Rather, the question is about the potential of the follower to shape the trajectory of technology –follower becomes leader– because its institutional structures form a coherent system with specific properties. The new leader has established distinct institutional arrangements, in order to meet the requirements of technological creativity (see above). These arrangements fit together and form even more elaborate **institutional clusters** that are self–reinforcing. A national system of innovation emerges.

VI.2 Catching up and the International Distribution of Innovative Capabilities in Europe

Such a model of catching up provides a rather optimistic view of the nature and characteristics of the process of technological accumulation in Europe, in particular if one takes into account the cases of the small less advanced countries and regions in Europe. For instance, the general statement of the FAST programme on globalisation is that on a European scale (including of course Southern Europe)

93

technological divergence has prevailed[29].

It seems to be true that many of the conditions laid down for catching up in Abramowitz argument are met in the case of the small, less advanced countries in Europe. These countries are involved to a greater and greater extent in European programs designed to support interactions in science and technology[30].

These countries meet the social capabilities requirements and are progressively building their own national systems of innovation. As suggested by J. Valls (1993) in the case of Spain: *"The national R&D Plan set new priorities, permitting publicly–owned research centres and universities to draw up and implement projects which were generally much larger than those hitherto. Moreover, the Plan provided incentives for collaboration between the private and public centres, particularly in the field of basic and applied research. The University Reform Law made university links to the outside world easier and more flexible, removing the legal obstacles to such collaboration and permitting teaching staff to receive payment for their work on joint private–public projects. This together with the subsequent creation of Technology Transfer Centers in the Universities, made company/university links in general and company/university links in particular far more dynamic and market oriented"*.

A rapid view of national technological specialization, however, gives a far less optimistic picture [Archibugi and Pianta (1992)]. In the case of Spain, a cluster of mechanical sectors emerges as the major area of Spanish technological specialization, including Special and General industrial machinery, Motor vehicles and Fabricated metal. Chemical and electronic classes show up as major weaknesses of the country. In the case of Portugal, the only sector where a high index of specialization is obtained is Organic Chemicals.

These countries are thus "locked in" to a narrow subsets of routines, goals and future growth trajectories in the medium tech. areas. The principal problem deals therefore with the transition from medium tech. to high tech. areas. Here resistance and obstacles are significant: European integration can re–inforce these specializations. For example, the participation of Spanish firms in Eureka are very much concerned with the mechanical sector.

---

[29] See volumes 20, 21 and 22 of the Prospective Dossier no. 2 of the FAST programme on Globalisation of Economy and Technology, which focus respectively on the cases of Ireland (O. Doherty and J. Mc.Devitt), Greece (T. Giannitsis) and Portugal (V. Corado-Simoes). On a general statement on convergence and divergence, see G. Dosi and C. Freeman (1992).

[30] For example, Spanish firms manage 253 links with other countries, in the framework of Eureka. Regarding the geographical nature of these links, only 23 are intra–peripheric (with Portugal, Ireland). See Eureka (1993).

VI.3 Increasing Value of On–Line Learning and Opportunity for Catching–up

If we look at the following table [Freeman and Hagedoorn (1992)], showing the (estimated) relative R&D efforts during the second half of the eighties for a large number of European countries, we get a rather alarming picture: the lesser developed countries in Europe spent less than 1% of their GNP on R&D during the second half of the eighties (Greece and Portugal spent less than 0,5%). Such a picture clearly suggests an inability of those countries to switch from a medium-tech. to a high–tech. regime of knowledge accumulation.

However, such a statement is based on a narrow perception of the knowledge generation system, as strictly located to the formal R&D institutions. In a world in which there is a strict division of labour between R&D institutions –which monopolize the activities of knowledge production– and other economic activities which are not thought of as generating knowledge, such an international distribution of R&D capabilities excludes some countries from the process of knowledge generation. Now, the advent of the new learning paradigm may greatly change our interpretation of the picture: in a world where on–line learning (learning by doing and learning by using) becomes more and more valuable in the process of knowledge generation, the importance of the international distribution of R&D capabilities could be over–estimated. We would expect that, since the value of on-line learning increases, the relative weight of countries with little R&D in the international distribution of innovative capabilities will increase, thanks to their manufacturing and consumption activities.

Table 1.– A Comparison of (Estimated) R&D Effort of Countries, R&D as % of GNP, Second Half of the Eighties

≥ 2 %		≥ 1 %		≥ 0.5 %	
France	2.3	Australia	1.1	Brazil	0.7
F.R. Germany	2.8	Austria	1.3	El Salvador	0.9
Israel	3.0	Belgium	1.5	Guatamala	0.5
Japan	2.9	Canada	1.4	Iceland	0.8
Netherlands	2.1	Denmark	1.3	India	0.9
Sweden	3.0	Finland	1.7	Ireland	0.8
Switzerland	2.3	Italy	1.5	Kuwait	0.9
United Kingdom	2.4	S. Korea	1.5	Mexico	0.6
United States	2.6	New Zealand	1.0	Singapore	0.5
		Norway	1.9	Spain	0.5
		Taiwan	1.1	Yugoslavia	0.8

Source: Freeman and Hagedoorn (1992).

Catching up has two aspects of course. The first is the simple increase of total factor productivity –using better techniques to produce the current spectrum of goods and services. The second is the introduction of new goods and services, which may already exist in leader countries, into the production economy. Both of these aspects involve knowledge use and generation: the first can be seen as process innovation; the second as product innovation. Thus the crux of the problem for a country trying to catch up lies in the production and use of knowledge and information. Of central importance, then, will be the national system of innovation [Dosi and Freeman (1992)].

We have argued above that as learning technologies evolve the importance of on–line learning will increase, and thus the importance of off–line learning will decrease. It follows that R&D spending similarly becomes less important. There are, however, two important caveats. The first is that there must be a line to learn on. The second is that there must be a critical mass of "knowledgeable people" to process the information generated on–line.

The critical mass issue is an important one as there are several different aspects of it: number of researchers; amount of data; number of users; and so on, each of which contributes to the creation of a community of active and efficient knowledge generation[31]. Futhermore, the kind of positive externalities associated with human capital arise not just from an absolute number of "knowledgeable people". Rather, they arise from the patterns of specialization and interaction which may occur; but whether these patterns of specialization and interaction do arise are matters of economic organization as well as agglomeration activities.

But the changes in learning technology may alleviate this problem considerably. First, the existence of electronic networks of researchers implies that the critical mass can more easily be geographically dispersed. Data can be accessed in other countries, and long–distance (inter or intra–national) research collaboration, and the attendant transmission of ideas, becomes much easier. This does not eliminate the need for personal contact, but it does increase significantly what can be done at a distance. The problem here, of course, is to create the networks. Second, in its most extreme form the collapse of the on–line/off–line dichotomy removes the distinction between researchers and users. This immediately increases the number of people with whom any agent can communicate ideas about the technology being used or developed. If the dichotomy and its functional assignment are strict, users interact with users and researchers interact with researchers: the separation occurs largely because of the type of information the two types of agents deal with. When the dichotomy breaks down, so does the difference in type of information, and so does the functional assignment. This increases the number of possible parties with whom any agent can usefully interact, and so alleviates the

---

[31] On this issue, see the seminal work of Rosenstein–Rodan (1943) about the industrialization of Eastern and Southeastern Europe; as well as more recent papers [Perez and Soete (1989), Stiglitz (1991), Dosi and Freeman (1992)].

critical mass problem.

The other caveat, the existence of a line to learn on, is perhaps more difficult from the point of view of the European periphery. This is not so in the sense of increasing total factor productivity –many lines of production certainly exist, and as on–line learning becomes more important, the smaller expenditures on R&D will be less of a bottleneck in increasing this productivity[32]. The problem arises in introducing new products. This is clearly a large part of the issue from the point of view of the periphery, as it currently tends to be specialized in medium tech manufacturing.

Starting up a new technology sector involves two types of difficulties. There are, first, fixed costs of manufacturing, and creating a market or capturing market share from incumbent firms. These problems exist in virtually any industry, and are not affected by the changes in the technology of learning that we have been discussing. There are also, though, knowledge problems: initial R&D; and rapidly capturing and deploying learning from production and use. These can be affected.

If more knowledge becomes codified and less tacit, we would expect a growing internationalization of the "public good" character of knowledge, i.e. its availability throughout the world to an ever larger group of companies[33]. This is particularly so if the knowledge used in the production process becomes codified and embedded in the technologies of production themselves. Adopting a strategy based solely on this fact does not permit one to be a leader, of course, though it may make it possible to be less far behind.

To be a leader it is necessary to generate new products and processes. The extent to which the change in learning technology affects this process will very much depend on the technology or industry in question. But for many industries, advances take place in a generational sort of way –a new product or process is not a radical departure from what has gone before, but rather builds on and extends knowledge and technology that was used in the production of the product or process it supersedes. When this is the case, the integration of learning on–line and learning off–line becomes extremely important. When this is difficult to do, only a small number of countries, those with the most advanced technologies, operating in the highest value–added industries, will be able to undertake it. If this becomes easier to do, more and more economies will find it possible.

---

[32] Off–line learning remains important of course, but as its relative importance declines, the inequalities in total learning activity that arise from the small expenditures on R&D among periphery countries becomes less extreme.

[33] On the implications of this tendency in the Information Technology Industries [see Steinmueller (1995)].

VI.4 Option Generation, Virtual Diversity and Opportunity for Catching–Up

Suppose part of the EC technology policy has been to (unwittingly perhaps) encourage diversity. Diversity is good because it keeps a stock of "genetic material" that can be used for future technical change; and it generates information of a certain kind. To date, diversity has been maintained by getting the periphery to do certain things that have kept it off the frontier. But the new technologies of learning suggest that this is no longer necessary. Diversity can be maintained "artificially" or "virtually". Policy can be re–directed so as to achieve this goal through means that do not have ill–effects on the periphery vis à vis the technological frontier.

Further, suppose we think of the technological frontier as the frontier between production technology and experimentation (i.e. technology that is tried and true in the production process, but the most recent vintage that meets that criterion). This means that the frontier lies at the frontier between diversity and standardization –if diversity is where learning is taking place, then it is learning about stuff beyond the frontier; if standardization is where learning is taking place, then it is learning about stuff on or inside the frontier. (It is probably this sense of the frontier that the periphery cares about). The collapse of the diversity/standardization dichotomy makes the notion of the frontier much harder to preserve. So "now there is no frontier"; technological advance becomes a much more fuzzy concept –it is harder to think of it as simply pushing out the frontier. If the idea of a frontier collapses, (or to put it another way, if there is experimentation going on "on the frontier") then there can be diversity among many producers who are all technologically advanced, since they are all experimenting. Thus the diversity comes, not from having people around using old and unusual technologies, but from having lots of people using cutting–edge technologies but experimenting with it in different ways.

However, while the collapse of the dichotomies provides a great opportunity of catching up (to the European periphery), it creates new problems as well: the most important policy orientation from the perspective of the emergence of the new learning paradigm deals with personnel training to support access to, and utilization of digital libraries. One might view the investment made in the formal education of the members of society, as required not simply to transmit what is presently thought to be useful knowledge, but to equip economic agents to retrive and utilize parts of knowledge stock that they may not perceive to be of present relevance but which have been stored for future retrieval in circumstances when it may become relevant. In other words, the accessibility of the extant codified knowledge stock may be indicated by the portion of population that has been trained to access and interpret it. This is a great policy challenge for the European periphery to be achieved, in order to exploit the opportunities of catching up, provided by the advent of the new learning paradigm.

REFERENCES

ABRAMOWITZ, M. (1986), "Catching up, forging ahead, and falling behind", **The Journal of Economic History**.

ABRAMOWITZ, M. (1991), "Following and leading" in Hanush (ed.), **Evolutionary Economics**, Cambridge University Press.

ARCHIBUGI, D.; PIANTA, M. (1992), **The technological specialization of advanced countries**, Kluwer Academic Publ.

ARORA, A.; GAMBARDELLA, A. (1993), **The changing technology of technological change**, Miméo.

ARROW, K. (1962), "Classificatory notes on the production and transmission of technological knowledge", **American Economic Review**.

ARROW, K. (1969), "The economic implications of learning by doing", **Review of Economic Studies**, vol. 29.

BROOKS, H. (1994), "The relationship between science and technology", **Research Policy**, Special issue in honor of N. Rosenberg.

COWAN, R. (1990), "Nuclear power: a study in technological lock in", **Journal of Economic History**.

COWAN, R. (1991), "Technological variety and competition: issues of diffusion and intervention" in **OECD**, Technology and Productivity, Paris.

DASGUPTA, P.; DAVID, P.A. (1994), "Towards a new economics of science", **Research Policy**, Special issue in honor of N. Rosenberg.

DAVID, P.A. (1993), "Knowledge, property and the system dynamic of technological change", **World Bank Annual Conference on Development Economics**, Washington D.C.

DAVID, P.A.; ROTHWELL, G. (1993), **Performance based measures of nuclear reactor standardization**, MERIT publication, 93–005.

DAVID, P.A.; FORAY, D. (1994), **Accessing and expanding the science and technology knowledge–base**, DSTI/STP, 94–4, OECD.

DAVID, P.A.; MOWERY, D.C.; STEINMUELLER, W.E. (1992), "Analysing the economic payoffs from basic research", **Economics of Innovation and New Technology**, 2(4).

DAVID, P.A.; ROTHWELL, G.; MAUDE–GRIFFIN, R. (1991), **Learning from disaster?: changes in the distribution of operating spell durations in U.S. nuclear power plants after Three Mile Island**, CEPR publication no. 248, Stanford University.

DOSI, G.; FREEMAN, C. (1992), **The diversity of developement patterns: on the processes of catching–up, forging ahead and falling behind**, International Economic Association Meeting, Varenna, October.

ERGAS, H. (1994a), **The new face of technological change and some of its consequences**, mimeo.

ERGAS, H. (1994b), **The new approach to science and technology policy and some of its implications**, Rand/CTI.

EUREKA (1993), **Evaluation of Eureka Industrial and Economic Effects**, Paris: ANVAR.

FORAY, D.; FREEMAN, C. (1992), **Technology and the Wealth of Nations**. Pinter.

FORAY, D.; RUTSUAERT, P.; SOETE, L. (1994), **The coherence of EU policies on trade, competition and industry. Case study: high technologies**, International Conference on EC policies, Louvain la Neuve, 27–28 October.

FREEMAN, C.; HAGEDOORN, J. (1994), "Catching up or falling behind: patterns in international interfirm technology partnering", **World Development**, Vol. 22, no. 5.

FREEMAN, C.; HAGEDOORN, J. (1995), "Convergence and divergence in the internationalisation of technology" in Hagedoorn (ed.), **Technical Change and the World Economy**, Edward Elgar.

HIRSHLEIFER, J. (1971), "The private and social value of information", **American Economic Review**.

PEREZ, C.; SOETE, L. (1989), "Catching up in technology: entry barriers and windows of opportunity" in Dosi et al. (eds.), **Economic Theory and Technical Change**, Pinter.

ROSENBERG, N. (1982), "Learning by using" in Rosenberg (ed.), **Inside the black box**, Cambridge University Press.

ROSENBERG, N. (1992), "Scientific research and university instrumentation", **Research Policy**.

ROSENSTEIN–RODAN, P. (1943), "Problems of industrialization of Eastern and

Southeastern Europe", **Economic Journal**, 53.

SCHELLING, T. (1994), **Research by accident**, IIASA WP., Laxenburg.

STEINMUELLER, E. (1995), "Technology infrastructure in information technology industries" in Teubal et al. (eds.), **Technological Infrastructure Policy: an International Perspective**, Kluwer Press.

STIGLITZ, J. (1991), **Social absorption capability and innovation.** CEPR publication no. 292, Stanford University.

TASSEY, G. (1992), **Technology infrastructure and competitive position**, Chapter 3, Kluwer Academic Publ.

VALLS, J. (1993), **Spanish national system of innovation, public *R&D aids and the internationalization of firms***, Eunetic Workshop, Strasbourg, March.

VON HIPPEL, E. (1988), **The sources of innovation**, Oxford University Press.

WEITZMAN, M. (1993), "What to preserve?. An application of diversity theory to crane conservation", **The Quaterly Journal of Economics**, February.

6 INDUSTRIAL DISTRICTS AND THE GLOBALIZATION OF INNOVATION: REGIONS AND NETWORKS IN THE NEW ECONOMIC SPACE

Richard Gordon
Silicon Valley Research Group
Center for the Study of Global Transformations
University of California, Santa Cruz

I. THE PROBLEMATIC OF INDUSTRIAL DISTRICTS

The emergence of a new "technico-economic paradigm" [Perez (1985), Freeman (1987, 1992)] and its association with "new industrial spaces" [Scott (1988a, 1988c)] has exposed certain critical inadequacies of traditional models of regional economic growth. Theories emphasizing the static efficiency factors of regional economies elaborate an entirely proper concern for local specialization and demand-supply interdependencies, but are weakened, in an age of perpetual innovation, by their disregard for the regional bases of dynamic technological change. Export-based growth models place an appropriate stress upon external demand, and its associated local multiplier effects, as central elements in regional economic development but fail to explain either the origins of initial regional specialization or, even more importantly, how to modify or shift modes of specialization over time. Models of territorial growth based upon inter-regional movements of production factors are particularly suspect in an era when qualitative transformations in the technological bases of production serve to reinforce and compound, rather than to remove, existing assymetries between localities at the leading edge of technological change and those already far from the forefront of even the prior industrial regime. Growth strategies oriented to the in-migration of firms from elsewhere appear invalidated not only by historical experience (the generation of branch-plant economies in which extensive intra-corporate external linkages supplant, rather than reinforce, local relationships) but also by the ostensibly indigenous nature of regional high technology-based economic development[34].

Recent perspectives on the problem of regional economic development, though divergent in their core arguments, all attempt to deal precisely with the dynamic emergence of innovation by focusing on new dimensions of the territorial

---

[34] For further detailed analysis of these theoretical perspectives see R. Camagni and M. Quévit, *Development Prospects of the Community's Lagging Regions and the Socio-Economic Consequences of the Completion of the Internal Market*, report prepared for the Commission of the European Communities (DG XVI), October, 1991.

X. Vence-Deza and J. S. Metcalfe (eds.), Wealth from Diversity, 103–134.
© 1996 *Kluwer Academic Publishers. Printed in the Netherlands.*

context. Individual-based theories center regional capacities for technology creation on the Schumpeterian entrepreneur, the extraordinary individual able to foresee new commercial possibilities and gamble on the risks of new technological development, or on specific types of cultural formation that tend to incubate entrepreneurial activities [Rogers and Larsen (1984), Weiss (1986)]. Place-based analyses assess geographic areas as autonomous reservoirs of "regional innovation potential" derived from specific locational attributes (research institutions, venture capital operations, "quality of life" amenities and so on) or as "industrial districts" in which inter-firm linkages and agglomeration economies combine to produce new technological possibilities [Rees and Stafford (1983), Boeckhout and Molle (1982), Piore and Sabel (1984), Pyke, Becattini and Sengenberger (1990)]. Advocates of spatial reconcentration suggest that location is a product of industrial organization rather than spatial attributes: the vertical disintegration of industry promotes spatial agglomeration as specialized producers achieve returns to scale through an external division of labor, locating in close proximity to each other in order to reduce the transaction costs of their unstable and non-standardized exchanges [Scott (1983; 1988a, 1988b)].

Despite their near-universal attractiveness to regional policy-makers around the world, these approaches exhibit a problematic understanding of the nature and trajectory of high technology economic development. First: the existence and character of local linkages remain largely presumptive. The efforts of individual (or corporate) entrepreneurs or, even more vaguely, the existence of local cultural predispositions towards innovative activity, may be more or less important to the creation of new technology, but neither the collective linkage networks inevitably underlying such efforts nor their explicitly territorial dimension are illuminated at all within this focus. The associated assumption that innovation tends to be concentrated in small firms ignores the fact that most small firms are not innovative and that innovation frequently emanates from large firms as well [Gordon (1989), Acs and Audretsch (1990)]. Similarly, the widespread failure of attempts to incubate high technology growth within specialized regional industrial complexes reveals that even the presence of favorable locational factors does not necessarily promote the inter-firm linkages necessary to the production of innovation: that is, locational proximity does not automatically generate inter-firm synergy. Nor does the existence of agglomerated firms demonstrate the existence of dynamic inter-relations among them: firms may locate in the same region for any number of reasons (the fact that founders often establish their businesses near their own residence is among the most prominent of these) and specialized linkages between firms frequently operate over long (not short) distances precisely because the need for technological complementarities outweighs the significance of transactions costs with respect to innovation [Gordon (1992, 1993a)].

A second problem with these approaches, inherent in the preceding observations, is that they lack an explicit theory of innovation (a weakness that mirrors innovation theory's general disregard for the territorial dimension of technological development). For entrepreneurial theory, the source of innovation remains necessarily exogenous, if not indeed, mysterious and essentially unfathomable. Neither the locational nor organizational approaches, even where they empirically observe certain institutional and inter-firm linkages, specify their relationship to the overall innovation process in individual firms (or groups of firms) or within a territorial agglomeration as a whole. In many studies, the relationships observed between firms within a localized territorial environment are divorced completely from any analysis of the totality of innovation linkages for these same firms, the latter remaining completely unexamined[35]. Consequently, it is impossible to assess the significance of localized inter-firm linkages in comparison with other types of linkage relations in which innovative firms are embedded.

A third, and equally fundamental problem is that, whether it adopts the guise of incubator seedbeds, regional innovation potential, specialized industrial districts or spatial reagglomeration, current theorization of industrial organization and spatial restructuring encourages belief in a singular mode of contemporary economic development: effectively quite distinct forms of regional economic development -Silicon Valley, Grenoble, Baden-Wurtemburg, the M4 Corridor, Orange County, the Third Italy- are assimilated to the same model. Policy-making in turn assumes the existence of a set of universal pre-requisites for high technology growth that can be implanted successfully in any given locality. Specialized industrial districts ostensibly multiply as discrete and self-contained instances of a single pattern: independent centers of comparative advantage emerge, apparently without regional inter-connections (except for market exchange or trade) or objective inter-relations within an international division of labor. As a consequence, these perspectives remain indifferent to several of the most potent forces transforming contemporary economies: spatial differentiation, inter-regional innovation systems and production linkages, and globalization.

This chapter suggests a more complex view of regional development -one that provides a basis for the comparative analysis of differentiated spatial systems of production- in order to situate strategic policy-making perspectives on regional innovation and economic development. Figure I elaborates a non-reductionist typology of innovative regions based on specific territorial configurations of six variables. Each factor retains a specific individual importance in structuring regional innovation processes, while the diverse configurations of factors constitute specific types of territorial innovation system[36].

First: the current fixation on regional innovation factors and localized inter-firm relations has obliterated the critical role of state action both in the initial establishment and in the long-term innovation dynamics of industrial regions. A

---

[35] Cf. the empirical researches of GREMI presented in Maillat and Perrin (1992), Maillat, Quévit and Senn (1993).

[36] For other recent attempts to develop similar typologies from which I have learned a great deal, see Gaffard et al. (1990), Storper and Harrison (1991), Perrin (1992). The present analysis builds on an earlier typology developed in Gordon (1991).

complete assessment of the government's role necessitates particular attention to the strategic nexus of state intervention and market processes established in each territorial context: is state intervention limited to an infrastructural role which presumes the predominance of market forces in the creation of a regional innovation dynamic? does the state, on the contrary, adopt a leadership role in the manner of traditional indicative planning? or is the architecture of state-market relations constructed as a co-ordinated whole in which "market–conforming" state actions operate to minimize market weaknesses and dysfunctionalities and to optimize the beneficial outcomes of market forces?.

Second: industrial structure exercises an important independent impact upon the possibilities for innovation within a territorial agglomeration. Vertically integrated firms tend either to be relatively self-sufficient enterprises or local units largely dependent upon external corporate parents: in either event, they tend (though not as a matter of necessity) to reduce their local linkages in a manner that attenuates regional innovation capacities. Horizontal integration involves the exploitation of external economies within a given sector and thus promotes the development of small firms and relatively dense inter-firm relations which may, under propitious circumstances and the favorable conjuncture of other factors, facilitate regional innovation processes. Horizontal disintegration, conversely, involves dense collections of firms that tend to lack any substantive connections with each other. Vertical disintegration tends to involve high levels of intra-firm specialization and inter-firm relations, though regional innovation capacities cannot be deduced from the structure of vertical disintegration alone.

Third: the governance structure of regional inter-firm relations is a factor of autonomous strategic importance which cannot be derived directly from industrial structure *per se*. Governance structure is critical to *the organization of inter-firm learning dynamics* which lie at the core of the regional accumulation and appropriation of knowledge regarding the creation and application of new technologies. Learning dynamics traverse the design, transfer and use of new technology[37]. Market-based dynamics, assuming that the knowledge relevant to innovation can be embodied principally in easily accessible information and capital goods, are oriented predominantly to composite (rather than integrated) design strategies. Transfer relations tend to be organized as permanently contingent and arms-length interactions mediated by price signals: inter-firm coordination is based principally upon information supplied by the market. Since production is treated as derivative of exchange and oriented to discovering the correct mix of commodity inputs (including labor) to minimize overall costs, market-based dynamics commonly rely upon production strategies focused on labor costs and the manipulation of external labor markets.

---

[37] For a more detailed treatment of learning dynamics as they apply to processes of technology design, transfer and use see Gordon and Krieger (1992).

106

Table I: Criteria for Assessment of Innovative Regions

Types of Innovative Region	State–Market Relations	Industrial Structure	Inter–Firm Linkage Structure	Extra–Regional Linkage Structure	Innovation Logic	Position in Global Division of Labor
High Technology Aggregation	state –or market-dominant	horizontal disintegration	market–based	market–based	breakthrough	subordinate/secondary
Science Ghetto	state–dominant	horizontal disintegration	market–based	market–based or hierarchical	breakthrough	subordinate
State–Facilitated Complex	market–dominant	horizontal disintegration	market–based or hierarchical	market–based	incremental	secondary
State–Dependent Complex	state–dominant	vertical integration	market–based or hierarchical	hierarchical	breakthrough	secondary/strongly competitive
Innovative Milieu	coordinated	horizontal integration & vertical disintegration	collaborative or market–based	market–based	incremental/ breakthrough	strongly competitive/leading edge
Network Region	coordinated	horizontal integration & vertical disintegration	collaborative	collaborative	fusion	leading–edge

Hierarchical dynamics tend to be driven by formalized, linear design strategies based principally upon hierarchical intra-organizational information processing. Technology transfer is premissed upon substituting internal coordination for external transactions and is organized through the bureaucratic integration of production stages within the firm. External relations, where it is necessary for them to exist, tend also to take a bureaucratic form, either as dependent subcontracting relations (suppliers) or as oligopolistic market relations (clients). Internalized cultures tend to promote the maintenance of command-control production systems, favoring internal labor markets for the provision of necessary skills on the one hand and, on the other hand, automation strategies based on self-regulating systems that reduce the role of human judgement or strip labor from the production process altogether.

Collaborative dynamics are based upon interactive design strategies, interdependent relations between firms based upon technical complementarities rather than transaction costs, reciprocal autonomy of production stages and the cooperative mobilization of the independent contributions to innovation and production of both internal and external labor groups. Collaborative interactions perceive work force capabilities, learning and flexibility as a positive competitive resource. The utilization of technology is oriented to extensive and high quality human intervention as central to flexible organizational adjustment to constant change. The key issue here is that the organization of learning dynamics independently affects both the possibilities for, and the types of, regional innovation.

Fourth: as will be seen in more detail below, the fiction of self-contained regional development processes encouraged by contemporary theory cannot be sustained. Regional clusters may provide collective learning processes essential to innovation, but these localized relationships are increasingly insufficient either to initiate or to sustain innovation trajectories. That is, the form in which a region's innovation system is integrated into relationships beyond the region is a critically independent aspect of territorial innovation. External linkages can be assessed in terms of market-oriented, hierarchical or collaborative dynamics in the same manner as inter-firm relations within the region.

Fifth: different types of innovation logic entail both different institutional presuppositions and different implications for industrial (and, therefore, regional) competitiveness. Existing analysis of this issue conceptualizes a basic division between radical and evolutionary technological change, that is, between fundamental breakthroughs in new technology on the one hand and path-dependent, incremental adaptations or extensions of existing technological knowledge and practice. However, while this distinction indeed refers to important differences in innovation logic, it might be argued that much, if not most, technological innovation falls precisely between these two poles, involving rather extensive technical and organizational change (i.e. beyond minor adaptations to existing practice) without necessarily presupposing fundamental changes in the prevailing technological paradigm (a form of innovation that I will call "social-organizational" or "fusion" innovation)[38].

Sixth: regional systems of innovation are not autonomous. Changes in the structural evolution of innovation and the spatial division of labor in an industry create a "complex global regional mosaic" [Gordon and Dilts (1988)] as each new round of adjustment and restructuring freezes some regional formations in place, induces adaptation in others, and brings entirely new regions into the ambit of global industrial organization. Thus, a region's position in the global spatial division of labor matters with respect to its innovation capabilities. Far from inducing uniformity of regional development, this global framework differentiates spatial organization around specific regional relationships to the changing historical conditions of innovation within an industry, each region indeed embodying, either in partial (i.e distorted) or complete form, particular aspects of historical innovation requirements[39]. The whole produces an integration of global and local levels in which not the homogenization, but precisely the differentiation, of spatial forms and dynamics is paramount.

II. A TYPOLOGY OF INNOVATIVE REGIONS

II.1 High Technology Aggregations

High technology industry has followed a growth path of simultaneous concentration and dispersion. A combination of factors -historically specific product characteristics, the dynamic growth and diversification of the information technology industrial complex, and its world-historical importance as a technical infrastructure- encourages a broad spatial scattering of high technology economic activity. Core high technology products are relatively indifferent to transport cost or production locale constraints, expanding the spatial ambit of the industry's locational possibilities in comparison with more traditional resource-dependent, historically rooted or cost-sensitive sectors. Low barriers to entry in many spheres of high technology production act as an incentive to extensive new firm formation. Information technology's utility as a control system linking geographically dispersed organizational units reduces the tyranny of distance and facilitates the spatial proliferation of high technology production. Its role as a core technology (the contemporary equivalent of railroads or electricity as an economic infrastructure) simultaneously animates national and regional concerns to develop autonomous production capabilities in high technology and spatially multiplies its linkages to other economic sectors and activities. The increasingly elaborate global division of labor within high technology industry extends its spatial incidence. High technology is a global industry of near-universal *local* significance.

---

[38] For more detailed analysis of social-organizational innovation, see section III below.

[39] Regional positions, however, are not necessarily frozen, but altering their position within the global division of labor presupposes a reconfiguration of the core elements of the regional innovation system.

Within this framework, isolated pockets of high technology activity, or *high technology aggregations,* are constructed by building upon particular dimensions of the multi-dimensional logic driving global high technology industrial growth as a whole. Entrepreneurial inertia (i.e. the impact of residential preference upon entrepreneurial locational decisions) can generate small localized complements of high technology enterprise. Localized and restricted spin-off effects may produce small semi-independent clusters of suppliers around existing technical branch plants. The global locational strategies of multinational enterprises can lead to the formation of territorially concentrated branch plant economies clustered around either particular phases of high technology production or specific sectors of high technology industry [Glasmeier (1985)]. Organizational decentralization within corporate headquarter complexes may exert strong local economic dominance, and even achieve global significance, without necessarily entailing a more diversified base of regional production activities[40]. Centralizing state policies, as in France, may attenuate the development of potentially promising sets of regionalized inter-relations [Ganne (1990)]. The proliferation of competitive state and local high technology economic development policies encourages spatial dispersal on the basis of individualized corporate responses to governmental incentive programs.

Even as high technology industry and employment tended over time to concentrate in a few privileged areas, therefore, it also revealed a strong propensity for geographical deconcentration in localized high technology aggregations. During the 1970s and 1980s, for example, greater total high technology employment growth was distributed in small, fragmented pockets of high technology activity throughout California than occurred in any established high technology location (outside Santa Clara County) [Bradshaw (1985)]. More than one-half of recent employment increase in the Los Angeles electronics complex is diffused in neighboring counties [Soja (1988)]. At the national level, the same tendencies prevail. In the U.S., all but 15 states have a minimum of 10,000 high technology jobs [Markusen et al. (1986)] and, in many instances, even a small share of total national high technology employment contributes significantly to overall manufacturing employment within the state itself. Keeble (1988) observes that in Britain "the geography of high technology development in the 1980s is one of widely dispersed growth in most regions... despite the large concentration in the south of England".

Whether initiated through state policy or by market forces, high technology aggregations invariably reproduce only particular facets of the total innovation requirements of the industry and their possibilities for dynamic growth are radically attenuated. Concentrations of firms and employment remain small (focused in a few large enterprises or disseminated in disaggregated groups of small firms) and generally lack the economic density to produce self-sustaining patterns of development: whatever the growth capabilities of particular companies, the significance of the aggregation as a whole is dwarfed by the dominion of the leading regional

---

[40] IBM in New York or Motorola in Phoenix constitute good examples of this phenomenon.

formations of high technology industry[41]. Internal linkages between firms within high technology aggregations are either non-existent (independent small producers follow discrete product market strategies with little orientation to other local firms; branch plant inputs and outputs are oriented to external parent companies) or they reproduce traditional non-synergistic relations of subcontracting dependence. Despite, or perhaps because of, the fact that they tend to represent a radical departure from existing regional modes of production, this group of firms commonly remains a small and disconnected enclave within the local economy as a whole. Economic diversification within such complexes is low and haphazard, and the division of labor, both within and between firms, is underdeveloped. External linkages tend to be truncated (as these areas tend to be marginalized from the mainstream of technological change in the industry) and consist primarily of exchange relations. The lack of significant dynamic local growth even in those aggregations centered on corporate headquarters which otherwise play a critical role in global high technology development testifies to the synergistic weaknesses of the high technology aggregation as a general form. Although individual firms might be innovative on the basis of their internal resources, regional innovation capabilities within these complexes are meager at best. The presence of particular locational factors (entre-preneurialism, external investment, headquarters location, regional development policies) may lead to the formation of a territorially localized collection of atomized producers -and collectively such aggregations comprise a substantial share of global high technology employment- but usually they will not generate more coherent centers of innovation without more systematic and more integrated direction.

II.2 Science Ghettoes

The notion that contemporary innovation is science-based, and that new firm creation can be incubated by an appropriate regional concentration of scientific knowledge, technical expertise and commercial acumen, has provided the rationale for a global frenzy of research/science park construction as the principal means of stimulating local economic development. In contrast with high technology aggregations (with which they otherwise share much in common), *science ghettoes* tend to result not from ad hoc causes but from deliberate initiatives on the part of local governments, university or research institutions, and development consortia. In each case, local implementation of this "universal" model of high technology development (the latter derived from the ostensible preconditions for Silicon Valley's emergence) is designed to incubate new forms of regional industrialization, either in virgin environments or as a substitute for traditional industrial activities now considered *dépassé*.

All of the guiding assumptions of the science park strategy are problematic

---

[41] Many of the world's illusory Silicon Valley's are formations of this type whose local (let alone, global) importance is magnified out of all proportion by public relations exercises.

in fact. The linear assumptions of the science park model -that innovation is applied scientific research and that commercialization involves conversion of research prototypes to marketable products through a fixed sequence of production stages- are at odds with the basic reality of technology creation, particularly in the information technology sector, but even in sectors like biotechnology that are far more closely derived from scientific research[42]. Though they are precisely the kind of enterprise science parks are designed to incubate, small and medium-sized firms particularly tend to lack the material or organizational resources required for sustained formal research contact with a university milieu and, where such relationships exist, they tend to be infrequent, contingent and purely instrumental in character on the part of the firms themselves. Finally, the idea that spatial proximity alone can incubate high technology synergies blindly ignores the salience of the material logic of the innovation process, prior regional investment patterns and existing industrial structure in determining innovative capabilities and locational patterns within high technology industry [Storper and Walker (1989)].

It is not surprising, therefore, that extensive research into the science park phenomenon reveals an absence of dynamic industry-university or inter-firm linkages in these sites. Macdonald's survey (1987) of British science parks indicates that their proliferation has not been in response to market demand and that they have almost universally failed both as incubators of technological innovation and as motors of regional development. Saxenian's examination (1989) of Cambridge Science Park, the most important in Britain, reveals an almost total lack of interaction among firms located in the zone, and little technical or informational exchange with Cambridge University itself. Enterprises in the Cambridge park remain small, failure rates are high, market penetration is negligible, and the park has contributed only minimal employment growth to the regional economy. A similar disarticulation of local firms and research institutions is apparent in the Zone for Innovation and Scientific and Technological projects (ZIRST) in the Grenoble-Meylan region of France [Boisgontier and de Bernardy (1986)]. The vast effort creating the technopolis at Sophia-Antipolis on the Cote d'Azur has generated only minor endogenous spin-off effects and little cooperation or synergy among the larger multinational branches that account for the large proportion of employment in the zone [Perrin (1988), Charbit et al. (1991)]. Nijkamp and his colleagues (1987) have found no correlation in the Netherlands between the regional incidence of innovation and the location of science parks. In the Japanese science City of Tsukuba linkages between research centers or with local firms are virtually non-existent [Tatsuno (1988)]. Stanford Industrial Park itself was a product, rather than a cause, of Silicon Valley's industrial prosperity [Braun and Macdonald (1982)] while recent research among innovative SME's in Silicon Valley indicates that only a negligible minority of such firms maintain any relationships with universities or public research institutions and that such relationships as exist tend to be contingent and not based upon long-term, genuinely

---

[42] For critiques of the linear model see Rosenberg (1976, 1982), Amendola and Gaffard (1988), Dosi et al. (1988, 1990).

112

collaborative basic research interaction [Gordon (1992)].

Thus, while individual companies may prosper in such locations, it is rare for science ghettoes to move beyond the status of formalized high technology aggregations, exhibiting minimal impact upon local economic development and occupying a quite subordinate position within the global structure of the industry. The role of formal scientific research, and the institutions in which it is based, in contemporary innovation is quite paradoxical since, as scientific knowledge becomes objectively more central to the internal structure of technological innovation, it plays an *increasingly peripheral* role in the *innovation process itself.* Accordingly, regional R&D policies require substantial re-contextualization: rather than utilizing research institutions as a fulcrum of radical shifts in regional technological regimes, R&D institutions would be better focused on specific existing regional capabilities and re-situated within the collaborative accumulation of scientific and technological know-how on a broad institutional front[43].

II.3 State–Facilitated High Technology Complexes

Extensive tracts of the global high technology spatial mosaic exist as *state-facilitated* or *state-induced high technology complexes*, designated development zones heavily subsidized and politically supported by state authority and constructed principally to attract substantial regional concentrations of foreign investment. Particularly when organized as specifically demarcated Export Processing Zones (EPZs), state policy is designed in the first instance merely to facilitate the spatial concentration and regional operations of external firms. Transnational capital in such areas is able to take advantage of subsidized land and utilities, investment allowances, promotional grants and subsidies, trade and tax incentives, and cheap, disciplined labor. In fact, transnational enterprises tend to determine the form of their insertion into the local region on the basis of an international strategy: the territorial specificity of the region is *utilized,* but a disconnection exists between the logic of industrial organization and the logic of territorial organization [Morgan and Sayer (1988)]. External companies dominate production, employment and exports in such regions. Enterprises tend to concentrate in a narrow range of production stages or operations. This disjuncture in turn accounts for the disarticulation between firms and with local markets generally found in these zones and for the lack of local economic development generated by these activities. Tight hierarchical external linkages thwart indigenous innovative relationships.

In zones like Tsukuba Science City, the North Carolina Research Triangle or the French technopolis Sophia-Antipolis, high technology research activities are

---

[43] Since universities and scientific research institutions focus on very different kinds and levels of innovation than commercial firms, the very justification usually elaborated for the likely success of science or research parks –viz. the leading-edge capabilities of local research institutions– is often, ironically, the most basic reason for the failure of the science park venture as a whole.

literally created *ab initio* by government initiative, in these instances the state (national and regional) not only planning and preparing the necessary infrastructure, but also directing the location of important national research centers, institutes and funding to the zone in order to attract commercial tenants. Although each of these zones has attracted a number of transnational corporate research laboratories, there has also been a marked lack of impact upon either local firm creation or the regional economy. At a broader level, the fact that EC R&D programs (such as ESPRIT, RACE, BRITE) *reinforce* the technological dominance of already privileged scientific regions and large multinational firms tends to exacerbate existing regional inequalities while doing little to foster regional potentials for dynamic innovation [Quévit (1992)].

A combination of government regional assistance policies and foreign investment has been instrumental in establishing zones of high technology activity in areas like Scotland's Silicon Glen and in South Wales. In both areas the high technology economy is oriented towards production activities as opposed to research and development and towards semi-skilled, rather than professional or technical labor: these regions operate essentially as branch plant manufacturing locations for foreign capital (particularly US and Japanese) establishing production platforms within the markets of the EC. In recent years, particularly in Scotland but also in Wales, some upgrading of activities has occurred towards higher R&D content, local technical and scientific input, more advanced products and indigenous firm formation. This expansion, however, represents a transformation of the regions' position within transnational strategies from basic assembly and manufacturing operations to the establishment of a regional European hub for re-design, marketing and service [Morgan and Sayer (1988), Morgan (1992)]. Both areas remain dominated by large external firms, employment in local companies incorporating only a small percentage (less than one-sixth in the case of Scotland and one-tenth for Wales) of total regional high technology jobs[44].

II.4 State–Dependent High Technology Complexes

The role of the state in state-facilitated complexes is relatively passive: the state builds the infrastructural conditions designed to attract the foreign investment that will ostensibly serve as the primary motor of employment creation, sectoral development or foreign exchange earnings. Characteristically, this strategy is pursued by national or regional authorities constrained by a lack of independent politico-economic leverage, either domestically with respect to indigenous or foreign capital or within the world economy as a whole. *State-dependent complexes* in contrast are established from the top down by states combining strong *independent political* capabilities for organizing domestic economic structure and policy with

---

[44] Local and regional governments, however, are learning to encourage multinationals to valorize, rather than simply exploit, local innovative capabilities: see Ashcroft (1993), Hill and Munday (1993), and Morgan (1993).

(commonly, though not invariably) a *subordinate economic* position within the prevailing structure of international competition.

State-dependent strategies attempt to leverage the autonomous power of the state to build an indigenous high technology industry capable of bridging the technology gap between leaders and followers and, reaping the "advantages of economic backwardness", of sustaining an independent competitive status within the global economy. Initially, state-dependent complexes are frequently considered defensive formations behind whose bulwarks the quest for international competitive parity can be prepared. However, the technological frontiers of high technology innovation are moving forward permanently, making it extremely difficult (though not impossible, as the Japanese and Korean cases demonstrate) for followers to catch leaders: as a result, these complexes often fail in their bid for competitive independence and remain permanently dependent upon the state for survival. This vicious cycle governs the essential character of state-dependent complexes which tend to adopt one of two overlapping modes of regional development: military high technology complexes whose regional economic rationale is justified in the name of national security and import-substitution complexes whose security is justified in the name of economic competitiveness[45].

As Markusen et al. (1986, 1991) have observed, state defense spending is the most important single determinant of high technology geography in the U.S.. In California, the electronics industry in the immediate postwar period congregated overwhelmingly around the Los Angeles Basin defense complex: as the aircraft industry, which had already substantially moved from the east to the west coast prior to WW II, expanded into aerospace and missile development, it shifted into avionics and around this complex grew a host of independent electronics manufacturers participating in the production of guidance and control systems for the new generation of spacecraft and weapons. This fundamental reorientation in the character of defense technology and military production promoted a corresponding relocation in the electronics industry's center of gravity from the East Coast to southern California. By the early 1960s, more than two-fifths of *total* employment in the Los Angeles-Long Beach region was directly or indirectly dependent upon defense spending, and in 1986 three-quarters of *all* electronics employment in southern California as a whole was in defense-related sectors (aircraft and parts, communications equipment, space vehicles and missiles).

South-east England's M4 Corridor was built upon the infrastructure of skilled labor, supply relations and marketing existing in the region as a result of the electrical industry's pre-WWII concentration in and around London [Breheny et al. (1986), Boddy et al. (1986)]. The state, most particularly in the form of Ministry of Defense research and procurement expenditures, played the critical role in decentralizing manufacture and stimulating the contemporary transition to high

---

[45] The latter, more prevalent in the NIC's and Third World countries, will not be discussed in the present context.

technology research and production to the western crescent on London's periphery. High technology activities along the M4 Corridor are predominantly defense-oriented. Research work is concentrated in government research establishments (GREs) and corporate defense-funded laboratories rather than in universities. The bulk of industrial employment, particularly of skilled labor, is deployed on defense-oriented custom production and small-batch manufacture. The leading sectors of private industry in the M4 Corridor are predominantly foreign-owned firms competing for government contracts or adapting foreign products to UK or European standards or market requirements. With the exception of traditional subcontracting arrangements, inter-firm linkages within the M4 region are virtually non-existent outside of state-sponsored contractual relations. Far from developing self-sustaining industrial and commercial synergies (even technology transfer from military to civilian projects within the same firm is low), the region's market power and innovative capabilities are attenuated and the complex continues its primary reliance upon state-provided infrastructure, subsidized markets and military expenditures.

A similar constriction upon local innovation and development possibilities is evident in the Paris Sud high technology complex of the Ile de France [Decoster and Tabaries (1986), Tabaries (1992)]. Like Tsukuba Science City, Paris Sud was in part a product of governmental decisions to concentrate scientific research institutions and high-level educational establishments within the same region. The area subsequently attracted leading industrial sectors and corporate research laboratories and headquarter functions. Yet, despite this vast centralization of innovative possibilities, linkages and cross-fertilization between and within research institutions and firms is minimal. Many of the region's activities are defense-related and are indifferent to other local enterprises or to regional supply and demand. Technology issuing from the Cité Scientifique's research apparatus is not commonly transferred to local firms: as Decoster and Tabaries (1986) observe, Paris Sud is essentially a national, not a regional, complex. Similarly, innovative small high technology firms in the region maintain no linkages with the Cité's research institutions.

II.5 Innovative Milieux

While entrepreneurialism, regional innovation factors or processes of vertical disintegration may (or may not) be present, neither the existence nor the innovative dynamics of innovative milieux can be reduced to mechanisms which treat the spatial dimension deterministically. On the contrary, innovative milieux are distinguished as a general category precisely on the basis of a specific articulation of territorial coordination and industrial organization: within these regions, the production system is constituted through spatial relations[46].

Innovative milieux utilize territorial organization as a basis for the establishment of the formal and informal collective processes essential to the production of perpetual innovation. As Camagni (1991) has observed, firms confront critical uncertainties regarding the complexity and cost of information, quality assessment of inputs and equipment, evaluation of alternative possible courses of strategic action, and the impact of decisions made by other firms. Neither internal decision-making mechanisms nor market processes are adequate to resolve these problems in a regime of continuous innovation and the institutional mechanisms of the milieu function in this context to reduce dynamic uncertainty and to mobilize interdependencies critical to the creation of new technologies. Formal knowledge embodied in customer-supplier relations, producer service linkages or skilled labor mobility, informal processes of information transfer and culturally-embedded inter-pretative schemas all reinforce territorially-specific collective learning processes[47]. Non-market relations with suppliers and clients, strategic interactions with research institutions, capital sources, and local authorities, and close proximity to competitors constitute important components of a firm's "support space" [Ratti and D'Ambrogio (1992)], actively mediating supply and demand. Agglomeration economies reinforce local advantages. In stark contrast with high technology agglomerations and science ghettoes, innovative milieux display strongly organized horizontal integration: they are constituted as territorial production systems. The entire set of localized relations generate a dynamic collective learning process which is fundamental to the regional production of innovation.

Three sub-types of innovative milieux may be distinguished, each exhibiting a specific developmental logic. First: innovativeness may be captured in old industrial regions by a largely internalized process fusing the mobilization of traditional strengths with the introduction of radical departures from established regional practice. This process of "filiation-rupture" [Aydalot (1986)] is exemplified in the textile, machine tool, leather and ceramics districts of the Third Italy [Pyke et al. (1990)], the Rhones-Alpes complex, the watch-making districts of the Swiss Jura [Maillat et al. (1992, 1993)], the Choletais region of France [Dubois and Linhart (1992)] and the engineering complex of Baden-Wurttemburg [Cooke and Morgan (1990), Herrigel (1993)][48]. Secondly, and in a radically different manner, innovative regions might appear *de novo* as a result of largely externalized processes contingently linking the adjustment difficulties of older industrial areas with the

---

[46] Extensive empirical research on innovative milieux produced by the GREMI group can be found in Aydalot and Keeble (1988), Camagni (1991), Maillat and Perrin (1992), Maillat, Quévit and Senn (1993).

[47] The circulation of information relevant to innovation may rest more decisively on one type of mechanism than another in a given case: for example, learning may rely more on traditional cultural institutions and practices (Third Italy, Swiss Jura), or on informal processes in large part produced by a new industrial formation (Silicon Valley), or they may be deliberately fostered as a result of public intervention (Japanese Technopolis program).

[48] As the contrast between the Third Italy (based predominantly, until recently at least, on small firms) and Baden-Wurttemburg (oriented around large engineering firms) makes clear, firm size is far less significant in the formation of such milieux than the collaborative linkages established within the territorial context.

competitive ability of newer regions to establish rapid predominance in new areas of expertise. Silicon Valley and Route 128 in Boston constitute prime examples of this phenomenon [Saxenian (1981), Gordon (1993b)]. It is possible that an innovative milieu is in the process of emerging in Orange County where a concentration of medical schools and research facilities around the University of California at Irvine form the core of new clusters of expertise in biotechnology and surgical instruments [Scott (1989, 1994)]. The formation of incipient biotechnology milieux around Trenton, New Jersey, Boston, San Francisco-Oakland, San Diego and Irvine in Southern California and Washington D.C. is intimately related to a direct relationship between the presence of university research institutions[49] and government funding (channelled principally through the National Science Foundation and the National Institutes of Health) which supports more than three-fifths of all research in this sphere [Blakely (1988, 1989), Castells (1989)]. Third: innovative milieux based upon the predominance of a single industrial sector should be differentiated from multi-sectoral regions with greater prospects for long-term viability. As will be seen below, perhaps the predominant characteristic differentiating Silicon Valley from competing regions like Route 128 or Los Angeles is the ability of Silicon Valley both to transform its logic of innovation over time and to diversify its industrial base.

II.6 Network Regions

Silicon Valley (ironically, since it has provided the basic model for the *self-contained* industrial district) constitutes the paradigmatic *network region*[50]. If we consider the trajectory of innovation in Silicon Valley's high technology industry it can be seen that in fact high technology development has not been driven by purely localist (or by purely global) dynamics but has been based upon *changing articulations of regional processes and external dynamics*. Each serial transformation in the structural preconditions of innovation, while absorbing preceding forms, has been dominated by a specific logic of economic governance, which has precipitated basic changes in the articulation of industrial and spatial organization.

In the first stage of its growth, extending broadly from the postwar invention of the transistor to the construction of manufacturing routines for integrated circuits by the early- to mid-1960s, the embryonic microelectronics industry emerged in the U.S. and in Silicon Valley not, as universally accepted mythology would have it, from localized entrepreneurial initiative, but from a conjunction of innovation in established firms and extensive state (i.e federal

---

[49] Until recently, biotechnology has been far more dependent upon basic scientific research than microelectronics. However, the relative failure of university-based, venture-capital funded entrepreneurialism to produce a broad range of marketable biotechnology products has begun to invalidate this approach. Increasingly, small research-based firms, formerly derived from and linked to, university research departments are being absorbed into the research operations of the large pharmaceutical and chemical companies.

[50] For more detailed treatments of innovation trajectories in Silicon Valley, from which the present account is drawn, see Gordon (1993b, forthcoming).

government) intervention.

The state was involved in every aspect of the microelectronics industry's emergence and early development. Advanced military and aerospace demand provided the principal market for microelectronics, established research priorities in product and process innovation, stabilized high profits for successful companies and underwrote the risks of new product development[51]. The vast majority of scientists, engineers and technicians in the microelectronics industry acquired state-of-the-art theoretical and practical knowledge in government-financed university or corporate research and development programs. The performance specifications of the world's hegemonic military power pushed development continually beyond existing technological frontiers into more advanced miniaturization and higher levels of performance. Government funding significantly extended to development of the manufacturing equipment and technologies facilitating the transition of device R&D into commercial production capability. Defense support was also critical in the establishment of the learning economies which governed the subsequent evolution of the industry: since market growth and expanded production volumes resulting from military sales accelerated cost reductions, U.S. firms were able to move faster and earlier than their foreign competitors down the learning curve towards more consistent quality, higher yields and lower prices. Government policy also provided a bridge between traditional industrial leadership and new entrants for, while established firms did receive substantial military funding for microelectronics R&D, defense contracts (including contracts for the construction of production facilities) were also awarded to smaller start-up companies, helping to diversify the structure of the emergent semiconductor industry. Military demand, therefore, was instrumental in creating and maintaining U.S. market leadership in electronics[52].

Silicon Valley's subsequent expansion did not proceed simply as an outgrowth of agglomeration economies associated with defense contracting (such as occurred in Los Angeles) or as a linear expansion of existing locational attributes, but necessitated a basic shift in the prevailing technological paradigm and a corresponding transformation in the character and integration of internal and external innovation relations.

The appearance of a novel logic of production can provide a critical basis for the consolidation of new regional centers of industrial growth [Storper (1985)]. New forms of production tend to seek locations beyond the obstacles posed by the structure of traditional commitments and practices in existing industrial areas. However, the new regions' superiority does not derive exclusively from internal sources of innovation. External linkages remain critical to regional growth but

---

[51] In 1962, military sales consumed 100 percent of integrated circuit, and almost 40 percent of overall semiconductor, production. By 1977, military sales constituted only 7 percent of IC, and 12 percent of total semiconductor, output.

[52] For more detail on the preceding summary, see Gordon and Krieger (1992), Gordon (1993a) and the numerous sources cited therein.

change their form: consolidation of the new region's dominance entailed, in the case of Silicon Valley, conversion of state-dependent relations to hegemonic external market linkages. Global market power is deployed as a source of internal growth, the region reaping the advantages of superprofits derived from capturing world markets against inferior competing technologies [Markusen (1985)]. Since the superiority of the new techniques or products is not established independently of their diffusion, market dominance constantly reinforces the initial comparative advantage of the innovative region and begins to restrict possibilities for alternative technical and social trajectories. As new investment swarms around the innovators [Schumpeter (1934)], escalating demand attracts regional factor supply, consolidating the process of spatial concentration as the innovative region is reconstituted around the specific requirements of the new industry [Walker (1988), Storper and Walker (1989)].

These synergies unfolded in Silicon Valley when the introduction of the integrated circuit (and, subsequently, the personal computer) transformed the institutional preconditions of leading-edge innovation from political interdependencies to *hegemonic market linkages.* Demand for the explosion of new applications resulting from the new technologies generated investment for new rounds of technical development which in turn generated markets for new products. Learning economy barriers simultaneously constrained competitors and allowed leaders to push technological frontiers forward. This transition from "state" to "market" provided the basis for a flourishing entrepreneurialism which was a *consequence,* not a cause, of the region's global hegemony.

State intervention and commercial hegemony mediated technological innovation in Silicon Valley to produce a process of regional agglomeration based principally upon evolutionary learning, industrial diversification, informal information flows and arms-length, market-based inter-firm transactions. Changing patterns of global competition, market organization and industrial structure subsequently threatened this particular articulation of external and internal dynamics. With the international diffusion of technical capabilities and the emergence of Japanese competition, Silicon Valley's "absolute advantage" [Chesnais (1986)], the result of superior levels of technical achievement, was substantially challenged in the 1980's. Market decentralization raised a multitude of new economic and cultural obstacles to production, marketing and distribution. Offshore labor platforms traded lower labor costs against a more methodical program of manufacturing process improvements, a strategy that proved increasingly misguided as competitive advantage shifted to manufacturing organization over product innovation. Firms can no longer simply provide standard devices and general-purpose systems but must adapt their offerings to specific end-user requirements: demand exists increasingly not as an abstract, external given, but as an anticipatory input into the design and development process. As international conditions of production moved away from Silicon Valley's traditional strengths, the weaknesses of the region's internal linkage structure were more starkly revealed.

The process of creating new forms of innovation presupposes that the boundaries between firms and their environment are no longer taken as given but, on the contrary, are subject to reconstruction as product and process technologies, as well as organizational structures, are reconfigured. The emergence of new objective material possibilities for technology creation may be compatible with the reinforcement or modification of existing organizational structures and strategies, but they also provide opportunities for a more fundamental restructuring that substitutes organizational embrace of the higher risks involved in forging new advantages for the inevitable shortcomings of more conservative approaches. Constructing a new mode of coordination for the existing system of innovation in these circumstances is not merely an option for the individual firm [Amendola and Bruno (1990)] but for industrial districts or regional innovative milieux collectively [Camagni (1991), Gordon (1992), Bruno and de Lellis (forthcoming)].

Silicon Valley has maintained its leadership as a global high technology headquarters in large part because substantial segments of regional industry have been able to meet each opportunity for "systemic innovation" [Imai and Yamazaki (1992)] with fundamentally new organizational linkages between local firms, new modes of innovative learning *and* most importantly, with new configurations in the structure of relations between the regional production system and the *extra-regional environment*. Detailed research into several different sectors within Silicon Valley's economy [Gordon (1992, 1993a, 1993b)] reveals two fundamental transformations in the current linkage structure of Silicon Valley companies: (1) the very high prevalence of collaborative strategic alliances with other firms, both inside and outside Silicon Valley, as the principal institutional basis for technological innovation (i.e. a shift away from the classically individualized entrepreneurial firm of the past) and (2) the increasing salience of *extra-regional* relationships precisely for the most significant inputs (technological and other) into the innovation process[53]. That is, over time, Silicon Valley has moved from "state" to "market" and, ultimately to "network" as the animating principle of regional innovation [see Gordon (forthcoming)].

The notion that this latest wave of collaborative innovation in Silicon Valley represents a reincarnation of the territorially agglomerated system of flexible specia-lization characteristic of the region's early development [Saxenian (1990, 1994)] is to misinterpret Silicon Valley's past, present and future as well as to misunderstand the contemporary logic of high technology innovation generally which increasingly demands *extra-regional* linkages as the essential precondition for technology creation in the contemporary world economy. If a region with all the agglomeration advantages of Silicon Valley can no longer produce innovation through localized relations within its own territorial framework, the policy implications of this latest transformation in Silicon Valley's innovation logic for all

---

[53] It is important to observe here that precisely those types of input that putatively engender territorial agglomeration in the suppositions of industrial district theorists are driving the formation of a dense network of *extra-regional linkages* in Silicon Valley.

other high technology regions are transparent.

III. CONCLUSION: THE GLOBAL MOSAIC OF REGIONS AND NETWORKS

Fundamental changes in the material logic of modern production -the appearance of new core technologies, the widespread diffusion of advanced technological capabilities, dramatic convulsions in long-standing patterns of international competition, global economic restructuring, the complexity and pace of technological change- have prompted both theoretical and practical re-assessment of the innovation process in advanced industrial societies[54]. R&D commitments transcend the resources even of investment leaders. Performance constraints imposed by firm-specific technology trajectories are exacerbated as international competition and technological convergence expand the range and complexity of the technologies a firm must monitor in order to produce innovation. The rapidity and complexity of technological change in all sectors necessitate closer and more flexible relationships with both suppliers and customers. The heightened importance of design specificity and operational complementarity reduces the efficacy of both market supply (the commodity status of inputs assuming less salience than their specific functional attributes) and vertical integration. Abbreviated product cycles and the globalization of markets expose inevitable shortcomings in corporate marketing expertise and distribution networks. New technology combinations alter the boundaries between sectors, plunging firms into unfamiliar activities outside the scope of their normal operations[55].

These developments, both severally and collectively, increasingly invalidate the traditional organizational alternatives of "market" and "hierarchy": explosive technical and economic change dramatically enhance the level of uncertainty and risk associated with market transactions, while the pace and cost of adaptability militate against a strategy of vertical integration. Firms are compelled to rely upon external transactions, yet markets cannot provide the necessary technological complementarities. Firms are coping with this contradiction in the material logic of production by elaborating new non-market forms of inter-firm coordination, focusing on their own core capabilities and delegating functions formerly produced in-house or purchased on the market to the complementary specialization of autonomous firms organized in an inter-dependent chain of production. Innovation in these industrial networks, therefore, is based primarily upon the *heightening of specialization* that occurs within each firm (as a result of the abandonment of the dysfunctionalities inherent in vertical integration) and within the network as a whole (as a consequence of collaboration which ensures the most efficient coordination of material, resource and knowledge flows throughout the network).

---

[54] Material in the next four paragraphs is derived from Gordon (1990).

[55] An incisive overview of these recent trends is provided in OECD (1992).

This *social-organizational model* of innovation, therefore, is based on a new relationship between the firm and its environment: the issue is not simply the management of established intra-firm innovation routines but the creation and use of new organizational and technological resources [Amendola and Gaffard (1988)]. In this model, innovation is endogenous neither to a single firm, nor to a statistical aggregate of firms, but to an integrated inter-firm *chain of production*. Firms are conceived as comprising an interdependent network in which the production strategies of individual actors are determined in part by their position within the chain and, hence, by the organization of relations between the constituent components. The reciprocal dependence of each stage of production upon the others means that the mode of organization adopted decisively affects the process of innovation. Innovation in this context is not simply a matter of adjusting to known environmental constraints or independently applying existing technical routines (static flexibility), but of modulating constantly uncertain environmental change through organizational creativity in cooperation with others (dynamic flexibility). Different forms of industrial organization will produce different processes of innovation. *In the social-organizational model, innovation is a product of the inter-organizational creation of a viable long-term innovation process.*

Thus, rather than confronting the invidious, and quite anachronistic, choices posed by mainstream analysis, contemporary firms, industrial sectors and regional/national economies in the era of permanent innovation are confronted with a far more synthetic problem. Technological innovation is increasingly *a product* of social innovation, of the ability not simply to react and adjust to technical and economic environmental pressures, but precisely to generate the organizational innovation capable of creating an environment in which continuous innovation is possible. Economic success appears to be geared increasingly to the ability to replace traditional organizational alternatives (market and hierarchy) with new logics of production organization: formation of these new production systems relies fundamentally on substantive innovation in the social (collaborative inter-firm relations) and spatial (inter-regional integration) organization of economic activity.

For less-favored regions (LFR's) in particular, the obstacles confronting pursuit of the self-contained district model are clear from the preceding analysis. The attempt to substitute high technology production *de novo* for traditional regional economic strengths is especially problematic given the new requirements of the innovation process, the rapidity of structural change within high technology industry and the structure of the existing global division of labor. Only the ability to engineer a shift in technological paradigms on the basis of distinctive regional breakthroughs is likely to produce successful economic development trajectories within this strategy. On the other hand, adherence to the incremental adaptation of existing regional forms of production is unlikely to alter the prevailing global division of labor and, on the contrary, threatens merely to widen the gap between LFR's and rapidly moving leading-edge regions. The reconstitution of industries based upon traditional regional know-how and institutions, and their integration with specialized high technology cores in a manner that permits rapid adaptation to continuous

123

change, facilitates technological modernization of regional industry and improves local product quality and differentiation, would appear to be a response to contemporary global economic change with the greatest likelihood of success, particularly in a context of expanding *intra-industrial* trade. That is, the application of new technological and organizational forms to traditional industrial strengths -the substitution of social-organizational, for linear and cumulative, innovation-constitutes the core of historical possibilities for LFR's[56].

Indeed, despite the dominion of the Silicon Valley model in regional policy-making over the past decades, examples of the success of this route to regional economic growth abound. In new industries like biotechnology and artificial intelligence, the vast bulk of innovative activity in Europe has been captured by *traditional* regions of industrial strength [Hilpert (1992)]. Moreover, the overwhelming majority of collaborative innovative relationships occur between these innovative regions, rather than between innovative regions and others. The revival of metropolitan areas in the 1980's (Milan, Grenoble, etc.) has commonly followed the strategy of "rupture-filiation", building on traditional territorial strengths while substantially moving in new directions at the same time. The Third Italy and Swiss Jura constitute merely particularly well-known examples of a more widespread process that has affected urban as well as rural regions. The ability of Japanese transplant automakers in the U.S. to operate facilities successfully in the same regions where the U.S. auto industry has failed [Bingham and Sunmono (1992)] testifies to the importance of new collaborative organizational practices accompanying strategies for technology creation at the regional level. The shift of segments of the coal and steel industry in the Kuhr into environmental technologies [Grabher (1993), Nordhaws–Janz and Rehfeld (1994)] and the growth of a micro-instruments industry in the Swiss watch districts [Maillat et al. (1994)] comprise important variations on this strategy.

A further lesson of the preceding analysis is that the kinds of linkages necessary for perpetual innovation in a territorial production system are increasingly unlikely to develop spontaneously: the creation of viable innovation processes and technology fusion presupposes extensive and deliberate coordination as the region seeks solutions beyond those ordained by prevailing allocation systems and cumulative technological trajectories. Coordination confronts substantial difficulties, however, since action by central government authorities is unlikely to amount to much in the absence of local initiatives towards the construction of collaborative modes of interaction while local initiatives, in a highly competitive global economy, necessitate strong support from central government interventions with respect to the construction of scientific and technological infrastructure[57], assistance with the

---

[56] The objective here is the development of specialized or niche market technologies out of the creative fusion of traditional and new technologies.

[57] This would involve such matters as the determination and provision of collective services, programs for the transformation and improvement of labor skills, information and incentives for cooperative interactions between suppliers, producers, and clients, diversification of technology sourcing, the clear specification of regional design and

establishment of linkages to global networks and, perhaps above all, the articulation of coherent cross-regional strategies in order to avoid destructive competition between regions for new technological activities. Moreover, at neither central nor local levels are state-facilitative or state-dependent strategies likely to engender the kinds of collaborative relations necessary to produce innovation under contemporary conditions: state policies must be *interstitial* in character, coordinating a new framework of state-market interactions to enhance and optimize potential regional gains from the impetus towards collaboration engendered by material transformations in the contemporary logic of innovation.

Finally, and perhaps most importantly, technological innovation can no longer be considered within a self-contained or purely localized spatial framework. Regional milieux provide collective learning processes essential to innovation: collaborative learning dynamics, collective services and agglomeration economies strongly link the creation and application of new technologies to regional frameworks of development. At the same time, these mechanisms alone are increasingly insufficient either to initiate or to sustain creative activity as technico-economic complementarities force production chains to incorporate extra-regional sources of innovation. The self-contained region, insofar as it existed at all, is now truly an anachronism.

On the one hand, industrial districts or innovative milieux are compelled to integrate extra-regional contributions as an essential component of the regional innovation process itself[58]. Technological complementarities enforce a broadening of the spatial arc of supply. The increasing application-specificity of products requires clients (predominantly located outside the region) to participate in the design of new technologies in an anticipatory manner. The global spread of innovative capabilities across the chain of production from R&D to marketing, in conjunction with the need for immediate access to diversified global markets, compels firms to enter a range of strategic alliances with other actors beyond the borders of the regional economy[59].

On the other hand, the globalization of industrial networks increasingly involves a transcendence of traditional approaches in which internationalization is propelled by the search for comparative factor advantage or internal transnational investment strategies. In fact, the preponderance of foreign direct investment is not directed towards low cost sites [Schoenberger (1991)] and the proportion of overseas investment in peripheral, as opposed to advanced industrial, locations has declined

---

development capabilities and so on.

[58] That is, in contrast with the argument in Storper (1991), the increasing uncertainty associated with the process of technology creation pushes firms outwards to make extra-regional connections (rather than reinforcing localization) and relations between regions are based upon the organizational dynamics of innovation (and not simply upon trade).

[59] Failure to achieve these linkages is one of the principal reasons for the present failure of many of the industrial regions in the Third Italy.

over time [OECD (1992)]. Rather, FDI is directed towards regions in which multi-national firms can *valorize,* rather than simply *exploit,* specific regional attributes in their attempt to forge new logics of innovation. Multinational networks increa-singly construct themselves not as centrally-coordinated global hierarchies but as a unified set of regionally interdependent, spatially dispersed and quasi-autonomous innovation centers, deploying precisely the specific innovation capabilities imbedded in regional production complexes [Petrella (1989), Julius (1990), OECD (1992)][60].

In this new global context, localized agglomeration, far from constituting an alternative to spatial dispersion, becomes the principal basis for participation in a global network of regional economies. At the same time, the viability of regional economies is a product of their ability to articulate a coherent organizational presence within a global milieu. Regions and networks in fact constitute interdependent poles within the new spatial mosaic of global innovation. Globalization in this context involves not the leavening impact of universal processes but, on the contrary, the calculated synthesis of cultural diversity in the form of differentiated regional innovation logics and capabilities[61].

---

[60] This development in particular accords new prominence in LFR's to traditional business attraction strategies, as long as they are strategically coordinated and integrated with the collaborative relations characteristic of the innovative milieu. The relationship between globalization and regions is discussed at length in Gordon (1995).

[61] It should be noted that, in this context, current EC innovation policy, which aims simultaneously to strengthen local networks of production and create trans-European or multinational industrial networks, operates fundamentally at cross-purposes: regionally introspective policies to build transportation and communications infrastructure, local entrepreneurship, the innovative capabilities of SME's and local collaboration lack any explicit extra-regional dimension, while efforts to create supra-national corporate networks are still guided primarily by traditional science-based innovation theory and lack explicitly regional dimensions. Fundamental reformulation and re-articulation of both strategies is required if they are to promote European competitiveness in a complementary, and not contradictory, manner. For pertinent observations on this problem, see Amin (1992), Bianchi and Miller (forthcoming), Quévit (1992), Perrons (1992), and Lehner et al. (forthcoming).

REFERENCES

ACS, Z.; D. AUDRETSCH, D. (1990), **Innovation and Small Firms**, MIT Press, Cambridge, MA.

AMENDOLA, M.; BRUNO, S. (1990), "The behavior of the innovative firm: relations to the environment", **Research Policy**, 19, 5, Oct., pp. 419–433.

AMENDOLA, M.; GAFFARD, J.–L. (1988), **The Innovative Choice: An Economic Analysis of the Dynamics of Technology**, Basil Blackwell, Oxford.

AMIN, A. (1992), "Big firms versus the regions in the Single European market" in **Cities and regions in the new Europe: The global–local interplay and spatial development strategies**, M. Dunford and G. Kafkalis (eds.), pp. 127–149, Belhaven Press, London.

ASHCROFT, B. (1993), "External control and regional development in an integrated Europe", presented to the **International Congress on The European Periphery Facing the New Century**, Santiago de Compostela, Spain, Sept. 30 – Oct. 2.

AYDALOT, P. (1986), **Milieux Innovateurs en Europe**, GREMI, Paris, 1986.

AYDALOT, P.; KEEBLE, D. (1988), **High Technology Industry and Innovative Environments**, Croom Helm, London.

BIANCHI, P. (1986), **Industrial Restructuring within an Italian Perspective**, Working Paper no. 2, Nomisma, September.

BIANCHI, P.; MILLER, L., "Systems of innovation and the EC policy–making approach" in **Working together for growth: systems of innovation**, P. Bianchi and M. Quéré (eds.), Kluwer, forthcoming.

BINGHAM, R.; SUNMONU, K. (1992), "The restructuring of the automobile industry in the USA", **Environment and Planning A**, Vol. 24, pp. 833–852.

BODDY, M., LOVERING, J.; BASSETT, K. (1986), **Sunbelt City?. A Study of Economic Change in Britain's M4 Growth Corridor**, Clarendon Press, Oxford.

BOECKHOUT, I.; MOLLE, W. (1982), "Technological Change, Location Patterns and Regional Development", **Fast Occasional Papers**, no. 16, Forecasting and Assessment in Science and Technology (FAST), Brussels, June.

BRADSHAW, T. (1985), **The Future of the Electronics Industry in California: Issues of Regional Growth**, Institute of Governmental Studies, University of California, Berkeley.

BRAUN, E.; MACDONALD, S. (1982), **Revolution in Miniature**, Cambridge University Press, Cambridge.

BREHENY, M.; HALL, P.; HART, D.; MCQUAID, R. (1986), **Western Sunrise: The Genesis and Growth of Britain's High Tech Corridor**, Allen and Unwin, London.

BRUNO, S.; DE LELLIS, A., "Innovative systems: the economics of ex ante coordination" in **Working together for growth: systems of innovation**, P. Bianchi and M. Quéré (eds.), Kluwer, forthcoming.

CAMAGNI, R. (1991), "local 'milieu', uncertainty and innovation networks: towards a new dynamic theory of economic space" in **Innovation Networks: A Spatial Perspective**, R. Camagni (ed.), Pinter, London, pp. 121–144.

CAMAGNI, R. (Ed.) (1991), **Innovation Networks: A Spatial Perspective**, Pinter, London.

CAMAGNI, R.; QUEVIT, M. (1991) **Development prospects of the Community's lagging regions and the socio–economic consequences of the completion of the internal market: final report**, Commission of the European Communities.

CHARBIT, C.; GAFFARD, J.-L. ET AL (1991), **Modes of usage and diffusion of new technologies and new knowledge: local systems of innovation in Europe**, 11, FAST Research Paper Series on Science, Technology and Social and Economic Cohesion in the Community, Commission of the European Communities, June.

CHESNAIS, F. (1988), "Multinational Enterprises and the International Diffusion of Technology" in **Technical Change and Economic Theory**, ed. G. Dosi, C. Freeman, R. Nelson, G. Silverberg and L. Soete, Frances Pinter, London, pp. 496–526.

CHESNAIS, FR. (1986), "Science, Technology and Competitiveness", **STI Review**, no. 1, Autumn, pp. 86–129.

COOKE, P.; MORGAN, K. (1990), **Learning through networking: regional innovation and the lessons of Baden–Wurttemburg**, Regional Industrial Research report no. 5, Cardiff, May.

DECOSTER, E.; TABARIES, M. (1986), "L'innovation dans une pole scientifique: le cas de la Cité Scientifique Ile de France Sud" in **Milieux Innovateurs en Europe**, P. Aydalot (ed.), pp. 79–100.

DOSI, G. (1984), **Technical Change and Industrial Transformation**, Macmillan, London.

DOSI, G.; PAVITT, K.; SOETE, L. (1990), **The Economics of Technical Change and International Trade**, Harvester, New York, 1990.

DOSI, G. (1988), "Technical Change and Industrial Transformation" in **Technical Change and Economic Theory**, Frances Pinter, London.

DUBOIS, P.; LINHART, D. (1992), **France: from local networks to a territory network**, Institut Arbeit und Technik, September.

ERGAS, H. (1987), "Does Technology Policy Matter?" in **Technology and Global Industry**, Bruce R. Guile and Harvey Brooks (eds.), National Academy Press, Washington D.C., pp. 191–245.

FREEMAN, C. (1987), **Technology Policy and Economic Performance**, Pinter, London.

GAFFARD, J.–L. ET AL (1990), **Coherence and diversité des systèmes d'innovation**, LATAPSES, Sophia–Antipolis.

GLASMEIER, A. (1985), **Spatial differentation of high technology industries: implications for planning**, Ph.D. dissertation, University of California, Berkeley.

GORDON, R. (1991), "Innovation, industrial networks and high technology regions" in **Innovation Networks: A Spatial Perspective**, R. Campagni (ed.), Pinter, London, pp. 174–195.

GORDON, R., "State, milieu, network: systems of innovation in Silicon Valley" in **Working together for growth: systems of innovation**, P. Bianchi and M. Quéré (eds.), Kluwer, forthcoming.

GORDON, R. (1989), "Les entrepreneurs, l'entreprise et les fondements sociaux de l'innovation", **Sociologie du Travail**, Vol. XXXI, no. 1, pp. 107–124.

GORDON, R. (1990), "Systèmes de Production, réseaux industriels et régions: les transformations dans l'organization sociale et spatiale de l'innovation", **Revue d'Economie Industrielle**, Vol. 51, no. 1, pp. 304–339.

GORDON, R. (1992), "PME, réseaux d'innovation et milieu technopolitain: la Silicon Valley" in **Enterprises innovatrices et developpement territorial**, D. Maillat and J.–C. Perrin (eds.), Neuchatel, pp. 195–220.

GORDON, R. (1993a), "Structural change, strategic alliances and the spatial reorganization of Silicon Valley's semiconductor industry" in **Reseaux d'Innovation et Milieux Innovateurs**, D. Maillat, M. Quévit and L. Senn (eds.), Neuchatel, pp. 51–72.

GORDON, R. (1993b), **Collaborative Linkages, Transnational Networks and New Structures of Innovation in Silicon Valley's High Technology Industry (IV): Industrial Supplier/Services in Silicon Valley**, Report prepared for DATAR, Paris.

GORDON, R. (1995), "Globalization and new production systems" in **The New Divisón of Labor: Emerging Forms of Work Organization in International Perspective**, W. Hittek and T. Charles (eds.), de Gruyter, Berlin.

GORDON, R.; DILTS, A. (1988), **High Technology Innovation and the Global Milieu**, Colloque GREMI II, Switzerland, April.

GORDON, R.; KRIEGER, J. (1992), **Anthropocentric production Systems and U.S. Manufacturing Models in the Machine Tool, Semiconductor and Auto Industries**, 18, FAST APS Research Paper Series, Commission of the European Communities, Brussels.

GRABHER, G. (1993), "The weakness of strong ties: the lock–in of regional development in the Kuhr area" in **The Embedded Firm**, G. Grabher (ed.), Routledge, London, pp. 255–277.

GRANOVETTER, M. (1985), "Economic Action and Social Structure: The Problem of Embeddedness", **American Journal of Sociology**, Vol. 91, no. 3, November, pp. 481–510.

HAGEY, M.; MALECKI, E. (1986), "Linkages in high technology industries: a Florida case study", **Environment and Planning A**, no. 18, pp. 1477–1498.

HILL, S.; MUNDAY, M. (1993), "Foreign direct investment and its role in the economic development of peripheral EC regions", presented to the **International Congress on The European Periphery Facing the New Century**, Santiago de Compostela, Spain, Sept. 30–Oct. 2.

IMAI, K.; YAMAZAKI, A. (1992), **Dynamics of the Japanese Industrial System from a Schumpeterian Perspective**, Working Paper no. 3, Stanford Japan Research Center, July.

JULIUS, D. (1990), **Global Companies and Public Policy**, Pinter, London.

KEEBLE, D. (1988), "High Technology Industry and Local Environments in the United Kingdom" in **High Technology Industry and Innovative Environments**, P. Aydalot and D. Keeble (eds.), Routledge, London, pp. 65–98.

KLINE, S.; ROSENBERG, N. (1986), "An Overview of Innovation" in **The Positive Sum Strategy**, R. Landau and N. Rosenberg (eds.), Washington, D.C., pp. 275–305.

LEHNER, F.; GORDON, R.; CHARLES, T.; HIROOKA, M.; NASCHOLD, F.; NIWA, F., **Global Production Systems: The Future of industry in Europe, Japan and the U.S.**, Commission of the European Communities, forthcoming.

LEVIN, R.C. (1982), "The Semiconductor Industry" in **Government and Technical**

Progress, Richard R. Nelson (ed.), Pergamon Press, New York.

LUNDVALL, B. (1988), "Innovation as an Interactive Process" in **Technical Change and Economic Theory**, G. Dosi, C. Freeman, R. Nelson, G. Silverberg and L. Soete (eds.), Frances Pinter, London, pp. 349–369.

MACDONALD, S. (1987), "British Science Parks: Reflections on the Politics of High Technology", **R&D Management**, vol. 17, no. 1, pp. 25–37.

MAILLAT, D.; QUEVIT, M.; SENN, L. (eds.) (1993), **Réseaux d'innovation et milieux innovateurs**, Neuchatel.

MAILLAT, D.; CREVOISIER, O.; LECOQ, B. (1993), "Réseaux d'innovation et dynamique territoriale: le cas de l'Arc jurassien" in **Réseaux d'innovation et milieux innovateurs**, D. Maillat, M. Quévit and L. Senn (eds.), Neuchatel, pp. 17–50.

MAILLAT, D.; CREVOISIER, O.; VASSEROT, J.-Y. (1992), "Innovation et district industriel: l'Arc jurassien suisse" in **Entreprises innovatrices et developpement territorial**, D. Maillat and J.-C. Perrin (eds.), Neuchatel, pp. 105-126.

MAILLAT, D.; PERRIN, J.-C. (eds.) (1992), **Enterprises innovatrices et developpement territorial**, Neuchatel.

MAILLAT, D.; LECHOT, G.; LECOQ, B.; PFISTER, M. (1994), "Analyse comparative de l'évolution structurelle des milieux: le cas de l'industrie horlagère dans el Arc Jurnssien suisse et français", presented to the Colleque de GREMI on **Les dynamiques d'ajustements structurals des milieux innovateurs**, Grenoble, June 10–11.

MARKUSEN, A.; HALL, P.; GLASMEIER, A. (1986) **High Tech America**, Alen & Unwin, London.

MARKUSEN, A. (1985), **Profit Cycles, Oligopoly and Regional Development**, MIT Press, Cambridge, Mass..

MARKUSEN, A.; CAMPBELL, S.; HALL, P.; DEITRICK, S. (1991), **The rise of the gunbelt: the military remapping of industrial America**, OUP, Oxford.

MORGAN, B. (1993), "Foreign direct investment and regional economic growth," presented to the **International Congress on The European Periphery Facing the New Century**, Santiago de Compostela, Spain, Sept. 30–Oct. 2.

MORGAN, K. (1992), "Innovating by networking: new models of corporate and regional development" in **Cities and regions in the new Europe: the global–local interplay and spatial development strategies**, M. Dunford and G. Kafkalis (eds.), Belhaven Press, London, pp. 150–169.

MORGAN, K.; SAYER, A. (1988), **Microcircuits of Capital**, Polity Press.

NELSON, R.R.; WINTER, S.G. (1982), **An Evolutionary Theory of Economic Growth**, Harvard University Press, Cambridge, MA..

NIJKAMP, P.; MOUWEN, A. (1987), "Knowledge Centers, Information Diffusion and Regional Development" in **The Spatial Impact of Technological Change**, J. Brotchie et al. (eds.), Croom Helm, London, pp. 254–270.

NORDHAWS–JANZ, J.; REHFOZD, D. (1994), **Ergebnisse einer Umfrage in der nordrhein–westfälischen Ummeltschutzwirtschaft**, Discussion Paper, Institut Arbeit und Technik, Gelsenkirchen, August.

OAKEY, R. (1984), **High Technology Small Firms: Regional Development in Britian and the United States**, Francis Pinter, London.

OECD (1992), **Technology and the Economy**, OECD Publications, Paris.

PERRIN, J.–C. (1988–89), **Nouvelles Technologies et Developpement Regional: l'Analyse des Milieux Innovateurs**, CER, Aix–en–Provence.

PERRIN, J.–C. (1992), "Dynamique industrielle et developpement local: un bilan en termes de milieux" in **Entreprises innovatrices et developpement territorial**, D. Maillat and J.–C. Perrin (eds.), Neuchatel. pp. 223–255.

PERRONS, D. (1992), "The regions and the Single Market" in **Cities and regions in the new Europe: the global–local interplay and spatial development strategies**, M. Dunford and G. Kafkalis (eds.), Belhaven Press, London, pp. 170–194.

PETRELLA, R. (1989), "Le mondialisation de la technologie et de l'économie", **Futuribles**, no. 135, pp. 3–33, Septembre.

PIORE, M.J.; SABEL, C. (1984), **The Second Industrial Divide: Possibilities for Prosperity**, Basic Books, New York.

PYKE, F.; BECATTINI, G.; SENGENBERGER, W. (1990), **Industrial Districts and Inter–Firm Cooperation in Italy**, ILO, Geneva.

QUEVIT, M. (1993), "Réseaux de partneriats technologiques et milieux innovateurs" in **Réseaux d'innovation et milieux innovateurs**, D. Maillat (ed.), M. Quévit and L. Senn, Neuchatel, pp. 119–148.

RATTI, R.; D'AMBROGIO, F. (1992), "Processus d'innovation et integration locale dans une zone péripherique" in **Enterprises innovatrices et developpement territorial**, D. Maillat and J.–C. Perrin (eds.), Neuchatel, pp. 167–192.

REES, J.; STAFFORD, H. (1983), **A Review of Regional Growth and Industrial Location Theory: Towards Understanding the Development of High Technology Complexes in the United States**, Report prepared for Office of Technology Assessment, U.S. Congress, April.

REHFELD, D. (1992), **Patterns of economic restructuring in an era of industrial decline: industrial development, change factors and regional policy in the Ruhrgebeit**, Institut Arbeit und Technik, September.

ROGERS, E.; LARSEN, J. (1984), **Silicon Valley Fever**, New York.

ROSENBERG, N. (1976), **Perspectives in Technology**, Cambridge University Press, Cambridge.

ROSENBERG, N. (1982), **Inside the Black Box: Technology and Economics**, Cambridge University Press, Cambridge.

SAXENIAN, A. (1989), "The Cheshire Cat's Grin: Innovation, Regional Development and the Cambridge Case", **Economy and Society**.

SAXENIAN, A. (1981), **Silicon Chips and Spatial Structure**, Institute of Urban and Regional Development, University of California, Berkeley.

SAXENIAN, A. (1990), "Regional networks and the resurgence of Silicon Valley", **California Management Review**, Vol. 33, no. 1, pp. 89–112.

SCHUMPETER, J. (1934), **The Theory of Economic Development**, Harvard University Press, Cambridge, Mass..

SCOTT, A.J. (1983), "Industrial organization and the logic of intrametropolitan location I: Theoretical Considerations", **Economic Geography**, Vol. 59, 1983..

SCOTT, A.J. (1988a), "Flexible Production Systems and Regional Development", **International Journal of Urban and Regional Research**, Vol. 12, no. 2, pp. 171-185.

SCOTT, A.J. (1988b), **Metropolis**, University of California Press, Berkeley.

SCOTT, A.J. (1988c), **New Industrial Spaces**, Pion, London.

STORPER, M. (1985), "Technology and new regional growth complexes: the economics of discontinuous spatial development" in **Symposium on Technical Change, Employment and Spatial Dynamics**, Zandvoort, Netherlands, April 1–3.

STORPER, M. (1992), "The limits to globalization: technology districts and international trade", **Economic Geography**, Vol. 68, no. 1, January, pp. 60–93.

133

STORPER, M.; HARRISON, B. (1991), "Flexibility, hierarchy and regional development: the changing structure of industrial production systems and their forms of governance in the 1990's", **Research Policy**, no. 20, pp. 407–422.

STORPER, M.; WALKER, R. (1989), **The Capitalist Imperative: Territory, Technology and Industrial Growth**, Basil Blackwell, New York.

TABARIES, M. (1992), "Nouvelles PME et cité scientifique en formation: Ile–de–France Sud" in **Entreprises innovatrices et developpement territorial**, D. Maillat and J.–C. Perrin (eds.), Neuchatel, pp. 23–40.

TATSUNO, S. (1988), **The Technopolis Strategy**, Prentice–Hall, New York.

WALKER, R. (1988), "The Geographical Organization of Production Systems", **Society and Space**.

WEISS, J. (ed.) (1988), **Regional Cultures, Managerial Behavior and Entrepreneurship: An International Perspective**, Quorum Books, New York.

WILLIAMSON, O.E. (1985), **The Economic Institutions of Capitalism**, The Free Press, New York.

WILLIAMSON, O.E. (1986), **Economic Organization: Firms, Markets and Policy Control**, Wheatsheaf Books, Brighton.

7 THE DIVERSITY OF EUROPEAN REGIONS AND THE CONDITIONS FOR A SUSTAINABLE ECONOMIC GROWTH

Jean–Luc Gaffard and Michel Quéré
Université de Nice–Sophia–Antipolis, CNRS–LATAPSES

I. INTRODUCTION

We intend in this paper to understand the problems underlying the obvious empirical diversity of economic growth within European regions. More especially, the focus will be on the institutional and organisational determinants of this territorial diversity in a global context that exhibits nevertheless a continuous and sustained economic growth. In a first part, we discuss some analytical implications of this diversity and focus on an analytical framework aiming at defining the conditions at which a sustainable economic growth can be obtained for European regions. In a second part, we provide some empirical illustration, based on recent research financed by the MONITOR/FAST European programme.

II. SOME ANALYTICAL INSIGHTS ABOUT A SUSTAINABLE EUROPEAN GROWTH

Recent contributions to growth theory, initiated by P. Romer's and R. Lucas's models, put into light the role of endogeneous sources of growth, namely the role of externalities and, over all, the role of increasing returns. Nevertheless, what is really analyzed are the conditions and the mechanisms for the exploitation of increasing returns. There is no reference to the conditions in which increasing returns are created. Thereby, what Austrian economists called the "production problem" is not taken into account in this analytical framework.

As a matter of fact, the exploitation of increasing returns implies for the economy to reach the path characterized by the maximum physical growth rate. This entails several implications. First, the obtention of this growth rate requires a strong mobility of production factors. Second, it requires, at the same time, a general adoption of, not only the optimal techniques, but also the optimal institutional rules and organizational forms. Finally, it entails to consider that some agglomeration effects, even when they are due to random factors, play a major role in the process of change.

From an "ex post" viewpoint, growth equilibrium shows itself as the best possible situation even if it is not Pareto–optimal. As a consequence, public

X. Vence-Deza and J. S. Metcalfe (eds.), Wealth from Diversity, 135–143.
© 1996 *Kluwer Academic Publishers. Printed in the Netherlands.*

intervention is aimed at strengthening the exploitation of given increasing returns in order to ensure an even stronger growth.

This type of analysis ignores transition issues. In fact, if we look at empirical evidence, it is obvious that, when the aforementioned strategy is put into practice, the very viability of the process of change is far from being assured. Two phenomena are addressed. The first one is that the mobility and the reallocation of resources imply an unavoidable fall of the growth rate that come before the development along the new path exhibiting the maximum growth rate. This fall is accompanied by sectoral and territorial imbalances, sources of tension and behaviours that might lead the economy far from the new equilibrium. Certainly, this phenomenon arises from a certain resource viscosity –the mobility and the adaptation to the new conditions are not instantaneous– and from the delays of adjustments it involves. The other phenomenom that evidently creates a problem of viability is the recurrence of shocks, i.e. the fact that the goal (the structuration of resources which assure a maximum growth rate) evolves under the effects of innovation in a broad sense (the opening of new markets, industrial reorganizations, the appearance of new products or new production processes).

In general, structural changes put the economy out of equilibrium and the conditions to maintain its viability are likely very different from case to case.

This questions the relevance of strategies aiming at exploiting increasing returns and searching a maximum growth rate. A conjecture can be emphasized according to which growth viability (its sustainability) depends on the way to adopt strategies with the objective of minimizing the distortions of the productive, sectoral, as well as the territorial structure. The truly optimal path is not the one that consists of reaching as quickly as possible the path characterized by the maximum physical growth rate. What really matters is to preserve a certain amount of diversity in the structures and institutions, not only in order to avoid disruptions inducing unviable evolutions but also to keep a maximum of opportunities.

In this perspective, one central concern lies in the understanding of the coordination mechanisms that characterize the economic growth. Specially, the hypothesis of substituability needs to be relaxed. In this inquiry about the irrelevance of the principle of substitution as central pinciple on the basis of which the working of the economy is explained, N. Kaldor expresses a very clear opinion. "This approach, he writes, ignores the essential complementarity between different factors of production (such as capital and labour) or different types of activities (such as between primary, secondary, and tertiary sectors of the economy) which is far more important for our understanding of the laws of change and development of the economy than the substitution aspect" [N. Kaldor (1975/.1978), p. 203]. Most of Kaldor's models of economic growth are equilibrium models. But many contributions (e.g. 1972, 1979, 1986) are dedicated to bring into light the essential features of an economy out–of–equilibrium. Considered jointly with contributions in the same topics by J.R. Hicks, they lead us to formulate a series of conjectures about the way

136

an economy works and evolves.

The first conjecture is that "the cumulative causation" (i.e. an endogeneous growth) cannot be explained but by the existence of complementarities between growth factors: the basic requirement of continued growth is that the various complementary sectors expand in due relationship with each other. The second one is that there is no growth without development, i.e. no quantitative change without qualitative or structural change which necessarily breaks the existing complementarities. The third one is that fluctuations and instability are the consequences of uncoordinated actions of economic agents facing structural changes and market desequilibria. The fourth one is that the required coordination is never complete, with the consequence that complementarity failures are never completely eliminated. Finally, the fifth conjecture is that viability of an economy involved in a process of change fundamentally depends on choices made by the agents as regards prices, stocks, and degree of utilization of productive capacity.

This last conjecture is particularly important. In fact, the main idea behind it is that the working of a fix–price system is very efficient. As Kaldor put it, "...there is no ultimate limit to industrial production in a world context, there is a limit to the rate at which output can grow, a limit given by the sequential character of production,...)". What does it really means?. The existence of a delay of gestation of investment coupled with that of a delay of transmission of information make it necessary not to have too strong fluctuations of prices. Such fluctuations would generate (and effectively generate) distortions in the temporal structure of productive capacity which hamper the mobility of the economy. Changes in prices are essentially temporary but they distort the structures of productive capacity for ever. What is at stake is the possibility for an economy (a system) to experience smooth quantitative adjustments in a context of incomplete information. This requires on the one hand price rigidities, on the other hand external interventions. Out–of–equilibrium, an economy cannot be viable but it is open and this openess is efficient whether its prevents too strong distortions of the productive capacity along the way.

Summing up. Complementarity between the different elements of a closed (or local) system —be it a sector or a region— is the actual source of a permanent growth. But such a complementarity is never completely realized because recurrent shocks break it systematically. As a consequence, the sustainability of economic growth essentially depends on the ability of economic agents to cope with complementarity failures. Price rigidities, and the existence of reserve assets are the main expressions of the required strategies. But, fundamentally, it is the organization of industry, i.e. the complex network of intra– and inter–system linkages, which favours (or not) the viability of the system.

137

III. AN EMPIRICAL ASSESSMENT OF GROWTH CONJECTURES: THE CASE OF INTRA AND INTER–REGIONAL LINKAGES IN EUROPE

A territory facing structural change can be defined as a set of economic agents implementing innovative strategies and of institutional structures aiming at favouring these innovative strategies. As such, innovative firms are not simply locating in accordance to some initial factors endowments, but they are also creating locally new specific opportunities through the learning effects stemming from local interactions. Companies need to use factors and more largely resources, internally and externally as well, to implement their innovative strategies. This dynamic and cumulative process implies to consider not only the exploitation of local resources but mainly the conditions by which a specific area is able to set up new resources. As a consequence, the viability of local economic systems depends on the existence of a institutional structure capable to favour the coherence of a complex set of internal and external relations. It is difficult to identify "ex ante" one such structure that could be seen as a model, according to some sectoral characteristics, to the size of companies, to the scale of production, and so on. For a territory, the viability of economic growth precisely requires a dynamic complementarity between internal and external resources under these institutional constraints.

Empirical studies of European regions allow to consider four different type of territorial strategies in order to face innovation or structural change along the line sketched out above. A first type of territorial agglomeration is one dominated by internal cooperative relationnships and by market relations with the external environment. This type of territorial development is common to very different territories, if considering the average size of companies and the concrete mechanisms implementing cooperative relationships. These regions (the Italian industrial districts, the Bade–Wurtemberg region, etc.) are places where the setting-up of new technological resources is the outcome of local learning among companies through specific partnerships. What can be named as a "cooperative institutional model" lies in this collective capability to learn which involves not only companies strategies but also incentives and information stemming from diverse local institutions (banks, business and trade associations, policy–makers, etc). The complementary effects stemming over time from innovative strategies are largely dependent on this collective capability to coordinate. Two different aspects are central: internally, learning and externalities need to be positively organized; externally, market information is shared and allows to avoid negative cumulative effects. Such a coordination of the production system implies means of exchanging information locally and learning about financial services, R&D activity, market functioning, technical devices, quality control, industrial relations, and so on. For policy–markers, favouring such exchanges of information can be thought of as the essential objective in order to implement structural change in a viable way.

However, this type of regions are also facing an external dependence. Their internal capability to face structural change is largely dependent on, on the one hand, their ability to attract and use external resources in order to enrich internal change

138

(in terms of financial investment, for instance) and, on the other hand, their ability to follow external market evolution. The latter can induce important internal restructuring to face changes in the external demand. This is a means of discussing the industrial district crisis where it seems that changes in the external demand largely involved internal reconfigurations based on a process of more explicit concentration and vertical integration.

A second type of European regions facing innovation can be characterized by the reverse strategy. There are territories where internal relationships are dominated by market agreements and where cooperation mainly occurs through external relationships. The aim of such areas is to ensure "balanced" transactions in the perspective of facing innovation processes without important price fluctuations or structural shocks or markets (concerning labour, intermediary goods, or final goods as well). This situation can be found in contexts such as Silicon Valley or South California in the United States. In Europe, this type of regions corresponds to metropolitan areas like "Ile–de–France" or "the London area". These territories are innovative places in so far as external relationships allow for cooperative agreements because internal relationships are largely dominated by market agreements. Therefore, the viability of these territories in facing innovation depends crucially on the capability to manage a relative equilibriated path (between external and internal relations) ensuring the existence and the disponibility of resources in the long run. This is certainly the best strategy to be implemented by local policy-market in order that no bottleneck can occur in the evolution of market relations in these areas.

A third type of regions is represented by areas that are dominated by external constraints. This means territories where internal components (and firms especially) have major relationships with external partners for all kinds of transactions. In other words, both market transactions and cooperative agreements are organized with the external environment, and even the local labour market is largely depending on the external supply. Therefore, those territories are very unstable in facing structural change because of the dependence from external decision centres. The positive effects on the local dynamics are weak and mainly embodied in phenomena like spin–offs, start–ups, local mobility of the human resources initially external to the region. Indeed, the possibility of policy–makers to diminish this intrinsic fragility of such territories is weak because they have difficulties to influence strategic behaviours that are decided outside their areas. Typical examples of such regions are the peripheral areas which largely depend on foreign investments. For instance, Scotland based its strategy on external American investments, and "Provence–Alpes–Côte d'Azur" has attracted external French and American large companies. These very different areas have in common the instability of these external locations and, as a consequence, the difficulty to set up a viable strategy locally. The internal complementarities expected from these external investments are difficulty to establish and this induces important bottlenecks in local economies (on the labour market or on the market for intermediary goods especially). Here, public intervention is certainly essential, even if difficult. Its objective should be to secure these local investments by favouring their participation in long term research or technological programs in view of creation of specific resources. In the most favourable case, one can expect a transition from these areas toward one or the other previous types (industrial district or metropolitan area).

A fourth type is represented by the European regions where internal and external relationships include in a significant way both types of relationships (market and cooperative agreements). Here, nobody can precisely anticipate what type of functioning will dominate the evolution of these areas. Therefore, we call it regions in transition. Different kinds of evolution can be discussed in this category. For instance, Wales belongs to this category when we consider the way in which this region succeeded in the restructuring of its steel industry because of a stronger partnership with other European regions that structurally modified the characteristics of its labour market. In another way, the Turin area is also developing a "comparable" strategy. Due to a change in the strategy of the major local firm, sub-contracting companies have tried to diversify and to find external markets in other European regions. This phenomenon allows the area to change drastically the characteristics of its competencies and of the skills of its internal labour market. However, one cannot be sure today that this evolution foresees the establishing of an industrial district mode of functioning. In the two last cases, the role of policy-makers is to allow for such changes by favouring the evolution they consider as the most viable strategy for the area.

IV. CONCLUSION

The viability of European regions in facing innovation and structural change largely depends on their ability to avoid too strong structural distortions. That means especially the capability to avoid strong price fluctuations on markets (concerning wages level, but also intermediary and final goods markets). European regions that benefit from a stable economic growth are those that are able to smooth price fluctuations in the long run. In that perspective, the interplay that exists between internal and external resources is crucial to implement such a strategy.

By considering empirical observation related to the ability of European regions to face innovation, one can discuss different strategies embodied in our taxonomy of regions. To a certain extent, this allows us to map and justify the existence of intra–european diversity of regions. The strategies of European regions in facing innovation are effective in so far as they are coherent. That means aiming at developing "equilibriated" transactions with the environment, as in the case of industrial districts or metropolitan areas in our taxonomy. The interest of this analysis is to consider that, because of the interdependecies among regions, no optimal model of local development exists; on the contrary, what is essential is not the search for such an optimal model but the establishing of a conceptual framework aiming at coping with the problem of viable strategies of local development.

REFERENCE

AMENDOLA, M.; BRUNO, S. (1990), "The Behaviour of the Innovative Firm: Relations to the Environment", **Research Policy**, 19, pp. 419–433.

AMENDOLA, M.; GAFFARD, J.L. (1988), **The Innovative Choice: an economic analysis of the dynamics of technology**, Basil Blackwell, Oxford.

AMENDOLA, M.; FROESCHLE, C.; GAFFARD, J.L. (1993), "Sustaining Structural Change: Malthus's Heritage", **Industrial Dynamics and Structural Change**, Forthcoming.

AOKI, M. (1988), **Information, Incentives and Bargaining in the Japanese Economy**, Cambridge Universtity Press.

ARTHUR, W.B. (1988), "Urban Systems and Historical Path Dependence" in Ausubel and Herman (eds.), **Cities and Their Vital System**, National Academy Press, Washington D.C..

ARTHUR, W.B. (1989), "Silicon Valley Locational Clusters: when do increasing returns imply monopoly", **Working Paper**, Santa Fe Institute.

ARTHUR, W.B. (1990), "Positive Feedbacks in the Economy", **Scientific American**, February, pp. 92–99.

BECATTINI, G. (1992), "Le district marshallien: une notion socio–économique" in Benko, G. et Lipietz, A., **Les régions qui gagnent**, PUF, Paris.

BRUNO, S.; DE LELLIS, A. (1992), **The Innovative Systems: The Economics of "ex ante" Coordination**, contribution to the International Conference on Innovation Systems, Bologna, October 5–6.

BRUSCO, S. (1990), "The idea of the industrial district: its genesis" in Pike, F. et al. (eds.), **Industrial Districts and Inter–firm co–operation in Italy**, Geneve, ILO.

CHARBIT, C.; GAFFARD, J.L. LONGHI, C.; PERRIN, J.C.; QUERE, M.; RAVIX, J.L. (1991), **Local Systems of Innovation in Europe**, Working paper of the European Comunity, (MONITOR/FAST Programme), Vol. XI, FOP 235.

GAFFARD, J.L. (1990), "Innovations et changements structurels", **Revue d'Economie Politique**, 100 (3).

GAFFARD, J.L. (1992), **Territoire et Innovation en Europe**, Miméo, CNRS-LATAPSES, 16 p.

GAFFARD, J.L.; BRUNO, S.; LONGHI, C.; QUERE, M. (1993), **Cohérence et diversité des systèmes d'innovation en Europe: éléments d'analyse et de politiques économiques pour les Communautés Européenes**, DG XII, Programme MONITOR/FAST, LATAPSES.

GAFFARD, J.L.; ROMANI, P. (1990), "A propos de la localisation des activités industrielles: le district marshallien", **Revue Francaise d'Economie**, Vol. V, 3, pp. 171–185.

GORDON, R. (1990), "Systémes de production, réseaux industriels et régions: les transformations dans l'organisation sociale et spatiale de l'innovation", **Revue d'Economie Industrielle**, no. 51, pp. 304–339.

HICKS, J.R. (1973), **Capital and Time**, Clarendon Press, Oxford.

KALDOR, N. (1966), **Causes on the Slow Rate of Economic Growth in the United Kingdom: An Inaugural Lecture**, Cambridge, Cambridge University Press.

KALDOR, N. (1967), **Strategic Factors in Economic Development**, New York, Cornell University Press.

KALDOR, N. (1972), "The Irrelevance of Equilibrium Economics", **Economic Journal**, Vol. 82, no. 4, December.

KALDOR, N. (1975), "What is wrong with economic theory?" in **Quarterly Journal of Economics**, reprinted in **Further essays in Economic Theory**, Duckworth, London, 1978.

KALDOR, N. (1979), "Equilibrium Theory and Growth Theory", dans Boskin (ed.), **Economics and Human Welfare**, New York, Academic Press.

KALDOR, N. (1986), "Limits on Growth", **Oxford Economic Papers**, Vol. 38, pp. 187–198.

KRUGMAN, P. (1991a), "Increasing Returns and Economic Geography", dans **Journal of Political Economy**, Vol. 99, no. 31, pp. 483–499.

KRUGMAN, P. (1991b), "History and industry location: the case of the manufacturing belt", **American Economic Review**, Mai, pp. 80–83.

LONGHI, C.; QUERE, M. (1993), "Systèmes de production et d'innovation, et dynamiques des territoires", **Revue Economique**, Vol. 44, Juillet, pp. 713–724.

LUCAS, R.E. (1988), "On the Mechanics of Economic Development", **Journal of Monetary Economics**, 22, pp. 3–42.

NELSON, R.; WINTER, S. (1982), **An Evolutionary Theory of Economic Change**, Belknap Press of Harvard University, Cambridge, Mass..

ROMER, P. (1986), "Increasing Returns and Long Run Growth", **Journal of Political Economy**, 94, pp. 1002–1037.

SCHUMPETER, J.A. (1934), **The Theory of Economic Development**, New York, Harvard University Press.

WILLIAMSON, O. (1985), **The Economic Institutions of Capitalism**, The Free Press, New York.

8 INNOVATION, REGIONAL DEVELOPMENT AND TECHNOLOGY POLICY: NEW SPATIAL TRENDS IN INDUSTRIALIZATION AND THE EMERGENCE OF REGIONALIZATION OF TECHNOLOGY POLICY

Xavier Vence-Deza[62]
Department of Applied Economics & IDEGA
University of Santiago de Compostela, Galicia

I. INTRODUCTION

The reemergence of the regions as promoters of economic development is a result of profound changes in the national and global economy and a deep movement to a broader regionalization. Many reasons contribute to this trend. The last decade is marked by great transformations in the structure of production systems; many manufacturing sectors have experienced declining production and employment due to the offshore location of traditional industries, the introduction of labor-saving technologies and the newly emerging sectors show a nuanced geography. Changes in the competitive conditions at European and world levels in a context of globalization have been accompanied, amongst other things, by huge transformations in production processes, changes in industrial organization, the development of new activities and products. Not all territories were able to adapt themselves to these changes at the same pace, nor did their reactions follow a single and uniform pattern. The geography of production suffered considerable changes, both at a world level and within each country. There is a great diversity in territorial dynamics which reflects the failure of governments' traditional regional policies (incentives and grants) and the need for a new approach, based on the invigorating capacity of the regions and the local networks. This diversity of dynamics also reveals the different and uneven capacity of each territory to assume innovation strategies and to incorporate the changes into technology and markets. A controversy exits about the real power of regional government to affect the level and growth rate of the economy and to support structural change. Both central and regional governments have different and specific ways of acting yet many goals require a close cooperation between different levels of governments. European and Spanish experiences can be a good reference to reflect on.

This article aims to revise the recent litterature on regional development,

---

[62]I would like to thank Nikos Kastrinos for his suggestions to the original version and to Xulia Guntin for her help in the statistical data handling.

X. Vence-Deza and J. S. Metcalfe (eds.), Wealth from Diversity, 145–197.

technology and policies pointing out the limits of standard approaches and current policies, mainly keeping peripheral regions in mind. The evaluation of regional impact of European Technology Programmes and the Spanish distribution of R&D activities provide empirical findings to support a new approach turned to the regional initiative.

II. TECHNOLOGY AND TERRITORY

The importance of technological change in economic development has been well established since the time of classical economists and some heterodox authors such as Marx or Schumpeter. The **neoclassical approach** paid less attention to this, in so far as it was excessively subjected to the general equilibrium model, which excludes structural change and dynamic evolution. For some time the unbearable weight of events imposed the need to incorporate technological change and innovation as the motor of economic development. So much so, that the general innovation capacity of a country has today become a key element for identifying its development power. However, at an analytical level technological change remained for a long time an exogenous variable in neoclassical growth models. In the last decade the situation has experienced a great stride in this field due to the developments in evolutionary approaches. Even the mainstream has contributed to this renewal with the "new (endogenous) growth theories".

Still less attention has been paid to the relationships between innovation and territories. A survey of the literature shows that the spatial dimension is a problem rarely incorporated in theories of innovation and technological change (Vence 1995). It is not easy to establish a convincing methodology to tackle the issue. As in other fields of economic theory, space (or territory) is a "disturbance" element for the "grandiose models". It is a difficult element to integrate into the abstract models precisely because from a territorial viewpoint innovation must be considered as something concrete, incorporating complete information from its economic environment. The problem also appears when we reflect on regional aspects of technology policy.

Consequently, it is vital to establish a conceptual framework to begin the study of the relationships between technological change and changes in spatial dynamics. Furthermore, it is also necessary to establish the mechanisms which allow spaces to develop their productive system throughout time, increasing productivity and capability for diversification and self-transformation. In this paper the aim is to systematize some aspects in order to understand **the dynamic innovation capability of a territory**. I will therefore try to establish conditions for an innovative and self-transforming activity within a **territorial productive system** and to sett up policy means to support such strategy.

The aforementioned issues cross the debates which have enriched the analysis on territory, industry and regional development in the last few years:

industrial districts, networks, technological districts, "industrial organization", endogenous development etc. Among the relatively few attempts to tackle the space/innovation relationship there are many different approaches according to one's concept of space and innovation, wich can esentially be summarized in two different paradigms (Aydalot 1986).

Space could be either seen as inert, accepting originally exogenous activities, or as a space with a specific history and structure that set up its own capabilities for development. The neoclasical theory of regional differences is based on uneven factor endowments. This identifies diferences in economic structure, and productivity, with differences in endowments of productive factors, on the assumption that regions are identical with respect to the preferences of the population and technology. The evolutionary perspective, for example, paid great importance to differences in technology (as a dynamic process of absortion of different technologies in an economic structure). In an evolutionary perspective the diversity of growth is a consequence of and a cause of evolutionary mechanisms (increasing returns and cumulative phenomena), in which innovation plays a major role (Metcalfe 1993). I would like to stress two important ideas on the subject. First, territorial externalities exist which are exploited by firms (increasing returns); such externalities derive not only from infrastructures but are also built by firms and institutions along the process of creating resources; territory is an active element in some important dynamic processes within firms (learning, exploring innovations, investement...) where all decisions imply sunk-costs, cumulativeness and irreversibility at territorial level. Second, regional clusters are not closed systems but their viability is a product of their ability to weave a consistent web of external relations with both their resource suppliers and their customers in a global market (Gaffard 1992; Gordon 1990).

Trying to conceptualize **innovation**, in the standard "linear model" one can differentiate between its genesis and its subsequent dynamics, that is to say, between a problem of creation of technology and a problem of selection techniques (Amendola&Gaffard 1988); the linear model sequence may suggest that R&D and the subsequent steps of the chain: innovation, production and economic growth, will take place within the same territory. This assumption can lead on policy makers to expect that policy measures aimed at increasing the level of R&D are enough to achieve a corresponding increase in technological innovation. Even more, since it is basic research from which innovation ultimately flows, basic research is a necesary object of policy (Malecki 1990: 100). The "chain linked model" from Kline & Rosenberg shows the complexity of multiple feed-backs between diverse activities that interact in a real innovation process; success in innovation depends not only on R&D capability but also on engineering capability and information about markets, customer's needs and competitors, among others. We should also bear in mind the differencies between product innovations and process innovations as well as the different aspects of current technological change (New Technologies, the microelectronic revolution, New Technologies of Information, flexible automation etc) that have a diverse role in the regional development. Summing up, to analyse

147

the conditions for innovation in a regional productive system, two approaches could be used which, on the whole, correspond to two different concepts of space. One consists in analysing the influence of New Technologies on industrial localization trends and on the spatial division of productive processes. A different approach stems from the observation of innovation and diffusion processes in concrete spaces and tries to establish the patterns to explain the uneven (and diverse) innovation capability of each space (Aydalot 1986; Gaffard 1992); here a territory is a **local system of innovation** which allows us to understand an innovation as a whole dynamic process, which includes more than one technology and also involves not strictly technological dimensions. From this perspective the key elements in defining the effectiveness of a region as an engine of growth are not only firms and markets but also the specific institutions, their goals, their decision-making processes and, in particular, their cooperation with firms and other institutions.

Empirical studies on geographical distribution of innovative activities show a very uneven regional distribution and a great concentration in some certain places (Feldman 1994; Audretsch&Feldman 1993). One obvious explanation why innovative activity clusters geographically is that the location of production is spatially concentred too. In this way, we can expect that the degree of innovative activities concentration in each industry depends on the degree of spatialy production concentration of each industry. In fact the relationship is much more complex. For example, the study by Feldman & Audretsch on the United States case concludes that the empirical evidence suggest that the propensity for innovative activity to cluster is more attributable to the role of knowledge spillovers and not merely to the geographic concentration of production (Audretsch&Feldman 1993). Moreover, industries in which knowledge spillovers are presumably the most prevalent, that is where industry R&D, university research and skilled labor are the most important, tend to have a greater propensity for innovative activity to cluster than in those industries where knowledge externalities are less important.

Technological change is a complex process of creation, development and difussion of new technologies, new knowledge, new forms of organization and new products. It is a complex process in which different types of agents participate, with different functions, at different levels and usually in different moments of time and also from different geographies. This broad set of economic and social factors constitutes the National/Regional System of Innovation and shape its patterns of evolution through time (Nelson 1993). The process is complex because innovation is not a instantaneous phenomenon but takes place through the time in a sequential, cumulative and irreversible fashion (Amendola&Gaffard 1988; Dosi et al 1988). And then again this process has a plain **spatial dimension**: global and local at the same time. Both its genesis and its industrialization take place in a global process where a number of agents take part, each one anchored in a concrete territory so that only territories capable to participate in this process can benefit of the dynamic effects of innovation and new markets opened by it. This activity of innovation-industrialization takes place in a territorialized system, creating interdependencies at local level, external economies and spillovers that conform specific conditions to development in each territory.

III. THE DIFFUSION OF INNOVATION AND THE TERRITORIAL ATTRACTION CAPABILITY: INDUSTRIAL ORGANIZATION AND LOCALIZATION

III.1 Diffusion and "Space-Point"

One common method of tackling the analysis of development is to consider space as accepting activities of exogenous origin. This is the prevailing perspective, for example, in studies of diffusion in which space is in fact inert, completely void of all significant characteristics (e.g. Griliches, Mansfield). Moreover, in most studies based on an evolutionary approach (Dosi's technological paradigms) or techno–economic paradigms (C. Freeman, C. Pérez), space don't exists or at most it seems to be taken into account only as a selection element (the environment), defined by resources, industrial structure and, in some cases, institutional system.

These approaches constitute an important part of current studies of industrial economy and, although different in other aspects, coincide in the importance attributed to the trends of the evolution of technical change and industrial organization. This question is studied using some of the variables which reflect the evolution of structural conditions of production, forms of competition, determination of profit rules, new forms of organization (integration–subcontracting). It is assumed that these aspects would determine the rhythm and characteristics of technical progress and the needs of firms in terms of work, skills and infrastructures.

In an increasing globalised economy, the prevailing standard approach considers that different spaces develop almost automatically according to the correspondence between structural characteristics of region and the general conditions of industrial change, and in particular according to its factor endowment. In other words, from this point of view, space is either absent or it is inert and passive, with an evolution derived from forces beyond the spatial field: if its configuration responds to the present needs of the most dynamic capital it will develop, otherwise it will fall.

To analyze spatial dynamics of innovation in the traditional "linear model" we can distinguish between the phase of creation of innovation and the later phase of diffusion amongst new users, new sectors and new spaces. In some standard theoretical approaches, the spatial dynamics of diffusion is established following more or less the base–ideas of the product cycle theory, and in this sequence: the first innovations appear in certain developed regions (countries), which have the suitable conditions in terms of resources, complexity of productive system, research capacity, market etc. Later, the process becomes internationalized/interregionalized, but only regions with developed areas can hope to take part in this global process. In this model, technology is fully defined from the start, so that it will be simply and

149

directly incorporated into the existing productive structure. Innovation created at an international level will slowly begin to spread all over the globe. This however will not be done evenly, but will occur according to the relative capability of each region to adopt these changes. Consequently, the regions will become hierarchical according to their capability to assume innovation, and their ability to incorporate it into the productive system. And, in standard analysis, it is assumed that this capability will be given mainly as a result of factor endowment.

III.2 Trends in the Evolution of Spatial Organization of Production and the Role of Technological Change

To study industrial change in a concrete spatial environment it is important to explain the method of tackling the relationships between technological change and change in spacial dynamics. One way to tackle these is to consider the patterns of the latter in the recent decades and try to discover the role played by new technologies, as just another element amongst others (Aydalot 1984: part I&II). More specifically, we can see this by means of the relationships between new technologies and changes in industrial localization (and industrial organization).

In the sixties and seventies a great process of territorial diffusion and dispersion of economic activities towards new areas had taken place. This is known as spatial division of production processes. At an international level this is seen in the internationalization of production, led by multinational firms, which, to a certain extent, breaks with the traditional polarization and concentrated organization of industrial space. In the eigthies new phenomena emerged whereby spatial division were joined by externalization of processes of production based on an organization of subcontracting and/or cooperation networks. As consequence a slow down in the spatial dissemination process of these activities could be observed, even accompanied by an inverse movement of recentring and "repatriation" of part of these activities, which consequently led to the protagonism of the urban metropolis.

During the delocalization period, which peaked in the years 1960–70, companies divided their activities into units-plants. This division was done as homogeneous as possible according to the type of work, and plants were set up in places where there was a suitable labour force available, in an attempt to lower labour costs (replacing traditional immigration towards the metropolis). This kind of production segmentation is particularly evident in Fordist processes (automobile, equipment, etc). Some activities were left in the central and urban offices (management, coordination, laboratories, some production units, etc) supporting agglomeration costs and higher salaries. Other activities were located in small or medium–sized towns taking advantage of the rural labour force and avoiding many other costs (rent, taxes, etc), unions etc. This movement occurred both within and across the borders.

In any case, this shift from sectorially polarized spaces (Perroux,

Boudeville) to spaces specialized in hierarchical funtions was only possible due to a series of circumstances in which a wide range of factors of different intensity and duration coincided (progressive reduction in transport costs, predominance of large companies, internalization and vertical integration of production and other services...).

Examining the movements of industrial localization within Galicia in the crisis period, we see that they are clearly marked by the supply of raw materials or basic inputs (energy) and also notice a slight movement away from the cities in some traditional sectors. For example, in the textile and food industries we see a fairly obvious process of re–ruralization (deurbanization). This at least is the case for manual activities although management and commercial activities remained in urban areas. In our case in particular, the trends described don't seem to be excessively influenced by reasons linked to technological change but rather by the close proximity of raw materials in these cases and the preservation of a flexible labour market, a cheap and obedient labourforce, etc, amongst other things. This movement away from urban areas would show an active policy of labourforce management directed towards recuperating an even greater capacity for reorganizing production (decentralization, workshops, putting-out system...) and the reduction of agglomeration diseconomies. Other factors such as the rise in the price of urban land and the growth of the city centre leads to the obsolescence of buildings and land speculation.

From a long term perspective we can make a distinction between two groups of elements which influence territorial tendencies of productive organization and localization in the postwar period: some follow steady trends and others are of a cyclical nature.

III.2.1 Tendencial Elements: New Technologies, organization and distance

Various important **tendencial elements** are emphasized in the literature:

a. A regular reduction in transport and communication costs as a consequence of continual technical progress in infrastructures and means of locomotion and transport (by sea and rail, lorry or aeroplane) and in the means of information transmission.

b. The trend towards concentration of capital is permanent suffering hardly any setbacks at all; crises are extremely short periods as far as concentration is concerned.

c. A steady lightening in techniques and production, with a small reduction in the relative importance of material production with respect to immaterial production (computerization of production or in other words " production intellectualization") which makes delocalization easier. This corresponds to an important increase in productive services, R&D and engineering activities.

d. Fiercer competition in the market favours a need to find a cheaper labourforce (in the periphery and even beyond market economy)

e. Every new technology introduced (since the Industrial Revolution) has always tried to make multispatialization easier, gradually reducing the importance of proximity and homogenizing space.

In contrast to the traditional localization theory (Weber and Marshall) which considers proximity as vital for creating economies to scale, agglomeration externalities and for reducing the transmission costs of goods and information, the aforementioned trends underline that technical change reduces distance-costs and proximity-advantages.

Based on this kind of argument, certain analyses upheld the hypothesis that telecommunications development in the eighties and improvement in transports lead to a strong shift in **comparative advantages, break the agglomeration economies** and allow a decentralization of activities moving out to the periphery and creating new growth poles (Planque 1983: chpt. 4; Piore&Sabel 1984). From this point of view the spatial system would be led to a higher degree of complexity altering previous spatial hierarchies, built since the first half of the 20th century and particularly in the postwar period (the so-called Fordist division of space). As B. Planque says, these changes "would tend to cause the emergence of multiple poles, interelated and interdependent in their diversity rather than to oppose growth-spaces, some of them encouraged by the decentralization of banal functions and other specialised in information and innovation (development poles). In the new model neither of multiple poles maintains the monopoly of innovation capability, and the newest, even being peripheral, have no less capacity to respond to the needs of adaptability, competitivity etc" (Planque 1983, 177). Even point at the possibility of a new tendence towards "decentralized invention", as a consequence of progress in "telematics".

Apart from this type of analysis, which could be called "optimistic" and which suggest that New Information Technologies allow decentralized development, in general more exhaustive studies on the subject tend to conclude that new technologies don't seem to be conclusive factors of a new territorial organization of industry[63]. New information technologies, merely, are not the explicative factor of the dynamics of spatial division of labour because they give rise, not only possibilities of decentralization, but also generate centripetal and cumulative local dynamics which lead to new forms of polarization.

---

[63] As shown by the works of "Groupe de Recherche Européen sur les Milieux Innovateurs", directed by P. Aydalot. [See Ayadlot, P. (1986) (ed), **Milieux innovateurs en Europe**, Paris, GREMI]. New approaches are also presented in the collective works of Benko, G.; Lipietz, A. (1992) (dir), **Les régions qui gagnent**, Paris, PUF.

In fact they should simply be considered as a means of decentralization or polarization, following the trend laid down by other types of social or economic variables. We could say that new information technologies reduce important traditional factors such as distances (considered as transport costs or comunication difficulties), and allow other fundamental economic variables to emerge. However, space is not neutralized. Rather, new forms of differentiation appear, leading to new forms of functional space hierarchies based on the existing regional/local labour market, infrastructures, telecommunication networks, and the dynamics created by public and private agents acting in this space (Amin 1992; Veltz 1992). For exemple, only a few cities are fully linked to the rest of the world by telephone, telecommunications and the interconnection of computerised networks. Progress in telecommunications enlarges the market but does not reduce the advantages of those areas with better communications. On the contrary. The effective availability of good infrastructures and telecommunication networks becomes more and more important, to the point of becoming a critical factor.

Moreover, distance/proximity affects other aspects that go beyond the impossibility of communication or transport costs and which become an ever-increasing priority, as studies of **industrial districts** show:

    – distance limits information exchanges such as direct contacts, dialogues etc. which are very important in certain activities, especially in more creative and less standardized activities.

    – distance reduces non–arranged or chance encounters which are often of great importance and profitability.

These two factors acquire greater relevance in knowledge and technology industries, particulary for small and medium–sized firms, where the synergies derived from these direct and chance exchanges are important for progress and for obtaining opportunities. In these types of technology–intensive activities, different activities such as conception/design of products and processes, organization, management, production and adaptation to usage are not separable nor strictly successive, but strongly intertwine, creating constant feed–back relationships. For this reason, agglomeration factors in these activities are not so much traditional localization factors (transport costs etc.) but instead other factors, related to the existence of local labour markets for engineers and technicians or to the defective spatial fluidity of the information and knowledge. This is a tacit and local ressource that frequently benefits from active cooperation which is better with the physical proximity of individuals (Marshalian "atmosphere").

To stress this view it is necessary to emphasize that concrete sectorial studies show a polarization process in the High-Tech Industries, concentrating growing parts of production in the same area. For example, in microelectronics, the segmentation shown by the Silicon Valley experience corresponds well to the old model of spatial division of labour.

153

Instead, in these studies emerge new forms of spatial organization in economic activities and new criteria for the space hierarchization which differs from the Fordist bipolarization (Veltz 1986). The introduction of new technologies combining computing and telecommunications permit a spatial deconcentration of automatized plants, combined with a strong functional concentration and a high centralization of the decisions even more important than in the Fordist model. This evolution does not extend to all industries but allows us to consider the existence of a logic of **functional** space hierarchization which in part substitutes the hierarchy based on **productive specializations** typical of the previous model. In contrast to a **zone space**, a **network space** with hierarchical functions would begin to emerge in some industries.

The Galician industry shows examples of industrial organization which adequately respond to this model, even in traditional activities as clothing industry. The production organization of the "Zara Group" shows this productive decentralization (in factories and workshops) and a concentration of conception, design and management activities. The implementation and importance of these organization methods in networks is largely related to the growing importance of logistics ("Just–in–time", "zero–stocks" supplies etc.), time management, the growing importance of conception and commercialization tasks as opposed to production itself, etc.

In this way some authors emphasize the appearance of contradictory trends both from the point of view of the centralization/decentralization of power and from the point of view of concentration/deconcentration of functions (Beckouche, Savy & Veltz 1986). There are three possible hypotheses:

a. new networks could bring about a reinforcement of decision–making centres to such an extent that they can stay in large cities.

b. on the other hand, information technologies could cause the decentralization of secondary power and functions linked to the management of manufacturing and distribution.

c. then again, computing and telecommunications do not seem to reduce the need for exchanges in the R&D and design phases because an ever-increasing density of networks of individuals will become necessary.

Although these recent changes are important, it is true that the spatial division of labor continues to intensify in some traditional sectors, whose dispersing and internationalizing trajectories have been observed for decades (automobiles, textiles, ...). Flexible automation and new production organization criteria (subcontracting, JIT...) are giving rise to a modification of the comparative advantages in favour of developed regions (and countries). In fact, in the activities where there is changing demand, flexible automation could be a substitute for active delocalization towards countries with low salaries (Mouhoud 1992).

154

Therefore, a survey of different studies on the subject shows that new technologies in themselves don't necessarily cause an aeceleration of spatial division of labor, although its appearance will coincide with a movement of delocalization of increasing segments of the productive processes towards geographically distant areas. So, although new technologies in themselves do not have to produce an evolution in one sole direction, new technologies can indeed be used by the logic of multispatialization and to emphasize spatial division of labour, as has been the case of all the developed technologies of transport–communications (boat, rail, automobile, aeroplane, telephone, etc). These can be exploited to make effective a tendency to spread in the space human activities and functions stemming from economic, social and even cultural factors. However, the complexity of the forces in hand, obliges us to introduce an additional point, in the sense that this trend could co–exist whith another of no lesser importance: in parallel to multispatialization, there is also a development of local initiatives which, when successful, cause cumulative dynamic effects and give rise to trends towards bipolarization. In this competitive game, urban or metropolitan areas have considerable advantages.

Summing up, from this perspective, new technologies of information don't mark a steady tendency of delocalization of industrial activities towards peripheral spaces. There is no characteristic in new technologies which allows to consider the peripheral and backward economic regions as potential beneficiaries if they maintain a passive behaviour. Incorporating new technologies and catching new industries requires an active strategy by institutions and private agents; all of them should cooperate to evolve the global strategy (which affects both the infrastructure for new technologies and institutional actions in the industrial, technological and social fields).

III.2.2 Cyclical Elements and the Importance of the Take-off Phase of a New Technological Paradigm

The cyclical factors in the evolution of spatial organization are linked to phases of technological innovation and the emergence of technological paradigms. The launch of a new technology or a new technological paradigm begins with a polarization phase around the new industries which produce the specific material support of these technologies. In emerging strategic industries, where products change quickly, firms are new and small without vertical integration and so they depend entirely on complementary industrial firms. This initial polarization is what can be observed for example in the birth of microelectronics in Silicon Valley, or in Highway 128 in Boston in the years 1960–70, which gave rise to a lot of literature on industrial districts.

This logic tends to disappear as the technological cycle progresses:

– with the development of these industries the entry barriers gradually rise

155

and the trend towards capital concentration grows. The activities become more and more capital intensive, the R&D becomes institutionalized...

– techniques and products stabilize: routines become established

– vertical integration develops

– large firms begin to acquire a dominating role and to gradually spread out their activity following the principles of spatial division of labour, transforming the pole into a relatively homogenous centre (services, management, R&D activities and knowledge–intensive industrial activities) and displacing other productive functions in the direction of the national or world periphery.

In this way, during the mature phases of the cycle, the regions are characterized not by their type of goods but by hierarchical functions and positions. They shift from a discontinous space formed by a juxtaposition of very autonomous territorial organizations ("a mosaic") to a structured, integrated space with hierarchical core-periphery relationships.

This excessively mechanistic vision is too close to the "product cycle" theory. The approach emphasizes that the mature phases of the cycle lead to a certain displacement of activities towards the periphery, while the beginning of a new cycle is expected to give rise to a new cluster. In fact, many reasons make real pathways more complicated:

– "Delocalization process" towards the periphery is not automatic and, above all, not indiscriminate. Even in the mature industries, the introduction of new technical and organizational systems which increase productivity could discourage them from moving out to the periphery and favour recentring.

– the decline in certain dominating core spaces does not have to benefit the periphery but it can benefit other core spaces.

Every new model of development presents new demands and could require a different environment so that the dominating industrial spaces may have to be moved; but the displacement takes place in the framework of the core economies themselves.

Rather than just describing these phases of the cycle, authors like Carlota Pérez emphasize the idea that the **transition periods between paradigms** are crucial because they can permit some spaces to catch-up with the technologically developed spaces. To achieve this aim requires being capable of carrying out a "massive process of social creativity" in the framework of the characteristics of the new paradigm; thus, it depends on the **social innovation capacity** to incorporate the new technological paradigm in the widest possible dimension.

Pérez's position is relatively optimistic in so far as it considers that, e.g., the capacity of developing countries to build an "imaginative and coherent strategy and the possibility of successfully putting it into practice", is determined mainly by the availability of "qualified human resources". On the other hand, in so far as the "new paradigm favours flexibility, adaptation to special conditions, the integration of activities and the exploitation of diversity" it seems that "making the most of the new paradigm depends on knowing how to value the specific characteristics of each country" (Pérez 1986: 85)[64].

The hypothesis of Pérez, in spite of its "cyclical" dimension, differs from ideas stemming from product cycle theories where the competitivdness of developing countries is said to be greater during the mature phases of products and technologies, when innovation and adaptation capabilities are no longer significant. From the perspective of the product cycle theory, industrialization would take place almost exclusively through technology transfer from developed countries during the mature phase of the paradigm. According to Pérez, a strategy based on new technologies would be possible in the early phases on condition that a certain amount of qualified human resources exist, giving rise to a learning process in which the non–existence of past routines does not have to be a problem but, in fact, an advantage. The experiences of Japan, Korea or Taiwan could support this point.

What could change the relative position of a country vis a vis leading countries are the changes introduced in national innovation systems. And the most appropriate periods are those of transition between paradigms, which are phases of "creative destruction" and discontinuity both at technological level and in the conditions for development. Therefore, there are periods when great changes in the strategies of each country can be made and the relative position of countries can take place. The capability of each country to assume and adapt socially, institutionally and politically to the conditions of the new paradigm are factors that determine the trajectory of each country in the new phase.

In order to design a regional strategy the most important is:

a. **to choose the entry points well**: the earlier the phase in which the new technology is found begins, the easier it will be (for example that woul be the nowadays case of biotechnology).

b. **to make a concentrated and sustained effort**: careful surveillance of the

---

[64] The position of this author seems excessively optimistic, as she does not consider any other restrictions except those related to human resources, e.g. infrastructures, capital, markets... which are considered in any development theory. In particular, this optimism goes too far when considering the capability of a country to adapt to the new paradigm in absolute terms without bearing in mind the capacity of others which, through competition and inter–dependency relationships, could make that strategy unviable in reality. In this case, it is necessary to approach this problem with an open economy, integrated in the world economy, fundamental for guaranteeing the actual conditions for development.

evolution of the international technological frontier is essential here.

       c. in this endogenous development process of technological capability, **State support** is, in general, vital.

       From this point of view the fundamental problem of regional development is how to adapt the socioeconomic configuration to the transformation of the development model and the requirements of each phase of the cycle. In this way, the capabilities for generating and acquiring innovations seem to be cumulatively concentrated in the most dynamic and industrialised urban areas, which are those that fulfil the greatest number of conditions to atract new localizations and favour the diffusion of innovation.

       This approach can show strong tendencies in innovation dynamics but it is also too generic and linear; it loses sight of the importance of temporal perspective and tends to overlook the "awakening effect" which the incorporation of a foreign technology can have on a local environment. In so far as it assumes that resource endowment and industrial structure are given, the dinamic feed–back effect does not fit well into this model. To incorporate it, is necesssary to carefully analyse what happens in peripheral regions across the time, taking account of endogenous dynamic factors and defining how ruptures can take place to the take–off of new dynamics.

       Summing up, from the perspective of spatial dynamics of industry it is not possible to establish a general rule about the role of technological change in this evolution. To do so, it would be necessary to tackle this question from a different approach, starting from the dynamics of each space and inquiring into the specific conditions of its development and the patterns in the dynamic relationships between different constitutive elements of this space. Particular attention should be paid to existing innovation structures, human resources and above all to the capacity to articulate and promote them. This will depend largely on the aid of public inititiatives and social and institutional innovation capabilities.

       An important finding of recent studies on the matter is that spaces are not currently defined sectorially but functionally, based on a skilled population (Veltz, 1992). So, regions and local communities should strive for a highly–skilled labour market, especially of engineers and graduates related to new technologies. This kind of labour market can make possible both self-developing local technological capability, if the appropriate support mechanisms are put into action, and also attracting large foreign firms needing highly qualified personnel.

158

IV. THE DIFFERENTIAL CAPABILITY OF EACH TERRITORY TO CREATE AND ADAPT TECHNOLOGY: DIVERSITY OF INNOVATION SYSTEMS

IV.1 An Alternative Approach

From an alternative approach, space is considered as "a local system of production" or as "a local system of innovation", with its own development capability. Here, the important concept is the territory as a concrete configuration of space. Territory is not a banal space. Even more, it is not defined by a given factor endowment but by a group of elements created through a long trajectory of regional development. This trajectory is not only reflected in its productive structure, infrastructures and the market, but also in its capacity to create new ressources, new firms, new markets. This capacity depends on skills, entrepreneurial initiative, innovative activities, organizational know-how, local patterns of industrial organization, governmental strategies, regional self-government etc, which together define the dynamic capabilities of territory as active player in its own development (Stöhr 1984; Aydalot and Keeble 1988).

This view of space can be easily related to an alternative way of considering technology, not as a given result, but rather as an open innovation process evolving in the space itself, based on specific resources and orientated towards solving concrete problems which arise for each productive system (Gaffard 1986)[65]. In this approach space is not inert, but instead can be seen as a means of generating a change process which is sequential and cumulative. Innovation is not the result from individual activities but from an active environment. The essential idea is not to consider this process as something leading to a definite end-result, but rather as a permanently open sequential process whose current results are the basis for future changes. It is a process which is constantly offering new possibilities, which allows agents to foresee at each step new productive processes and increase the available range of products and techniques. An efficient management of this sequence can permit territory to increase its **dynamic productive flexibility**[66] and adaptative capacity (Amendola & Gaffard 1988).

Empirical studies reveal the heterogeneity of innovations in relation to space and sectors. As a consequence, diffusion processes are also heterogenous. Diversity in the configuration of innovation systems implies that no single model of diffusion

---

[65] J.L.Gaffard, "Restructuration de l'espace économique et trajectoires technologiques" in P. Aydalot's **Milieux innovateurs en Europe**, op. cit., pp. 17–27. The concept of specific resources contrasts with that of generic resources, especially in the sense that specific resources are not given a priori nor are they presented as abstract elements, but instead are treated as resources produced throughout a productive process and which therefore present specific capacities. See M. Amendola, J.L. Gaffard, **Dinamique économique de l'innovation**, op. cit., chpt. 1.

[66] Note that here the concept of productive flexibility is radically different to the standard use; it deals with a dynamic flexibility which derives from the continual creation of a diversity of productive alternatives which in turn give rise to other new ones.

nor of localization patterns exist[67].

The association of progress with innovation should also be stressed. The relevance of the innovation process does not so much depend on technical brilliance, or its more or less radical character, but on its capacity to irrigate the local productive system. The progress that an innovation could mean at general level does not necessarily have to be automatic in a concrete region, but will depend on the involved industries and their relative importance in the region. Innovations originating in a region do not always benefit the economy of that region. This often depends on the unpredictable formation of new local industry or the capability to incorporate findings into the existing firms. It also depends on whether the benefits are retained by the region, or whether they are drained through interindustrial relationships or financial circuits. As Malecki and Sweeney pointed out, from a schumpeterian perspective, this passage from innovation to regional economic change critically depends on entrepreneurship which develops through a long process of local industrial experience and the creation of an appropriate local socioeconomic environment. Entrepreneurship is the main factor to permit the regional incorporation of innovation because it is a local phenomenon, not transferable from place to place and is dependent upon the economic structure as well as the attitudes and culture of the region (Sweeney, 1987; Malecki, 1991).

Precisely as a consequence of the diversity of the patterns of local impact of innovation, one should insist on the idea that regional innovation capacity is not a substantial enough condition for its economic development. Despite being a crucial element, we should however value their ability to merge into the background of local productive system. Thus, innovation should be evaluated on the basis of concrete short–term consequences (above all in employment and income), also taking into account long–term dynamic effects.

Finally, we should also take into account the external relationships of each territory. The right emphasis on local determinants often leads to an underestimation of the relationships with the whole economic system and of the determining effects of structural change and mutations in strategies of multinational firms. This is a decisive question not only since the development of many peripheral spaces depends on their capability to selectively attract innovative and productive functions belonging to external firms (Wales or Sophia-Antipolis can be examples), or because of the aforementioned drain of technical results; but because external relationships and outside sources of innovation are vital for all dynamic production systems in the context of the globalization process of innovation activities (Gordon 1990). Currently, local cooperation is necesary but not enough to support internal

---

[67]   This approach is analytically supported by Amendola–Gaffard's models and also has a notable empirical component which was built up starting from a monographic study of concrete innovations in concrete territories, picking up their characteristics, internal force lines and evolutive trends observed. The aforementioned studies by GREMI, the studies on industrial districts and technological districts in Italy and the United States, studies carried out on technopolis and scientific cities in Japan, etc.

160

innovation, the complexity of which increases more and more. An innovative region should have antennes to participate in different global processes of innovation.

IV.2 A typology of Spatial Patterns of Innovation

In order to establish the relationship between innovation and territory we can make a simple taxonomy of spatial patterns of innovation. Different sources and ways of technological development are possible depending on characteristics of regions, technologies and time. Each one of these typological forms is more or less related to a conception of innovation, reflecting a different logic. This simple taxonomy is built according to the origin of technologies from a spatial and sectoral point of view: on one hand, technology can be created by an active performance of regional agents (firms, universities etc) or by foreign agents, in the later regional agents only adapt technologies and produce an operative knowledge; on the other hand, the degree to which technology used in a sector is generated whithin the sector or comes from outside through the purchase differs enormously (Pavitt, 1984). In each case the key territorial factors which could boost the whole innovation process will be different. (See figure A)

Figure 8A. TYPOLOGY OF THE SECTORAL AND SPATIAL ORIGIN OF INNOVATION

	External to Space	Internal to Space
External to Industry	Use of New Technologies (reconversion)	University, Public and Industrial Research (not industrie refered)
Internal to Industry	Decentralization of "high–tech" firms (spatial division of labour)	Emergence of "high–tech" firms (Internal research)

In fact, this formal typology can be summarized in four relevant patterns, two spatially endogenous and two spatially exogenous:

a) innovation in a territory as a consequence of the dynamics of delocalization of the activity by external innovative firms.

b) innovation as a consequence of the adoption of techniques generated in the exterior by the local industrial system.

c) innovation generated within the research bodies of the space itself (which always implies a certain connection between various sectors or techno-scientific institutions).

d) innovation generated within innovative firms that can diversify activities

161

in the region.

Of course, this is a typology but does not mean that each of these logics are independent from the others nor does it imply that each territory experiences a single type of them. Some dynamic territories can benefit of a complex combination of different processes. Other regions should concentrate their efforts in a specific way to catch their own opportunities.

Type A: Innovation Through Decentralization: this type of innovation is introduced by large firms from sectors linked to new technologies as a consequence of a decentralization strategy of its activities. As we saw earlier, this territorial dynamic of innovation is linked to another, more general one, related to the changes in spatial division of labour of the large firms in these sectors (encouraged by new information technologies) or associated with a diversification strategy of firms creating new plants elsewhere (spin-out).

In this case, knowledge is created within the firm, in its own research laboratories with specialized personnel placed beside production centres or detached in an autonomous Division. A typical decentralization strategy could be represented by firms which enter in the domain of a new (or relatively new) field whithin the same industry, where there is a certain continuity between old and new technologies. Firms can create new plants to new products anywhere in the world and territories have to find the way of influencing the location behaviour of firms and attract them. As it's about "high tech" activities but in new fields, the potential market will be impportant and also the Public support.

The main territorial factors involved in this type of innovation dynamics are above all the **attraction factors** of each space, creating external economies and agglomeration economies:

– transport, communications and telecommunications infrastructures

– qualified personnel and training systems (in some cases, research)

– industrial structure and possibilities of subcontracting

– standard of living

Therefore, the viability of an innovative strategy requires a certain complexity of regional innovation system (and regional productive system) and a suitable institutional configuration. In certain cases an active and suitable strategy of regions can remove important obstacles and lacks derived from the past (certain technology parks could be an exemple); but the limited number of succesful

162

experiences reveals the strong weight of the past.

Type B: Innovation Through Reconversion/ Adoption: in this case, a territory incorpores technology through a process of internal reconversion of existing industries by adopting and adapting foreign inventions.

Here, there can be two different sources of technological knowledge, both in a variable proportion: on one hand, the main technological knowledge is generated in other (various) environments although a certain internal knowledge of local firms is necessary. This internal knowledge can be created within the production process and its main agents are engineers and technicians from the production departments of firms.

From a territorial point of view, this type usually involves a reconversion of an industrial system constituted and diversified in advance. The reconversion stems from a consolidated industrial structure, with a group of highly qualified Small and Medium Firms, which allows some of them to take advantage of the relationship between old and new technologies (mechanics, micromechanics, electronics, etc.), through reconverting process of their production processes.

As for territorial factors relevant to type "b" pattern of innovation the importance of the aspects related to the type of local industrial system should be highlight:

    − sectoral structure (main specializations and diversification)

    − industrial structure (size of firms, type of interfirm relationships, clusters)

    − degree of local dependence with respect to exterior agents

    − importance of local R&D, especially process and product engineering.

Type "C": Endogenous Innovation and Dynamic Synergies. This model of industrial technological development could be based on the bifurcation of existing firms towards new activities or on the endogenous creation of small and medium firms stemming from research and constituting an innovation pole or scientific pole. The main characteristic of this kind of configuration is that it involves an active process of creation of resources at local level. Obviously, the relevance of the local activities doesn't imply ignoring the importance of external relations to move ahead the innovation dynamic in a context of increasing globalisation of innovation processes, but what should be pointed out is the central role played by local agents in this global process and their capacity to root it territorialy.

In this last type, knowledge is created mainly outside the firm (in Universities or public laboratories) and incorporated directly into the firm by researchers who either set up their own company or are contracted for research by existing companies. In either case, innovation proceeds from science rather than from industry; the initiative to develop and commercialize the innovation also comes usually from the world of research although the business world can also act as a promoter.

An specific case is the science–based innovation, created "ex nihilo" from scientific laboratories (spin-off). When this happens the complete new characteristics of the products exclude links with previously existing products and demands instead a tight link with basic and applied research. These type of innovation dynamics tend to take place round local Universities, public research centres, etc with whom they have permanent feed–back (formal and informal). Of course the degree of exteriority between innovation and previously existing industry will not normally be so extreme. On the contrary, usually it is a result of research combined with business experience in related activities which allows a diversification or bifurcation in the local industry.

For "endogenous innovation", local **synergy** factors between different agents and institutions will be important:

– research infrastructures (University, public and industrial laboratories)

- cooperation and interface structures between Research and Industry

– interfirm cooperation

- availability of skilled resources and higher education

– personal interchanges of qualified information from different agents ('face-to-face communication in a information rich environment')

– financial support: availability of venture capital, credit policy, subsidies

– availability of productive services and telecommunications structures. This can be provide by active policies creating business incubators, consulting services and others.

IV.3 General Tendencies of These Three Innovation Patterns

Recent experience on innovation regional dynamics shows a very rich diversity in terms of patterns and successful results. The successful experiences are not a pure type but a mix of diverse configurations and diverse strategies. This complexity permits strong contrasts among different analysis even refering to the same case (Aydalot, 1986, 1988; Malecki 1991; Amin, 1992; Gordon, 1991; Scott and Storper, 1990). In this circumstances it is not easy to advance too general hipothesis.

The difficulties encountered by many industrial regions in decline to reconvert towards new activities are a clear example of a relatively important fall in the second type of innovation dynamic. The close specialization and routines on specific industrial activities do not facilitate the rapid adaptation of workers and entrepreneurs to new skills, new business, new markets etc. In any case, only selected companies develop activities which could allow them to take steps in this direction, on the condition that they count on financial and institutional support. In fact these usually constitute two important additional limitations. In addition, the disconnection between University R&D and the needs of companies can be just as limiting because the research fields usually have little to do with the technological paradigms of such industrial system.

Furthermore, this trend is linked to the Long Term trend to a larger concentration of capital and an increasing spatial division of labour observed in mature industries. In the last decades a process has emerged whereby SME's are taken over by large groups which organize themselves in terms of spatial division of labour, concentrating their research activities on the metropolis and breaking largely with local innovative synergies. This does not mean to say that traditional industrialized regions cannot begin a new innovation trajectory using the infrastructures, high qualifications, management and business know–how, support services, capital etc developed in the past. But it seems clear that the concentration of capital reduces the type "b" technological capability of local environments in benefit of two other patterns of innovation:

– the internalization of innovation by large multi-spatial firms.

– the externalization of research by knowledge-specialized institutions, which are becoming more and more involved in productive activity.

Some remarks should be made specially from the view of peripheral regions. Peripheral territories have shown little capacity to promoting large firms or to attracting them, and even less in the case of attracting functions of higher technological intensity. Taking Galicia as an example, we observe that its share in Spanish foreing investment in the last decade was only 0.5%. A study on regional share of investment in high technology industries through the crisis period[68] shows that "expansion investments" were concentrated in very few provinces (Madrid, Barcelona, Valencia, Vizcaya), in a steady degree of concentration from previous period. A certain delocalization could be seen in "new investments" towards provinces bordering with the aforementioned (especially in the area of Madrid: Toledo, Segovia, Avila) but didnot reach the most peripheral provinces. Deficiencies in road and telecommunications infrastructures increase the costs arising from a peripheral geographic position. The proximity of Universities, the standard of living, the existence of airports (which is very important for internationalization) were not enough to counteract other dynamic elements. Probably the lack of support from public institutions, the absence of a regional policy of valorization of potential in each territory, the non–existence/non–functioning of local organizations to encourage and promote industry are important factors which enable us to understand the absence of large projects in spanish peripheral regions. The Education and Training System was historically not capable of offering a wide enough range of specialities which were needed with the present technological revolution. In this way, the traditional limitations for endogenous development and the traditional lack of factors to attract outside investment, are added to the evidently obvious lack of scientific and technical specialities in the fields of new branches and industries which have emerged since the sixties. Of course technological change does not rely wholly on technical professions but on the gradual incorporation of economists and other graduates into the organizing structure of firms giving them a dynamic factor; however, the significant incorporation of the latter was very late on (on mass in the mid eighties).

On the other hand, due the importance attributed to the interface research/industry it is necesary to stress that there is an obvious coordination problem between the research and industrial systems, both in type "b" and type "c" patterns. In the case "b", it could be difficult for firms to use the knowledge from research system given the weakness of their R&D structures and their orientation towards traditional activities. In case "c" the step from research to production is a complex process (both from the technical and engineering aspects and also from the point of view of financing and manangement). Thus, few projects really start and many of them fall on the way. Either of the two cases reveals the complexity of weaving the heterogeneous segments of a "large collective worker", a complexity which increases when it coincides with a period of change in general conditions of production and market.

---

[68] Giraldez, E. (1988), "Comportamiento inversor de los sectores de alrta tecnología 1975–1985. Tendencias spaciales", **Papeles de Economia Española**, no. 34, pp. 431–453. The sectors of high technology considered in this study were: "office machinery and computers", "telecommunications machinery and equipment", "electromedical machinery and apparatus" "signalling and control electronic apparatus and equipment" and" electronic components and integrated circuits".

The compatibility and integration of diverse activities taking place in the complex innovation process shows the importance of a systemic approach which harmonises and gives coherence to diverse strategies of local agents. For example, I consider that mobilizing synergies between University research and productive activity requires a harmonization strategy of behaviour and rules of both poles, which nowadays are quite disparate (Vence 1995: ch. 11). Market mechanisms lead to a predominance of short–term purposes and are an imperfect means of valueting strategic resources which, like basic or applied research, are only profitable in the long–term. Even more, their social utility can be not profitable under market conditions. At the same time researchers want to maximize their own curricula on the basis of exclusive scientific community criteria (evaluated depending on the international scientific frontier) and not from its utility for its social environment. Such behaviour will lead to an abysmal gap with respect to the needs of productive system, specially in the case of a backward productive system. These aspects among others offer the deep challenge of designing alternative forms of cooperation and social appropriation which would allow us to value activities which nowadays the market is unable to value. Obviously, these changes are not feasible at regional level and the goals should be less ambitious. But, in fact, at regional level it is possible to weave a close cooperation by different and specific means that can remove important obstacles. The most difficult step is to start a cooperation and this can be better promoted through local and specific engagements. But this requires an active strategy because experience indicates that the simple contiguity of research and industry does not engender innovative synergies.

V. TECHNOLOGY POLICIES AND REGIONAL DEVELOPMENT IN THE AGE OF GLOBALIZATION: CENTRALIZATION, DIVERSITY AND COHESION

Standard Technology Policy has been based on neo-classical concerns (appropriability, externalities, uncertainty etc...). The territorial dimension has been underdeveloped, consisting predominantly of external economies, spill-overs, agglomeration economies within a static framework. As a consequence policy recommendations emphasize global benefits from concentration of innovation activities. In fact, government concern on the regional aspects of technology policy is quite recent. The aforementioned changes and the perception of the growing role of Science and Technology in the regional economic growth and competitiveness justify the growing importance attributed to regional aspects since the last decade. Moreover, the above refered theoretical renewal also contributes to this.

Some regional experiences show how a development strategy centred on technology can be carried out to build up a solid base directly linked to the world economy, bypassing to a certain extent the state level. This does not mean that the State structure and its institutions are insignificant. Quite the reverse. State policy, to a certain extent, prepares the ground for regional action which can enormously differ between contries. In particular, international comparisons show that decentralized structures can favour regional development.

From an evolutionary point of view the aim of technology policy is to improve the characteristics of National/Regional System of Innovation knowing well that changes in this field are not fast but show great inertia. From a narrow point of view, technology policy can be considered as a group of public activities which affect the promotion of technology, its diffusion, technical innovation etc, carried out by the different Governmental levels (R&D, infrastructures, incubators etc). But, in fact, technology policy involves a wide-range of industrial aspects that characterize a National/Regional System of Innovation as a whole:

- research organization and technology infrastructure
- training activities
- company organization
- type of relationships between producers
- user-producer interaction
- relationships between firms and the financial system
- institutional arrangements

There are three different levels (or aspects) in the analysis of relationships between Technology Policy and Regional Development:

a. Regional Dimension of Global Technology Policy efects
b. Regional Technology Policy of the States
c. Technology Policies of the Regions

Obviously, there is a strong interrelationship between the three levels and there is a debate about which is the most adequate level to develop each type of activity. Amongst other factors, adequacy depends on the desired aims and **the characteristics and innovative behaviour of each region**.The advantages of assigning funds for regional technological policies are the creation of local companies and the improvement in human capital; the cost of these regionalizing policies is the decrease of the economies of scale (or agglomeration economies).

Both the European Commission and the Central Government, as well as the Regions, have specific means of technological policy, in some cases set up and financed at their own expenses and in other cases negotiating and managing programmes from higher Administrations. The effectiveness of the measures of each Administration depends on the design and the adequacy of the instruments for the goals in mind but above all the efficiency of this institutional packet depends on deeper mechanisms such as the technological system of each territory and the

168

infrastructure for research and teaching. Therefore, the effectiveness of a policy depends on how well it adapts to the conditions of the concrete productive system. This is precisely the problem of European technology policies with respect to backward regions.

We will lead our reflection from the point of view of the peripheral regions and the way to acheive **cohesion** within the national (or European) system.

Peripheral regions usually present a weak industrial structure, traditional sectors, mostly very small firms, a weak Science-Technology-Industry system, a virtually undynamic social structure and low political bargaining power. Regional development in these type of regions does not stem from attracting large firms nor from attracting high technologies but instead from the progressive transformation of the local industrial system. Even more, as consequence of its weak entrepreneurship they will have great problems in benefitting from the generic Technology Policies, where the support goes towards new technology sectors and the projects selection is realized according to the top technological level.

V.1. The discussion on the regional dimension of technology policy. The Spanish experience

Nowadays authors agree that technology policies have direct or indirect regional effects. There are very contrasted arguments and recommendations on the degree of centralization/decentralization of R&D activities or the adequate level to design technology policies. A first set points out the importance of economies in centralization R&D activities. On the opposite side there are some important elements which support the decentralization of R&D activities and Technology Policy. A second set of arguments is related to the limits of the regional governmental actions and in favour of a technology policy designed at the central level. Obviously, the relevance of different arguments depends on the type of technologies take into account and the specific polical configuration of each country.

V.1.1. Arguments in favour and against centralization of R&D activities

One possible way of discussing the effectiveness of regional technological policies is to focus on the evaluation of the decentralization of R&D activities. The question arising is important: does a net social benefit exist in a regional decentralization of R&D policy, whether it might be due to decentralization of university research, public research centres or through grants for private firms?

169

The standard evaluation models establish a balance between the costs and advantages of evaluating the net benefit in social terms, based on pareto optimality (Martin 1986).

The main argument against a R&D decentralization policy towards the peripheral regions is the **decrease of economies of scale and agglomeration economies.** Consider that a positive relationship exists between the dimension of the R&D laboratories and the productivity of the researchers who work there, deriving from work division, shared information and common use of infrastructures. **Agglomeration economies** also exist linked to the concentration of other research activities, industries or services. Existing concentrations of R&D activity attract additional R&D activity, reinforcing the pool of professional and technical workers. Information, contact networks and technical progressiveness are standard in areas where entrepreneurship is common. Moreover, we can consider that while the costs of transmitting **information** may be invariant to distance, presumably the cost of transmitting **knowledge** rises along with the distance. For example, the aforementioned study by Feldman & Audretsch on the United States case concludes that the empirical evidence suggests that industries in which knowledge spillovers are presumably the most prevalent, that is where industry R&D, university research and skilled labor are the most important, tend to have a greater propensity for innovative activity to cluster than in those industries where knowledge externalities are less important (Audretsch & Feldman 1993).

An additional argument is that regional development is less dependent on local generation of innovation than on **the access** to new technologies. Local production of knowledge does not guarantee the local industrial use. To support this idea a contradiction between University Research and its use as a motor for local development is underlined: university research is above all fundamental and usually its use is not appropriate for the region.

From another point of view R&D decentralization is encouraged. The decentralization arguments mainly focus on the systemic, dynamic and cumulative aspects wich characterize the innovation process.

1. A discretionary localization of R&D activities in a region has a multiplying effect, directly and indirectly, on the employment and economic activity of this region. It raises qualifications of human capital of this region and could allow the reduction of brain drain[69].

2. Innovation is a learning process and the knowledge is to a great extent

---

[69] It is also true, in the opposite sense, that if there is a dominance of research training and it is not possible to include researchers in the productive system later because it could give rise to a brain drain at a higher level.

tacit (institution or person-embodied), cumulative and **localised** in its nature and effects. Local capability for creating or merely assimilating innovations is a requirement to the technological development of a region. The development of R&D activites is a cumulative phenomenon and, in the long-term, the volume of research results produced in this region will tend to increase.

3. Its contribution to the modernization of the local industrial structure as a consequence of **spin-offs** generated by R&D activities (university, public or private). Furthermore, the decentralization of R&D activities allows to strengthen links with the **industrial environment** and so it can lead to fostering the applied-orientation of the academic research.

4. The search for economies of scale in R&D activities should not automatically be associated with spatial concentration. A strong spatial concentration could coexist with a large part of the small research teams, and on the other hand, the geographic decentralization could be done setting up adequately large laboratories.

5. According to evolutionary approaches, decentralization can favour the diversification of research projects, giving rise to an increase in the amount of learning processes which in the long term benefit the economy as a whole.

V.1.2. Arguments in favour of designing policy at central level and limits for decentralization policies

Another way to discuss the decentralization policies is to focus on the competencies and capabilities, at each level of Government, to move programmes capable of changing a given situation. We can review the main arguments against decentralization policies(Hilper 1991; OCDE 1991):

1. Regional governments do not have the capacity to modify the initial conditions in order to achieve regional development. Peripheral regions lack the industry and research structures to raise critical mass required to undertake structural changes. Usually policies designed at a regional level tend to focus on the role of an innovative entrepreneur but policies cannot create it.

Though this argument is partially valid I consider that changes take place step by step and even when "initial conditions" cannot change radically, they are at the basis of a chain of very imperceptible events. This process can be set in motion by qualitative actions from government but such actions do not necesary imply a great budget or, rather, the budget amount it is not the main aspect.

2. Regional governments are unable to cope with regional disparities. Decentralization can even give rise to the emergence of different regional patterns determining divergent developments and so increasing disparities. Most of the existing high-tech regions continue to strengthen their capability and, on the other hand, disadvantageous regions have very reduced oportunities to participate in national trends of socioeconomic development, to catch up with the leading regions or to introduce a high-tech-based industrial recovery. Differences in traditional innovation models are added to the effect of new technologies.

Nevertheless, not too many cases can ilustrate the real reduction of regional disparities from centralized initiatives. The breakdown of the traditional regional policies implemented by central governments is an experience to have in mind.

3. A regional government cannot substitute national innovation and industrial modernization policies. The generation of innovation is not a fast process and needs a lot of resources. The regional government can support the existing innovative capacity or try to attract foreign firms, but it cannot introduce appropriate development programmes for such high-tech industries (telecommunications, biotechnology, energy or environmental technologies ...).

Effectively, some general objectives require a wide cordination not only at the national level but also at the european level. We are mainly concerned here with finding the adequate level of coordination for each technology involved and for different economic goals.

4. "Central government and the strategic bodies see themselves as responsible for a general interest which is greater than the sum of the particular interests of each region; they also feel resposible for making the allocation of resources as effective as possible, which often involves efforts being concentrated and existing strong points being reinforced. For their part the regions are obviusly primarily concerned with their own interests" (OCDE 1991, 59). However, this statement should be confronted with empirical experience because central governments might not be territorialy neutral, mainly when there are nationalist claims behind.

5. As regions are generally smaller, the standing bodies and formal mechanisms are often less developed than at national level, and ad hoc procedures, informal consultations and personal relations play an even more important role. Nevertheless, this behaviour is not exclusive at the regional level but it is also quite common at centralized levels.

6. From a centralizing point of view the role of regional governments may favour a redefinition of national policies more than a political decentralization. What

it can do is creating the appropriate facilities (research institutions, for instance) or activities in order to attract initiatives carried out by the central government. The creation of Technology Parks for innovative small and medium firms could encourage the location of R&D departments of larger firms in these parks and also to foster the links between firms, universities and R&D institutes. In this way, the regional government can promote the participation of regional industry in the modernization of national industry.

The inevitable conclusion, therefore, is that it is not a matter of setting up centralization against decentralization but to search an adequate balance between the two different stances. That balance varies from technology to technology and from policy instrument to policy instrument. The role of Supraregional Technological Policies can produce positive effects on regional development and cohesion:

   a)if it serves to create the infrastructure conditions for new activities and innovation.

   b)if it can guide with territorial criteria the location of industries and services.

   c)if it can also support the innovation in traditional industries.

   d)if it can promote interregional/transnational cooperation

V.2. Regional dimension of Spanish R&D policy. The doubtful efficacity of concentration of R&D resources

Studying the Spanish experience might help to elucidate the relative importance of each of the preceeding statements. The first step in order to analyze the matter will lead us to examine the regional concentration of inputs in R&D activities (expenses and researchers) and to compare them with the R&D output (patents and publications). Finally, we will consider the regional distribution of technology policy aids (grants for scientific research and grants for firm innovation).

From 1986 to 1991 the total expenditure on R&D in Spain grew by 84.2%, starting from a very low level. As a consequence the effort in R&D (Expenditure R&D/GDP) increased from 0.6 % in 1986 to 0.95 % in 1991. The increase was the same in the business sector and in the Universities and Centres dependent on Administrations. The increase in scientific production was slightly smaller. Furthermore, the technological deficit increased from 0.27 % of the GDP to 0.39%.

At a regional level the degree of concentration of expenditure was maintained : Madrid (44 %), Cataluña (20 %), Andalucía and País Vasco (8%). Only three regions show an above average effort in R&D : Madrid, País Vasco and Navarra. The sectoral distribution of R&D expenditure shows a reduced importance of business R&D in most regions. Only in País Vasco, Cataluña and Madrid did the business sector show a porcentage above the national average (58%). Cataluña is the sole region to present a balanced structure in R&D distribution, more similar to developed countries (70% business- 30% Universities&Public centres).

The Regional concentration Gini Index has high values and has increased from 1986 (0.62) to 1991 (0.64) which shows that concentration was maintained, even increased. This concentration is no higher as consequence of the effort realized by the regional governments; if we leave apart funds from the Regional Governments itself the Gini Index would show an important increase from 6.5% in 1987 to 7.1% in 1990.

Analyzing the concentration of researchers there are some important results. First, the region of Madrid alone concentrates 46.5% in the business sector, 56.7% in Public Administrations and 20.8% in Universities. Second, the relationship between Researchers/Active population(%) in Spain as a whole is 2.5% but it is 7.3% in Madrid, 3.5 % in Navarra and 2.9% in País Vasco. That concentration in the cases of Madrid and Navarra, is particularly high in Public Research Centres. Third, the expenditure by researcher is also much higher in Madrid: 8.98 million pesetas/researcher, doubling other important regions as País Vasco, for example. This difference is especially obvious in the Public Administration sector where Madrid receives 47% more than the average. This occurs in spite of the fact that the Regional funds have been also included here (and Madrid Regional government does not expend any fund at all).

This brings us to the question of whether a correlation between high regional concentration of R&D inputs and the regional concentration of R&D output exists. Against the hypothesis that geographical concentration of R&D is justified because it increases the productivity of researchers and generates economies of scale and economies of agglomeration, the empirical evidence of Spanish case does not support this statement. On the contrary, the high R&D concentration in the region of Madrid offers a low output measured by both the number of patents or the number of publications. This unexpected finding can be observed in both the industrial research or the scientific research. The relationship between business R&D and patents shows a lower productivity of industrial research in Madrid than in other regions. That result could be easily attributed to the different sectoral specialization of R&D activities and the uneven value of patents. Nevertheless, it is more difficult to explain the relationship between public R&D (University and Public Administration) and publications, which also shows less scientific productivity by the Madrid researchers.

174

Figure 8.1.R&D BUSINESS EXPENDITURE VS NUMBER OF
PATENT APPLICATIONS. SPANISH REGIONAL DATA

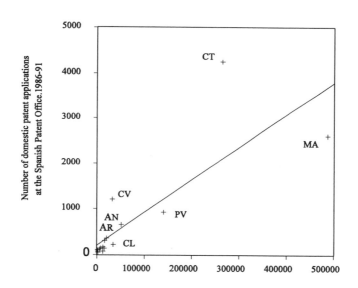

Cumulative business expenditure on R&D, 1986-91.
Million ptas.

Figure 8.2. R&D UNIVERSITY AND PUBLIC ADMINISTRATIONS VS SCIENTIFIC PUBLICATIONS. SPANISH REGIONAL DATA

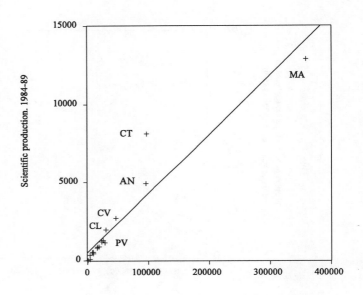

Cumulative public expenditure on R&D, 1986-91.
Million ptas

Figure 8.3. R&D STATE AIDS VS NUMBER OF PATENT APPLICATIONS. SPANISH REGIONAL DATA

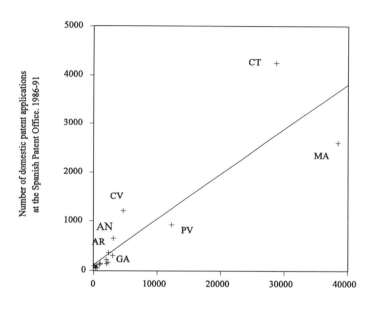

States aids. 1988-92. Million pts

The Spanish R&D policy carried out in these period, has served to shorten the differences in this field with other European countries. However, it has not been able to reduce territorial centralization, but on the contrary it has reproduced differencies. The industrial technology policy has been guided by the idea of making Madrid the high-tech pole of Spain (with the deliberate support from the state-public companies); policy has been focused on high-tech sectors, neglecting innovation in traditional sectors. Furthermore, R&D policy has been based on criteria of competitivity of applications without regional corrective mechanisms. There has been no political awareness on the part of different political forces nor on the part of those responsible of Administrations, to take into consideration the regional variable when putting the R&D policy into practice at a State level. For that reason the dialogue initiated in the eighties between the different Administrations to distribute the roles has not continued.

We have analysed the impact of funds from CICYT and CDTI-Ministry of Industry in the regional redistribution of funds to carry out R&D. Empirical evidence prove that the territorial redistribution effect does not exist. The same occurs with funds from the EC to carry out R&D projects (Framework Programmes), given that they are also competitive and regional criteria are not included.

Willing to counterbalance that situation many regions designed technology policies but their implementation was not exempt from difficulties, and the use of funds was far from efficienty. Regions tend to copy the behaviour of the Central Government. The regional plans do not try to complement the national plans and do not select clearly areas of priority to act upon. They usually reproduce the objectives of the National Plan. Secondly, there is no coordination between the policies of the different regions. Thirdly, within each region there is a very deficient (if it exists at all) coordination among the Science Policy, the Technology Policy and the Industry Policy.

V.3. On the Regional Dimension of European R&D

V.3.1. Some preliminar remarks

Several studies released by the European Commission (CEC-STRIDE 1986; CEC-ARCHIPELAGO EUROPE 1992; CEC-ERSTI 1994) show a very uneven regional distribution of R&D activities.

Disparities on R&D can be considered among countries or among regions. Between States of European Union the ratio of highest to lowest intensity of R&D

178

(GERD/GDP) is about 5:1 and for GERD per capita it is about 11:1. Obviously differences grow when the comparison is estabished at a very desintegrated level. The "Archipelago Europe" study shows that in each of the countries there appear to be a small number of 'islands'. Those 'islands' are relatively small with a high concentration of research laboratories both from firms and public bodies (Universities etc), working intensively together in a very exclusive cooperative networks of "islanders". Very few cooperation opportunities exist for laboratories and entreprises from regions outside these major 'islands', in fact only 5% of all cooperation partners in Europe are located in peripheral regions. The ten 'major European islands'[70] shelter 80% of the all research laboratories and firms that participate in transnational R&D cooperation networks in Europe. Even more, the study shows that a relationship between the quality of cooperative links and the peripherality of the region exists.

The observed tendencies show no convergence. Disparities remain. The emergence of new technologies does not change the current situation but, on the contrary, they are used more efficiently by the core regions to strengthen their competitive positions and to restructure traditional sectors.

Other studies emphasize the diversity of regional patterns of innovation using the concept of Regional Systems of Innovation (Gaffard 1994). Differences from region to region are not only quantitative but also qualitative. The articulation of the different constitutive elements of each Regional System of Innovation varies significantly, above all the quality of internal cooperation and external links with other regions.

V.3.2. European tecnology policy: networks and cooperation

An important feature of European Technology Policy is represented by the preeminent role of cooperation. Cooperation is the basic principle articulating the whole of the R&D initiatives of European Commission. Cooperation is essential for European integration taking into account the great initial fragmentation of the research community. But cooperation is also important as a new concept to organize and to optimize research resources of firms or Public Administrations. Moreover, cooperation seems very useful in the framework of changes in the new generic technologies (more complex and integrative).

---

[70]South East England, Ile de France, Frankfurt, München, Turin, Rotterdam/Amsterdam, Rhein/Rhur, Stuttgart, Lyon/Grenoble, Milan.

179

Figure 8.4. PARTICIPATION ON 3RD FRAMEWORK PROGRAMMME BY COUNTRIES, in %

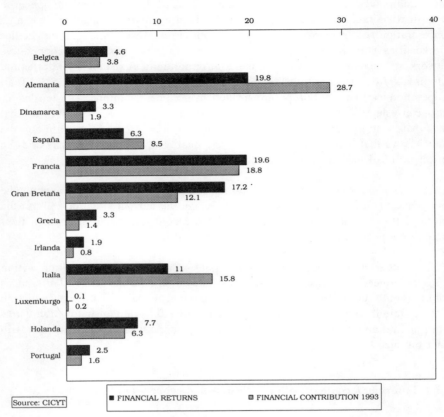

Source: CICYT

Figure 8.5. PARTICIPATION OF SPANISH REGIONS ON THE 3RD FRAMEWORK PROGRAMME (in %).

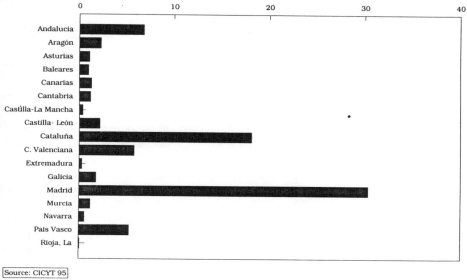

Source: CICYT 95

181

As all R&D activity supportted by European Commission is cooperative either at transnational level or at local level, any study on European technology policy is implicitly an analysis of the cooperation on R&D. Participating in most important programmes (i.e. Framework Programme) requires transnational cooperation among different type of agents (entreprises, Universities and other). Beyond the programmes explicitly devoted to create networks and favour mobility and other ways of increasing exchanges and interfaces, the main programme (FP) only supports transnational cooperative projects. Ten years after the first Framework Programme we can realize that its efficacity in creating an integrated research community at European level has been achieved. Nowadays the European cooperation is an usual and increasing dimension of activities of scientists and enginers throughout the Union either in Universities or in entreprises. But the oportunities or capabilities to participate in cooperation networks are not the same for everybody. In fact, the data analysis shows a very uneven participation according to the type of agents or regions.

Some characteristics of Community cooperation agreements established with a view to participating in shared-cost research actions are common, others are radically different from those of inter-firm technological cooperation agreements. Technological cooperation agreements established through the Community programmes have an institutional framework developed in contrast to those implemented by firms on a contractual, private basis which are not subject to national regulations in the large majority of countries. The limits established are justified by Community objectives and by atempts to achieve efficiency and even-handedness and, finally, by the desire to reduce any negative impact of agreements. These limits initially make themselves felt in the composition of European research consortia and the area of research is limited to precompetitive activities, in the fields priorized by Commission.

Due to the potential advantages of cooperation we can expect favourable effects on the regional dynamism of R&D and innovation. Transnational cooperation could permit breaking gaps and isolation of partners from the less favoured regions; so, entreprises and researchers could participate in large projects, unthinkable from their own forces.

V.3.3. The uneven participation of regions in European networks

The cooperative links set up between European countries through their participation in Community programmes permits a first evaluation of the qualitative participation of different countries and regions. The official data on III Framework Programme (CICYT 1994) show that the first three countries (U.K., France and

Germany) participate in 50% of proposals and they are also leaders in 52% of total supported projects. The analysis of the relationship between proposals and supported projects reveals an unequal success of countries: Netherland, France, UK and Denmark are higher (more than 24%) but proposals leadered from Italy, Spain or Germany have worse results.

Another significant data is related on the first partner (behind the leader): UK appears in first place seven times and France appears five time; the second partner of a large majority of countries (8) is Germany, UK three times and France one time (ERSTI 1994, 249). The positioning of countries in partnerships reveals a greatly uneven capability of research but also other factors such as complementarities within national systems of innovation, national incentives and policies, linguistic factors, traditional cooperation relationship, bargaining power etc.

That uneven participation provokes a contrasted partipation in financial returns by different countries and, more significantly, no proportional to the share in the funding of the programmes. The situation is far to the fair-return principle. Only three countries have lower returns (in %) than their financial contribution: Spain, Italy and Germany. UK has the most favorable proportion(17% against 12%).

The data analysis shows a very uneven participation according to the type of agents or regions (ERSTI 1994). Features of agents (as size, sector of activity etc), technology opportunities and local environments are factors influencing participation on European Programmes due to the priorities established by European Commission. From all this it follows that a very uneven regional distribution of benefits from cooperation can take place, either in short term (financial returns) or in the long term (results and effects on innovation capability).

Although there is no clear information about the regionalization of funds from the EC, it has been considered that in EC countries the three most developed regions of each country obtain 80% of the projects of the Framework Programme in which they participate. In Spain, Madrid receives at least 40% of funds (higher than % of partipants); followed by Cataluña (18%) and Pais Vasco (9%). However a clear regional redistribution effect is showed by the EC Structural Funds which Spain has been receiving since 1990. Taking into account the Spanish case, Madrid only receives 5.7% while Andalucía receives 19.4%, Galicia 10% and Castilla-La Mancha 6.4%. To the peripheral regions the Structural Funds received from the EC are as large as those received from the Framework Programme, in spite of the large difference of the respective amount.

In my opinion, the benefits from these funds lessen for the following reasons. Firstly, because of the lack of dialogue between different Administrations which consequently leads to a lack of investment planning both at a State level

(R&D Institutes which are the same in different Regions but no have relation with one another) and within the same Regions (R&D infrastructures which are carried out when the scientific and technological priorities have not been defined). Secondly, the R&D infrastructure endownment, being a necessary condition, is not enough to decrease significantly the territorial differences; these territorial differences only manage to be reduced with adequate variations in the distribution of real capability of research, which implies for example research personnel, access to the international networks etc.

Thus, a superficial analysis of European cooperation shows that peripheral regions have difficulties in participating in European networks. The expected advantages of cooperation to the regional dynamism of R&D and innovation do not take place automatically as a result of the existence of Community mechanisms and incentives. On the contrary, the hypothesis of this paper is that improving the regional technological environment and the local cooperation are prerequisites for success in transnational cooperation. It seems that local cooperation can constitute a usefull learning process towards more ambitious tasks in cooperation.

Progress in regional participation and cohesion may require to pay more attention to the regional dimension of policies and programmes in order to take into account regional features to really strenghten the Regional Systems of Innovation.

V.4. The cohesion challenge for European technology policy: from infrastructure towards improving R&D cooperation by strengthing Regional Systems of Innovation

The aim of increasing European industrial competitiveness and, at the same time, reducing regional disparities seems to be partlly coincidental but also partlly contradictory. It is coincidental in so far increasing competitiveness can permit to increase income and to strengthen the production networks that contitute the European production system in which firms from peripheral regions also take part. Nevertheless, it is also contradictory because external competitivity is mainly led by large firms: the European leaders in the strategic sectors, usually concentrated in a selected number of regions or metropolitan areas. Thus, the pre-eminence of competitiveness goal above any other objectives can concentrate efforts and finantial support in the "best" placed firms and regions (e.g. in strategic technologies), contributing to reproduce disparities in regional development. In fact, as shows the FAST Report 1992, just ten cities-regions absorb the two thirds of R&D Public expenditures and this proportion is higher in private expenditures. As innovation capability is not suddenly created but requires a certain time to start whatever objective founded in short term competitivity leads to strengthening centres of innovation in previous technologies, sectors and regions in the detriment of creating new capabilities in other technologies, sectors and regions. Otherwise, that logic changes when we are placed in the medium and long term strategy. In such a situation competitiveness can be much more solid on the basis of diversified production systems, allowing higher posibilities to face future changes in the main technologies, organizational forms and markets. In the last two centuries, industrial development shows that changes take place creating a shifting geography of industrialization, derived from the uneven capability of territories to create new industries in new sectors, adopt and adapt external innovations, and adapt to variable conditions in markets (Storper & Walker 1989). So, a certain investment to guarantee and invigorate the diversity of innovation and industrialization patterns, in places territorially diversified, can constitute a solid strategy to maintain the medium and long term competitiveness of European industry as a whole.

In a globalization framework the competitiveness of European industry depends to a large extent on its capability to innovate and reduce the length of time required for maturing innovation and products. That is, it has to innovate and do it fast. It requires a great capability to assimilate information, technologies and knowledge. Thus, technology policy must create conditions to assimilate rapidly new technologies and to introduce new goods and services along the production system.

The first condition to achieve this innovativeness is the permanent improvement of the telecommunication and transport infrastructures providing a fast and efficient circulation of goods, services and information around the world. Development of such infrastrutures constitutes itself the main way to promote certain high-tech activities and some other related to them, as shows the experience of the telecommunication sector. Obviously, the availability of this infrastructure creates new comparative advantages in well connected regions and raises the interregional unevenness. The aim of European Commission to incorporate the whole European territory to the communication and telecommunication networks faces the scarcity of financial resources and the necessity of time to do it; the time gap itself (eventually several years) is enough to create comparative advantages and evolve cumulative benefits to the first connected regions.

The second pack of mesures to push innovation and diffusion aims to create and strengthen the Innovation System, constituted by Public Research Centres, Universities, Firm and sectoral research centres and a broad set of cooperation networks. The efficacity of such system not only depends on the individual Science and Technology capability but from the cooperation capability to creating and diffusing new results and to tranfer them quickly towards its industrialization and commercialization. Building up an appropriate environment for innovation requires homologation and simultaneity between a wide-range of agents, with logics and purposes frequently disparate. It involves a continuous effort of orientation, flexibilization and coordination. Obviously, the chain of creation and diffusion requires to connect agents from different places of the European geography but the

185

differential capacity of territories to benefit from this innovation process depends on the density and strength of local cooperation links. Thus, European policy to promote cooperation in research and innovation can't ignore its geography and must support partners from lagging regions on the industrialization of results or the promotion of dispersing plants from large partners. In the long term, regional diversity and geographical distribution of production activities seems the only way to promote cohesion and reduce regional disparities.

Diversifying innovation centres can still constitute the way to accelerating changes by means of competence between different centres. At the same time it permits to select alternatives to the same problem, avoiding the "lock-in" in a worse technology. To this effect the **Maastricht memorandum** suggests that "the goal for policy is to maximize the amount of innovative experimentation with various technologies while at the same time minimizing the length of time required for experimentation. The amount of experimentation can be increase through policy support for simultaneous experiments with different technical solutions... Public goals could be pursued through policies to support experimentation by different research institutes located in separate geographical regions, though this must be followed by economic and social evaluations to select the best solution and complementary policies to encourage users to adopt it"(CE 1993, 72). From European perspective this strategy does not have to hinder the standarization of techniques or, in any case, their connectivity. As Gaffard points out "only the existence of conversion technologies providing compatibility between different standars can reconcile the explotation of increasing returns and the preservation of diversity" (Gaffard 1994, 35). But the main problem would be that national governments tend to adopt and promote national solutions regardless of their technical performance.

The Community Policy has included different programmes aiming to strengthen scientific and technological infrastructure in peripheral regions and to promote linkages and networks in order to raise cooperation in innovation and to favour knowledge diffusion and assimilation. Nevertheless, because of the link weakness in the innovation chain, peripheral regions can bear a perverse effect: peripheral researchers can contribute to develop innovations in core regions and even facilitate the 'brain drain'. The same phenomenon can take place within one given country. Breaking this vicious circle must be an important objective of the Community Policy and the regions themselves. It requires means and incentives working in the opposite sense: creation of local conditions to value results and this implies a close coordination between technology policy, industrial policy and investment recruitement policy.

Thus, EU regional cohesion challenge requires progresses in designing specific policies devoted to this aim. The risk exists of designing common policies, programmes and rules to implement it in every region, following some successful cases. Such starting point can lead to a disaster. A successful policy in a concrete region doesn't guarantee success in another one. Just the main idea of the approach proposed here is to pay much more attention to the region specificity and its concrete needs in order to design technology policies. In fact, studies on regional patterns of innovation during the last two decades show the existence of great originality and deep contrasts among them (Aydalot 1986; Gaffard and al.(FAST) 1993). Though strategic European competitiveness in the world market must be settled in high-tech industries it doesn't imply that all regions must compete in this sector. This isn't the way to guarantee cohesion, neither are core regions the sole candidates to lead the process. In a long term strategy it is perfectly possible to build up a specialized competitive region if adequate means are employed. Obviously, it is not possible that all the regions compete in the same range of activities. The process of Single Market creation itself leads to a growing regional specialization and Community Technology Policy must strengthen a sectoral oriented Regional System of Innovation -well adapted to the needs of the strategic clusters of regions- in order to settle the basis to the cohesion challenge. That strategy implies a primordial role of regional agents and authorities in order to design a regional technology policy involving the real needs and capabilities of the Regional Production System.

Insofar as this is in the right road we can consider that a Community Technology Policy guided by the regional cohesion principle must be based on diversity. Support to develop diversity can avoid irreversible excesses of industrial polarization around a reduced number of metropoles. Although some Community funds are devoted to peripheral regions by programmes as STAR (1986-91), SPRINT (1989-93) or STRIDE (1990-93) their effectiveness is limited. Besides their amounts being not in the mesure of challenges, the great problem is that they are mainly focused on infrastructures, but even when infrastructure is a condition for a dynamic of innovation, however the simple built of that is quite far from activities of innovation. In fact, usually, the main problem is to make a good use of that infrastructures. The main technology programmes (the four R&D Framework Programmes or the extracommunity Eureka) absorb most of the Science&Technology funds but no regional criterion is considered in their implementation. What consequences derive from this statement for the design and implementation of the R&D Framework Programme?. I consider that FP Funds musn't be fully distributed in a sole centralized selection of applications excluding regional criteria. Taking into account the subsidiarity principle, central evaluation must decide which programmes and projects are **really** strategic to the European Union as a whole and other innovation activities must be selected on the basis of criteria established by regions. Obviously, no clear frontier exits between strategic and non-strategic goals. Otherwise risks and difficulties can be adduced againts this regionalization of the non-strategic part of the FP. Certainly, besides the high complexity of administration, there can be tendencies to duplicate projects or inefficient management from unexperienced regions. In order to avoid lack of coordination, duplication of effort or an unsuitable use of funds a "contract-programme" could be introduced, signed by regions and supervised by the European Commission. Central coordination can identify overlaps and recommend cooperation among regions in order to reduce costs and increase returns. Another different or complementary way would be to attribute competencies to the **Regions Council** to coordinate and to deal with this part of the FP.

Such a proposal to regionalize technology policy does not imply a reduction in cooperation at European level but a reinforcement of tendencies to create large networks dispersed on the territory. Coordination is settled on complementarities revealed by a comparison of different programmes and projects proposed by each region. So regions should establish priorities and specific agencies may articulate cooperation to develop joint projects or coordinate projects. The mechanism should not be really different from the coordination rules of Transborder Programmes, but generalized to all regions. In any case, the regionalization may permit changes in current rules to define the "community character" of projects by agreement of private-individual agents (firms, universities...), which allows leaderships of large firms face to the Small and Medium Firms. Furthermore the current criterion on multinationality of partners (three or more) can favour cooperation among a restrained number of high-tech partners from different countries but leaving aside most of the european regions. Network models as the "four motors of Europe" are far from the cohesion strategy of the Maastricht Treaty but, on the contrary, allow the emergence of the so-called "Archipelago Europe".

VI. CONCLUDING REMARKS: CONDITIONS FOR INCREASING COOPERATION AND DEVELOPING INNOVATIVE ACTIVITIES IN PERIPHERAL REGIONS

In so far as the capacity to attract outside investments to stimulate the industrial and technological system remains reduced -as the experience of most European periphery regions shows (NEI 1991)- it is necessary to focus our efforts and resources on a development strategy based on the capability of the production system to adapt and to develop innovations. Regional development goes through the transformation of the production system with the aim of increasing its competitiveness, as market conditions become stricter and it is not possible to hope to maintain this competitive position based on a low-salaries strategy. To do so, it is necessary to develop a local productive capability stemming from the endogenous potential, but integrating all the external factors affecting its incorporation in the world market and providing complementary elements of innovation processes. Local companies from peripheral regions face more difficulties than core-regions, not so much because of geographical reasons but for reasons related to economic structure and agglomeration economies (size of local market, availability of credit and venture capital, accessibility to suppliers and customers, availability of technical labour force etc.). Taking into account the lacks of peripheral regions any innovation strategy must articulate endogenous and external factors.

Concluding remarks on decentralization and diversity

1. The decentralization of technology policy and R&D activities can favour at the same time cohesion and diversity. From an evolutionary perspective the purposes of technology policy are to increase the rate of innovative experimentation in the economy and to ensure that the market selection process is open and works to diffuse superior product and process technologies. The rate of experimentation depends on the conectivity of the science and technology system; the policy-makers can increase conectivity by promoting the interaction between the higher education and industrial components. The distribution of research funds between institutions, the existence of bringing mechanisms to connect industry to the public science and engineering base, and the degree of public investment in skill formation all play their role in raising creativity in general. In terms of static efficiency letting a number of competing experiments run simultaneously involves duplication and appears wasteful. From an evolutionary perspective this is not so (Metcalfe 1993). Therefore, R&D activities decentralization is not an economically inefficient decision when we think of the long term. Regional Technology policy can constitute an important element in raising the technological capability, productivity and competitiveness of firms located in less favoured regions. That is precisely a very important contribution to the aim of **regional cohesion.**

2. The main problem is not only distributing regionally R&D funds, but also designing different policies in order to strengthen Local Innovation Systems that are different to one another. To choose priorities at regional level is necesary. An important reason, among others, is that a single market implies greater specialisation within Europe, and therefore regional cohesion policies should seek to built on regional diversity and the strengths of each region (CE 1993, 81). I agree with Gaffard's proposal which enfasizes that "Community or other (national or regional) programmes must meet the requirement of maintaining and strengthening diversity: this means that, in terms of their conception and organization, they must be sufficiently diversified to be able to correspond to (sectoral, territorial9 innovation systems which are themselves diversified, and thus help make them viable" (Gaffard 1994, 38). In our recent experience we can observe a great mimicry between the regional programmes on technology and the national (or European) programmes. All indicates that the results are quite different and disappointing. Thus policies to support regional cohesion and convergence within the EC must avoid the error of

189

using identical policies to encourage all regions to move in the same direction. It's not enough to decentralize the policy executing bodies. A diversity of bodies has to exists, so as to design the appropriate means and programmes to the needs of each regional production system.

Factors dealing with the capacity of a territory to incorporate innovation into its local industrial system and basis to orientate a technology policy at regional level:

1. The existence of a **local demand for innovation**: a voluntarist intent to catch up advanced technologies is a political temptation which runs the risk of failing, as shown by the multiple failures of science parks and technology parks all over the world, but especially in peripheral regions; this could be an exorbitant expense with no real effects on the technological dynamism of the territory. The same could be said of industrial parks since broader international experience shows that they are not a prior condition to localize firms; if they do not form part of a development strategy of the local productive system, they run the risk of becoming "cathedrals on the desert". New technologies can be gradually incorporated when the configuration of the production system based on old schemes no longer provides solutions to the new problems arising from the evolution of markets and competition. This can occur when local firms recognize concrete needs. The dynamics of continuous change in innovative areas constitutes in itself the key factor that constantly demands new answers and sets new challenges to innovation potential in well identified fields. Thus, technology policy must pay more attention to technology demands from regional in rooted sectors. Otherwise, when innovation is institutionalized in its own local structures, those structures could create new needs because usually each innovation asks for other innovations in a sequence of positive feed–backs.

2. The existence of **technology infrastructure and innovation capability**. It is difficult for a region to participate in the diffusion of new technologies without an adequate technical capacity for adaptation or innovation. To acquire an innovation produced outside, regional firms must be able to transfer innovation and incorporate it efficiently into their productive structure. All incorporation of technical change implies a certain innovation capability and a tacit knowledge that proceed from the body of companies and local innovation system. The achievement of this aim requires a strengthening strategy of R&D activities within firms, complemented by a network of sectoral centres of innovation (especially interesting for small and medium firms) and a fluent network of cooperation between firms, Universities and other public research centres. There is no way to technology without a solid **R&D infrastructure** and an organizational ability to build real capacity to innovation. To build up basic technology infrastructure can be considered the first step to develop a regional system of innovation and, of course, it does not imply only physical infrastructure but also at least technical and scientific resources and organizational

knowledge.

3. The capacity to **remunerate the innovation**: the market must guarantee the profitability of investment in innovation. This depends on the **entrepreneurship, financing** and **markets**. Entrepreneurship is the critical factor to value innovations through existing firms or new firms formation; unfortunately, it is just the less manipulable factor from policy institutions. In some cases, despite liberal dogmatism, an active and direct role of government creating innovative firms can be an adequate way to appropriate economic benefits of innovation efforts and to start up a clustering process of activities related to the first one settled. To do this regional governments need a high level of competencies on the matter.

A basic condition to develop innovation and start up creation resource processes is the existence of an National/Regional Financing System able to promote new projects of technological investment. The major problem is the difficulty to maintain a regional commitment of Financing Institutions in the era of deregulation and globalisation of financial markets.

As for the markets, except in the case of internationalized companies, profitability depends on the solvency of the regional or national markets to absorb new merchandises, particulary in the case of equipment, intermediate goods or technical services. Since regional markets are small, especially in peripheral regions, policy measures must be introduced to favour the competitiveness of local firms in internationalized markets, which allows them to value investments and enlarge the production scale. Conventional policies to promote exports are today inefficient and also confront international agreements on the matter; thus, new ways must be founded to create stable commercialization networks and to adapt to the market trends, variety, quality and after–sales services[71]. Furthermore, an important way to create market in certain high-tech activities (e.g. information technology) is the purchase policy of regional government. This can constitute a safe market that permits firms to learn and build up the economies of scale necessary to compete in new markets.

4. **Complementarities** and **cooperation** to introduce innovation technically and socially: as new technologies become more and more systemic, efficient implantation depends on the actual local existence of infrastructural and interindustrial complementarities, skills and technical labour force, financial support etc. Therefore, an industrial development strategy should closely coordinate efforts in the different areas attending to the specific needs of each local productive system.

---

[71]As an intersting example, some traditional migrant regions try to weave great networks of customers or distributers of merchandises exploiting cultural and friendly links.

On the other hand, increasing cooperation between local firms along the different phases of production process and commercialization permits to internalize external economies. Particularly, cumulative feed–backs into the innovation activities can be created through **innovation networks,** which favour sunk-cost reductions and increasing returns from enlarged information and externalities of research.

Beyond the aforementioned static advantages most important dynamic advantages can be created leaning on complementarities and cooperation. First, local cooperation can be the first step in "learning to cooperate" and can facilitate further and faster cooperation. Second, the capability of a territory (e.g. Industrial district) to face qualitative changes in the economic environment and to create new technological opportunities depends on the amount of **diversity** and **complementarities** in products, technologies and competencies developed in this territory. As Amendola and Gaffard pointed out diversity is the source of future qualitative flexibility, thus diversity must be preserved even if this could imply a suboptimal allocation in the short term; but, in fact, diversity can be no less productive even in the short term when complementarities are well valued. Nevertheless, a nuance should be made because diversity is a worthwhile concept to explain, above all, benefits from complementarities between different regional innovation systems rather than from internal heterogeneity. In any case, policies must face diversity as something to be encouraged and not only as an inevitable heritage. In order to balance potential losses in economies of scale, complementarities should be carefully worked out.

5. **The political–administrative competencies of regions**: the degree of centralization-decentralization of decisions at political–administrative level greatly influences the region capability to incorporate innovations for many reasons. Among them we can point out three. First, due to the Institutional's previous influence on the historical configuration of production structures of the country at spatial level. Second, because decisions at a centralized level also tend to cluster capability. In a decentralized system, each region acts concurrently with respect to the others, adding additional means to acquiring innovation. The Spanish experience, where industry policy rests as an exclusive competence of the Central Government, is a good example of the perverse and shortcircuited effects of centralization; nearly 45% of Spanish R&D is concentrated in Madrid and it receives also nearly half of public aid for innovation. In fact, the Spanish Technology Policy is only contrebalanced by Tecnology Policy of regions and by the Structural Funds of the Community Technology Policy. Third, the degree of self-government determines the wide-range of undertaken means.

6. The configuration of **socioecomic dynamics** and the type of wage relations ("relation salariale") also affects the capability to introducing innovations, as many "régulationistes" emphasize. A hierarchical society may be more opposed

to the introduction of innovations, in so far as these could unbalance the relative positions of power between social groups or socioprofessional categories. On the other hand, a horizontal integration of society makes communication and innovation easier, especially if this becomes the means to social mobility. Although there are negotiated mechanisms for distributing the benefits of innovation, this is much easier when all agents collaborate in its introduction. Thus, the authoritarian or hierarchichal models of labour relations constitute a factor of disintegration and conflict which makes it difficult to introduce any changes and also makes future efficent profitability more difficult.

REFERENCES

AMENDOLA, M.; GAFFARD, J.L. (1988), **Dinamique économique de l'innovation**. Paris, Economica.

AUDRETSCH, D.B., FELDMAN, M.P. (1993). "The geography of innovation and production", CEPR Conference on the "Location of economic activity: new theories and new evidence", december, Vigo (Galicia).

AYDALOT, P. (Ed) (1984), **Crise & Space**. Paris, Economica.

AYDALOT, P. (Ed) (1986), **Milieux innovateurs en Europe**. Paris, GREMI.

AYDALOT, P.; Keble (Ed) (1988), **High Technology Industry and Innovative evironnements: The European Experience**. London, Routledge.

BAILY, M.I.; CHAKRABARTI (1988), **Innovation and the productivity crisis**. Washington: Brookings Institution.

BANDT, J. DE; FORAY, D. (1991), **L'evaluation économique de la recherche et du changement technique**. Paris, CNRS.

BECKOUCHE, P.; SAVY, M.; VELTZ, P. (1986), "Nouvelle économie, nouveaux territories", **Colloque Economie et Territoire**. Paris.

BENKO, G.; LIPIETZ, A. (Dir) (1992), **Les régions qui gagnent**. Paris, PUF.

BERGMAN, E.; MAIER, G.; TÖDTLING, F. (eds.) (1991), **Region reconsidered**. London, Mansell P.L.

BERSTEIN, J. (1988), "Costs of production intrafirm and inter–industry R&D spillovers: Canadian evidence", **Revue Canadienne d'Economie**, no. 21, pp. 324-347.

BURSTALL; DUNNING; LAKE (1981), **Multinational enterprises governments and technology: Pharmaceutical industry**. París, OCDE.

COMISION INTERMINISTERIAL DE CIENCIA Y TECNOLOGIA (cicyt) (1995), **Memoria de actividades del Plan Nacional de I+D en 1993**, Madrid, CICYT.

COMMISSION OF THE EUROPEAN COMMUNITIES (1985), **Science and Technology for Regional Innovation and Development in Europe**. Luxemburg.

194

COMMISSION OF THE EUROPEAN COMMUNITIES (1993), **An Integrated Approach to European Innovation and Technology Diffusion Policy a Maastricht Memorandum**. Editors: L. Soete and A. Arundel. EEC, Brussels.

COMMISSION OF THE EUROPEAN COMMUNITIES (1993), **New location factors for mobile investment in Europe**, Regional Development Studies, n. 6, Luxemburg.

COMMISSION OF THE EUROPEAN COMMUNITIES (1994), **The European Report on Science and Technology Indicators 1994**, EUR 15897, EC, Brussels.

COOPER, C. (1973), "Science technology and development". London, Frank Cass.

G. DOSI ET AL. (1988), **Technical Change and Economic Theory**. London, Printer Pub.

FELDMAN, M.P. (1994). **The geography of innovation**, Boston, Kluwer Academic Publishers.

GAFFARD, J.L., "Restructuration de l'espace économique et trajectoires technologiques" in P. Aydalot (ed.): **Milieux innovateurs en Europe**, pp. 17–27.

GAFFARD, J-L; Longhi, G; Quéré, M. (1994), "Coherence and diversity of innovation systems in Europe", Draft Paper. IDEFI-Université de Nice.

GIRÁLDEZ, E. (1988), "Comportamiento inversor de los sectores de alta tecnología, 1975–1985. Tendencias espaciales", **Papeles de Economía Española**, no. 34, pp. 431–453.

GÓMEZ, M.; SÁNCHEZ, M.; DE LA PUERTA, E. (1992), **El cambio tecnológico hacia el nuevo milenio. Problemas debates y nuevas teorias**. Madrid, Fuhem/Icaria.

GORDON, R. (1990), "Global Networks and the innovation process in high technology SMEs: The case of Silicon Valley", Working Paper, Silicon Valley Research Group, University of California, Santa Cruz.

GRILICHES, Z. (1984), **R&D patents and productivity**. Chicago, U. Chicago Press.

HILPERT, U. (Ed) (1991), **Regional innovation and decentralization**, Routledge, London.

HILPERT, U. (1992), ARCHIPELAGO EUROPE - Islands of innovation, FAST Dossier: Science, Technology and Community Cohesion Vol. 18, European

195

Commission, Brussels.

HIPPEL, E. VON (1988), **The Sources of Innovation**. Oxford U.P.

MALECKI, E.J. (1991), **Technology and Economic Development**. New York, Longman S.T.

MARTIN, F. (1986), **Une évaluation des politiques gouvernamentales de décentralisation régionale des activités de R&D**. Colloque Tecnologies Nouvelles et Developpement Régional. París.

METCALFE, J.S. (1993), "The Economic Foundations ofTechnology policy: Equilibrium and evolutionary perspectives", Draft Paper. PREST-Manchester Unversity.

MOUHOUD, E.M. (1992), **Changement Technique et Division International du Travail**. Paris, Economica.

MOWERY, D.; ROSENBERG, N. (1982), "The commercial aircraft industry" in R.R. Nelson (ed.), **Government and technical progress**. New York, Pergamon Press.

NELSON, R.R. ed. (1993), **National Innovation Systems. A comparative analysis**, Oxford, Oxford University Press.

NOBLE, D.F. (1986), **El diseño de América. La ciencia la tecnología y la aparición del capitalismo monopolista**. Madrid: Ministerio de Trabajo y S.S.

O.C.D.E. (1989), **Universidad industria y desarrollo**. Madrid, Ministerio de Trabajo y S.S.

O.C.D.E. (1991), **Choosing Priorities in Science and Technology**. Paris.

O.C.D.E. (1992), **Technology and the Economy. The key relationship**. Paris.

PAVITT, K. (1984), "Sectoral patterns of technical change: Towards a taxonomy and a theory", **Research Policy**, 13, pp.343-373.

PLANQUE, B. (1983), **Innovation et développement régional**. Paris, Economica.

PÉREZ, C. (1986), "Las nuevas tecnologías: una visión de conjunto" in C. Ominami, **La tercera revolución industrial**, p. 85. Buenos Aires, RIAL–GEL.

SCHMANDT, J; WILSON, R. EDS. (1990), **Growth policy in the age of high**

technology. The role of regions and states, Boston, Unwin Hyman.

STORPER, M., WALKER, R. (1989), **The capitalist imperative**, Oxford, Blackwell.

STÖHR, W.B. (1984), "La crise économique demande–t–elle de nouvelles stratégies de developpement régional?" in P. Aydalot (ed.), **Crise & espace**. Paris, Economica.

VELTZ, P. (1986), "L'espace des industries électriques et électroniques", **Les Annales de la Recherche Urbaine**, no. 29.

VELTZ, P. (1992), "D'une geographie des coûts à une geographie de l'organisation. Quelques thèses sur l'evolution des rapports entreprises/territoires", **Revue Économique**, vol. 44, 4, 1992.

VENCE, X. (1989), **Potencial innovador e cambio tecnolóxico na industria. Estudio da área sur da provincia de A Coruña**. Santiago de Compostela, Cámara de Comercio de Santiago/Banco Pastor.

VENCE, X. (1991), **Innovación e cambio tecnolóxico na industria: o potencial científico–técnico e as transformacións nos procesos de producción na industria en Galicia durante o período de crise**. Tese de Doutoramento, U. de Santiago (microficha).

VENCE, X. (1992), "O sistema de ciencia–tecnoloxía e as asimetrías nas relacións Universidade–Industria", **Revista Galega de Economía**, 1, pp. 25–48.

VENCE, X. (1995), **Economía de la innovación y del cambio tecnológico**. Madrid, Siglo XXI.

9 STRATEGIC TECHNOLOGY POLICY FOR NEW INDUSTRIAL INFRASTRUCTURE: CREATING CAPABILITIES AND BUILDING NEW MARKETS

Moshe Justman and Morris Teubal
The Industrial Development Policy Group
The Jerusalem Institute for Israel Studies

I. INTRODUCTION

A new area of technology policy has emerged in the last decade, focusing on the development of technological infrastructure for industry and aimed expressly at promoting economic growth. We refer to this area as "strategic" technology policy, to distinguish it from policies that offer direct, "tactical" support for individual R&D projects They are strategic in the sense that they must be targeted explicitly, requiring a measure of coordination that can only be achieved through the active involvement of business and political leaders at the highest level. This represents a marked departure from earlier programs which were either targeted at non-economic goals such as defense or space exploration ["mission-oriented" technology policies in Ergas's (1986) terms]; or aimed at supporting market-based technological progress on a wide front through "neutral" subsidies or broadly-based tax relief for R&D activities or for venture capital. These latter policies were generally devoid of industry bias and hence could be administered by an impartial professional bureaucracy. It also represents a departure from the policy approach frequently followed in the newly-industrialized countries, of targeting specific industries. Strategic technology policy targets technological infrastructure rather than industries; it is characterized by "mild" rather than "strong" selectivity [Justman and Teubal (1990)].

Far-reaching changes in the underlying conditions of the world economy -notably rapid technological change stemming from the revolution in information technology, and political changes that have spurred a dramatic trend towards economic liberalization- have signaled the need for structural change throughout the industrialized world. The frequent emergence of new products and processes and their rapid diffusion in a shrinking globe have greatly accelerated the rate of obsolescence of industrial capacity, creating a recurring need for industrial renewal. With increasing frequency, economies are arriving at nodes of structural change where it is evident that new infrastructure is needed to pave the way for further

X. Vence-Deza and J. S. Metcalfe (eds.), Wealth from Diversity, 199–227.

growth, though the optimal direction of growth can often only be guessed at. These nodes require a very different kind of policy effort than that required at periods of "routine" growth [Justman and Teubal (1990, 1991)]. Strategic policies are aimed at meeting this challenge.

It is a challenge that is all the more formidable because elements of infrastructure needed in effecting these changes go beyond "traditional" infrastructure such as roads, power, and water, and involve the generation of new technological capabilities. The more sophisticated of these capabilities, required by leading-edge industries, are located somewhere along the spectrum between pure science and firm-specific applied research; they are associated with "generic research" [Nelson, Peck and Kalachek (1967)] and "strategic science" [Irvine and Martin (1989), J. Senker (1991)]. Other industries may require technological infrastructure based on an existing, foreign knowledge base that must be imported and adapted to local conditions. This is particularly true for the second tier of industrialized countries such as South Korea, Singapore, and Israel, and may also apply to the industrial development of peripheral or declining regions in the most advanced economies. The characterization of technological capabilities suggested by Westphal et al. (1984) and Dahlman et al. (1985) regarding the assimilation and utilization of foreign technology in the context of traditional targeting policy can also be applied to sector-based (rather than firm-based) technological infrastructure serving firms in mid-tech and low-tech sectors of advanced countries

Japan, which pioneered the well-known VLSI program in the late seventies, is the main origin of this new type of technology policy, as well as the premier example of its success. As Ergas (1986) has pointed out, in the early 1980s Japan was the only country to have in place "mission-oriented" policies aimed at promoting economic growth (rather than defence, health, or the environment, objectives which prevailed in other countries). Since then, and due in no small measure to the success of the Japanese model, strategic technology policies have been implemented by other countries or groups of countries, such as the Alvey program in the UK, Esprit and Jessi in the EEC, and the MCC and Sematech programs in the US. [On information technology policy, see E. Arnold & K. Guy (1986 and 1987), and K. Guy & E. Arnold (1991)]. Moreover, a large number of sectoral and regional programs were established for mid-tech/low-tech sectors in countries such as Germany and Italy [Malerba (1991), Ergas (1986)].

Nonetheless, despite the widening application of strategic technology policy in practice, we still lack a well-formed theory which might provide a systematic analytical framework for formulating, implementing, and evaluating such policies. The type of project-oriented "anatomy of market failure" which served as a framework for the previous generation of technology policies -e.g., broad-based subsidies and tax advantages for R&D- cannot provide an analytical basis for this new generation of policies; the key issues that define strategic technology policy are fundamentally different.

The distinction emerges most sharply when the economy is at a node of structural change and must choose from among a set of discrete, mutually exclusive, alternative paths for growth. In such circumstances market failure is endemic: not only is there no reason to assume that the market will choose the best path -or even identify the elements of infrastructure that are necessary for growth when they are needed [Nelson (1987), Justman and Teubal (1991)]- but market failure analysis alone cannot identify a "preferred" path of growth. Additional analysis is required. Consequently, in identifying the potential for policy intervention, the economic analysis of strategic policy places greater emphasis on economies of scale and scope, rather than on the non-pecuniary externalities that play a central role in justifying R&D subsidies [Justman and Teubal (1986)].

But there is more to it than that. Even when the appropriate path for growth has been determined, getting on this path may take more than just the price-based incentives -subsidies, tax credits- on which technology policy has mostly relied in the past. It may require coordination and cooperation among potential users of the new technological infrastructure in determining which elements are necessary and should be supported. And in some cases, coordination of supply and demand will be needed.

This implies that the government may have a positive role to play in promoting information exchange among the parties, financing early background studies and pilot projects, and supporting cooperative efforts aimed at defining future needs and developing new capabilities. This is true also for 'conventional' infrastructure when local absorption of foreign technology depends on developing local capabilities. The complexity of these needs may be such that it is not possible to generate a supply of the necessary infrastructure services without enlisting the active participation of intended users, some -or all- of whom may not yet be active in the market. Hence the emphasis which the present paper places on market building. These policies represent a new type of intervention, requiring new skills, organizational forms, decision-making processes and feedback mechanisms. Their design and implementation represents a formidable challenge. Much additional work must be done -within the context of a specific institutional framework- before practical conclusions can be drawn from the general analytical considerations discussed here. The present paper can only hope to lay the groundwork for such an effort.

The next section describes what we see as the main characteristics of technological infrastructure. We then examine how new technological infrastructure is generated and why the market might fail to do this efficiently, considering separately two types of infrastructure: one that links a conventional industry to foreign sources of technology, and one that requires the creation of new capabilities in support of a leading-edge industry. A simple formal model of critical mass in infrastructure development highlights some of the points raised here. Issues regarding the formulation and implementation of strategic technology policy are then addressed, and we conclude with a brief summary.

II. EXAMPLES AND CHARACTERIZATION OF TECHNOLOGICAL INFRASTRUCTURE

Technological infrastructure for a group of firms -we may sometimes call it an industry- comprises a set of capabilities and market links which these firms need in order to function on an efficient competitive basis, yet which transcend the needs of any individual firm. It manifests itself in one of two main forms: technological support services, e.g., in product design or quality control; and collaborative R&D projects that support the development, by individual firms, of new products or processes. Correspondingly, we may talk of two main classes of technological infrastructure - sector-oriented and functional.

Sector-oriented technological infrastructure -which may serve sectors within the mid-tech and low-tech groups- includes a number of distinct areas:

A quality control, testing and analysis capability for conventional firms, particularly small and medium sized ones. Between the mid-fifties and mid-sixties Japan created several Engineering Research Associations (ERAs) to house capabilities of this kind; some of these served the auto parts industry [Sigurdson (1986)].

A product design capability for industries such as textiles, plastics, metalworking, furniture and footwear. These capabilities are usually absorbed at the sector level and subsequently diffused to individual firms, sometimes with the help of consultants. Israel successfully developed such a capability for the clothing industry; it currently needs one for the plastics sector.

A capability for identifying, selecting and absorbing novel production technology, equipment and raw materials. [See Malerba (1991), for examples in Italy]. This is increasingly important given the exponential growth in the options available. Small and medium sized firms -which are particularly good at providing specialized products for niche markets- find it difficult to scan the technology horizon effectively and select what they need. While consultants can do part of the job, markets for consultancy services associated with new technology are imperfect and require stimulation from the organization absorbing and housing the new capabilities.

202

A capability for solving contamination and ecological problems facing firms in a particular sector or region (for example, from the tanning of leather). A consortium of firms may get together or the local industry association may initiate a program for tapping foreign sources of knowledge that may bear on a common problem. A second stage of assessing this information and arriving at a collective solution may be required, prior to effective implementation. The same may apply to the conservation of natural resources, such as water or energy.

These examples show that the notion of technological infrastructure is related to information and implies a need for collective behaviour of some kind rather than simply accessing world technology on a firm-by-firm basis. Indeed, in many cases it may be a precondition for accessing world technology, especially for SMEs. Moreover, the above examples make it clear that conventional industry infrastructure may involve little, if any, R&D, and often deals with the "bread and butter" problems facing firms. With the emergence of a new technological paradigm of more flexible manufacturing we can expect the need for these elements of sectoral technological infrastructure to grow dramatically.

The best-known example of functional technological infrastructure is the development, in Japan, of design and production capabilities for 1 megabit DRAMs through the MITI-orchestrated VLSI project in the late seventies, capabilities that enabled the participating electronics firms to launch a series of innovative semiconductor devices in the decade that followed. The specific capabilities were associated with crystal technology (how to avoid the bending of silicon crystals); fine processing technology (the electron beam delineator); testing and evaluation technologies; and design capabilities [Sigurdson (1986), p. 46]. In biotechnology, the specific capabilities that comprise a functional infrastructure might include fermentation technology, genetic engineering, biosensors, protein engineering and downstream processing [J. Senker (1991)].

In general, functional capabilities stem from cutting-edge generic technology in areas such as information technology (including microelectronics), biotechnology, new materials, etc. Their function is also to "mediate" between technology needs of firms and other domestic users) and sources of supply. However, while "conventional capabilities" are intended to link firms with foreign technology, the development of advanced capabilities focuses, in addition, on harnessing the results of scientific research (domestic and foreign) to industrial use. This involves much more than "dissemination" and "transfer" of scientific knowledge, since it may also require the development of "enabling" engineering capabilities (e.g., downstream processing capabilities in biotechnology).

The basic building blocks of technological infrastructure are the elements of human capital embodied in individuals, comprising their formal education, training and work-experience. Capabilities can be thought of as combinations of these elements of human capital linked together in purpose-built groups, generally within firms. For example, one might think of a bank's consumer-loan granting capability as deriving from the individual abilities of the loan officer, the commercial assessment of the loan, the information technicians who provide access to data on the prospective lender's credit-worthiness and process the paperwork, the bank's lawyers who draw up the loan contract, and the branch manager who provides the physical conditions in which the transaction takes place, all linked together by routines and procedures as well as mutual confidence and trust. Individual elements of human capital may originally be developed within a university research institute or an industrial technological center and spin off, at a later stage, to industrial firms where they can become integrated into purpose-built groups.

These capabilities offer their services in markets -broader networks of buyers and sellers that trade in definite services or goods. Participants in the market know what they can expect from one another on the basis of the institutional arrangements that govern its operation (laws, regulations, contracts) and tacit understandings that derive from "membership" in the market and reflect its advantages. In the subsequent discussion we will distinguish between the capability-creation phase of infrastructure generation and its market-building phase.

A number of features characterize technological infrastructure. First, it has a generic quality: the capabilities involved are intended for different uses by a number of firms, they are not geared to individual innovations or firm-specific R&D projects. Although generic research is an important component of some technological infrastructures, not all technological infrastructure involves research, and not all scientific research generates technological infrastructure. It is generally aimed at reducing the distance between the results of scientific research and their ultimate utilization, often involving a multi-disciplinary effort that combines scientific and engineering skills. Thus development of technological infrastructure in bio-technology requires the combined effort of micro-biologists, bio-chemists and bio-engineers; and the VLSI project placed great importance in the development of production capabilities in conjunction with design capabilities.

A second salient feature is its pre-competitiveness, i.e. its indirect economic value, and the general absence of a market for its output. This provides possible justification for some form of non-market intervention, as individual firms in the private sector may not have an adequate economic incentive to undertake such activities (we shall have more to say about this in the following sections). At the same time, it increases the chances for successful cooperation among competing firms that need to protect their trade secrets from each other.

Economies of scope are a third feature of technological infrastructure which

204

distinguish it from conventional infrastructure such as roads, power, or water. Whereas the latter exhibit economies of scale in the production and supply of a standard commodity, the critical mass associated with technological infrastructure derives from the need to provide a spectrum of linked but specialized and distinct capabilities, on which the firms in the industry (or user group) draw in a variety of patterns.

Thus, for example, there might be seven potential elements of infrastructure -A, B, C, D, E, F and G, each representing a distinct capability, and four firms that might be able to participate in the industry. Firm 1 needs A and B and may need D to achieve competitive production; firm 2 needs C, D and E, firm 3 needs B and possibly C, and firm 4 needs E, F and G. Some of the firms -and some of the elements of infrastructure- may not be commercially viable. Successful development under these circumstances requires more than simply ensuring that a critical mass of demand materializes. A cooperative effort will be needed to determine which elements of infrastructure should be developed, who should cooperate with whom, and in what sequence. It is this need for cooperation among prospective users of the technological–infrastructure that sets it apart from conventional infrastructure.

We next consider the process of building technological infrastructure. For this purpose it will be convenient to distinguish between two basic types of technological infrastructure: one geared to serving conventional industry by linking it to existing foreign sources of technological know-how and by building a market for these technological services; the other aimed at creating new capabilites for a leading-edge industry. In both cases, there are good reasons why a market solution, without intervention, might not lead to the most efficient development path, and we consider what might be an appropriate role for government intervention in reaching closer to this path.

III. MARKET BUILDING BY LINKING LOCAL NEEDS TO FOREIGN SOURCES OF TECHNOLOGY

Consider first establishing a technological infrastructure for conventional industry. Its role is to mediate between the technological needs of the industry and potential -mostly foreign- sources of supply. The role of infrastructure services in this regard is, in the first instance, one of promoting static efficiency by providing information and advising local industry regarding the availability of foreign technology. But it should also serve the function of promoting dynamic efficiency through a twofold action: stimulating local demand for foreign technology by helping local industry redefine its needs in terms of the possibilities that the new technology offers (user need determination -see below); and increasing the effective supply of technology inputs by stimulating investment in adapting them to local needs and promoting local sources of supply (including technical consultants).

Some firms may have the capability both to define their needs in terms of the foreign technology and to access it directly without the intermediation of a specialized, multifirm oriented technological infrastructure. In these cases neither capability creation nor market building is relevant. However, in an increasing number of sectors in mid-tech and low-tech areas, especially those populated by small and medium sized enterprises, this is not possible. Enabling these firms to access foreign technology (including developing their awareness of its relevance to their activities) may necessitate both establishing local capabilities and building a market for the services flowing from them (see Figure 1). In what follows we assume that the prior capability requisite has already been developed so we focus exclusively on market building.

Initially, there is little or no domestic market for these services, i.e. there is neither supply nor demand: they cannot be obtained locally, nor is there a well-articulated need for them. Market building relates to a number of things. First, the local market for imported technological inputs must be developed; next there should emerge a derived market for local linking or intermediation services; finally, these should stimulate the creation of a market for local substitutes for foreign technology, as the domestic economy is able to develop a competitive advantage in an increasingly mature foreign technology.

Consider business software as an example of such infrastructure services. After establishing a core of basic software capabilities relevant to the needs of a target group of users, the process of building markets might begin with an import agent acquiring distribution rights to a specific software package and advertising its availability to potential users. This might lead to additional investment, either by the import agent or by other parties (including the organization housing the core capability), in translating operation manuals, translating the software, modifying the software to accomodate special local needs, etc. At the same time, potential users might be modifying their operations to take advantage of the new software, e.g. computerizing their "accounts payable" files or building an automated warehouse. This might lead, in turn, to other agents importing other, complementary software packages that enhance the utility of the first package, further increasing the user base. Eventually, the local user base might be large enough to support locally produced business software for domestic use, and in some cases this might even lead to the export of domestically-produced software, if an advantageous wage differential could be maintained after the technology was mastered.

From this example it should be clear that market building could proceed without public intervention. Nonetheless, there are a number of reasons why a purely market-driven process might not succeed when it was needed:

Externalities among users learning about the application of the new technology. New users may hesitate to adopt the new technology because they are uncertain about its potential for meeting local needs, and information that could reduce this uncertainty may be costly to obtain and difficult to appropriate. However, if a group of potential users can arrange a coordinated effort to study common aspects of the new technology based on a free sharing of information, lear-ning costs can be reduced, risks shared, and external effects internalized (see below for possible market failures in generating potentially profitable collaborations).

Codification and standardization. By this we mean transforming individual experience about the adaptability of the new technology to local conditions into a codified body of knowledge that allows distinct user types and product types to be identified and effectively linked. Teubal and Zuscovitch (1991) refer to this as "general discriminating capabilities" and argue that it requires an explicit allocation of resources, over and above tacit experience. This activity may have enormous social value in promoting rapid diffusion of the new technology, but individual entrepeneurs may be reluctant to invest the necessary resources because of the difficulty in appropriating the benefits from such an effort. This holds especially true for small-scale entrepeneurs who may well fear that any early effort on their part to expand the market through standardization will be co-opted by larger established firms with stronger complementary assets.

Network externalities. These arise when late adopters of a new technology derive inappropriable benefits from the prior existence of a large user base. Telephone networks are the core example, but such effects can arise in a variety of contexts, e.g., late adopters of widely used computer hardware may benefit from a pre-existing supply of specialized software, maintenance and repair technicians, expert consulting services, etc. Such external effects will benefit the entrepeneur supplier of the new technology -the local distributor of the computer hardware in our example- who may indeed want to subsidize early adopters in some way. But without some measure of coordination this may be a very risky investment, especially if there are a number of competing entrepeneurs. If such private efforts fall short of inducing the socially desirable rate of diffusion, some measure of neutral public support for early adopters of the new technology, from whatever source, may be warranted.

Expert consulting services. A new technology in the early stages of its diffusion may require the support of a pool of local consultants able to advise users about which products to select and how to make the best use of their selection. There may be market failure in the training of such a pool. Early experts in the technology may find that it is in their best interests to limit the spread of knowledge, e.g. by extracting a high price for its transmission, while offering consulting services themselves, rather than work for a wider dissemination of knowledge, in line with

207

social welfare. A dynamically efficient market must, of course, pay a premium for the early acquisition of valuable knowledge. Nonetheless, in some cases adequate compensation may be compatible with a wider dissemination of knowledge than is achieved in practice.

Figure 1

Technology Infrastructure Policy for Conventional Industry

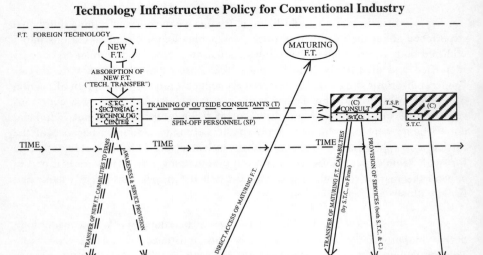

In each of these cases, an increased level of private entrepeneurship can work to overcome some of the distortions that may arise. Public policy should be geared in the first instance to promoting such entrepeneurship where it can, reserving more direct forms of intervention for the gaps that are left by the private sector. But it is important to bear in mind that this type of government support is catalytic in nature: it has a role to play in the early stages of structural change, after which it should be phased out, and its resources directed to new challenges.

IV. CREATING NEW CAPABLITIES FOR A LEADING-EDGE INDUSTRY

In developing technological infrastructure for a leading-edge industry the initial conditions are such that necessary elements of infrastructure -specific capabilities which will enable the industry's participation in the market- are neither available domestically nor can they be imported. Therefore creating these capabilities is a necessary first stage in the process of establishing technological infrastructure. Even though the set of firms that are likely to participate in the new industry may be known (though in most cases, with some uncertainty) they have only a vague understanding of their needs -certainly nothing that can be quantified with any precision. Much of this knowledge is tacit.

These capabilities could be generated through regular market channels. Individual industry firms could develop the know-how they need in-house, possibly spinning it off as an independent business unit at a later stage. Rosenberg (1976) describes just such a process in his account of technological convergence and vertical disintegration in the U.S. machine tool industry in its early stages. Or, two or more firms could form a commercial joint venture for the same purpose. There are a number of such examples in the aircraft industry. Alternatively, an external entrepeneur, not one of the industry firms, might undertake to develop the necessary capabilities and sell the industry the services it needs.

Each of these solutions can work under the right conditions, but there clearly are circumstances that might undermine their feasibility. Individual firms in the industry might hesitate to invest in capability creation if the needs of any single firm could not by itself justify such an investment and if the results of such an effort were unappropriable or difficult to commercialize. And adversarial relations between potential industry participants may deter any one of them from placing itself in the difficult position of relying on a competitor for infrastructure services. At the same time, while it may be clear that cooperation among firms is likely to be beneficial, the early vagueness of needs and possibilities may undermine their ability to form business partnerships for infrastructure development on a commercial basis. And an outside firm might hesitate to make the necessary investment because it lacks

209

sufficient knowledge of industry conditions that would enable it to identify commercially viable elements of infrastructure, or it might be deterred by its poor bargaining position vis-a-vis industry firms with superior complementary assets" in technological knowhow, product lines, marketing skills, distribution networks, etc. [Teece (1988)].

When the market cannot provide for the timely development of the necessary elements of technological infrastructure for a new industry, it may be possible for potential industry participants to undertake a pre-competitive, extra-market cooperative effort to create the required capabilities (a sort of consortium-led vertical integration). Cooperation offers the obvious advantage of distributing development costs over a broad base of users as well as allowing a division of labor in developing the new capabilities that hopefully capitalizes on the respective strengths of the participating firms. But its most important contribution may lie in facilitating user-need determination and coordinating future supply and demand.

Fundamental uncertainty regarding the new technology, and the absence of a market for the services it makes possible, may preclude the separation of user-need determination from the capability development process In this case it is essential that potential industry firms work together to identify a common set of user needs as they develop these capabilities. Moreover, a collective specification effort may be more efficient than a sequence of bilateral interactions. This is especially important if the critical mass for a viable infrastructure is large, requiring the support of a large number of firms with varied needs.

The presumed absence of a futures market for the technological capabilities which comprise the industry's infrastructure indicates a potential need for coordinating their supply and demand. Such a need may arise wherever there are interdependent investment decisions [cf. Chenery (1959), Heller and Starret (1976), Murphy, Shleifer and Vishny (1989), Justman and Teubal (1991)]. Investment in farmland, for example, may depend on access to rail transportation, while the profitability of the railroad depends on the extent of adjoining settlement. In some cases the railroad may act unilaterally, setting down track in the anticipation of future settlement, but in others the uncertainty regarding future settlement patterns may be too great for the railroad to act without explicit coordination with other factors involved in the settlement process: farmers, water resources, distribution facilities, etc. The situation is more difficult with technological infrastructure, since users may have to cooperate directly not only to supply the service but to assure that demand for the service materializes. There is high uncertainty regarding the "needs" which the infrastructure is supposed to satisfy (see user need determination, above) and the ability of the technological infrastructure to adapt itself to meet these needs. And potential suppliers of the specialized capabilities may anticipate a weak bargaining position vis-a-vis a narrow user base, prejudicing the outcome of any

210

future negotiations that may arise. Thus, user cooperation and joint ownership of the capabilities may be critical. The uncertainty surrounding early collaborations of this type (especially when cooperation is not yet routinized in the economy) may call for temporary, catalytic government support. Ultimately, mature capabilities should be privatized, i.e. spun off as marketable services offered on a commercial basis

This type of user cooperation and coordination is essentially an industry activity, the success of which depends first and foremost on the efforts of the firms involved. Yet because it is a new form of cooperation among firms, government has a role to play in defining its form and institutional context. This comes out most clearly in the United States, where cooperation among firms often hinges on the waiving of antitrust regulations. But even where such barriers do not exist, cooperative efforts to build technological infrastructure must still be exploratory in large measure. Hence the lessons that can be learned from an early effort are an external benefit for future attempts in other industries, even if the effort itself fails; indeed, failed efforts are often the most instructive. Government can play a fruitful role in this process by encouraging and supporting private intiatives in this regard, while such efforts are experimental. It can participate, in one way or another, in the systematic collection of information; in developing multidisciplinary skills that can help foster cooperation in building technological infrastructure; and in modifying the institutional framework where necessary. In return it can require firms to share with others what they have learned about the process of building a cooperative effort of this type.

We shall have more to say about specific forms of government intervention in Section 5, below. But first we would like to discuss the issue of critical mass in new capability development in the more formal context of a simple model.

IV.1 A Simple Conceptual Model

Consider an industry defined in terms of the technology it uses (such as the specialty steel industry) and potentially serving a set $N = \{1,...,n\}$ of separate but structurally identical markets. And assume that each of these market segments can sustain at most one supplier firm Firms can enter these markets by developing suitable products at a cost that will later be defrayed from profits earned on sales.

The cost of developing new products and their value for buyers depend on the domestic availability of industry-specific technological services, purchased from individual entrepreneurs -we will call them "consultants"- each offering a particular capability. Let $M = \{1,...,m\}$ be the set of these capabilities. The mix of capabili-ties which each firm needs is different, though there is substantial overlapping among firms; indeed this overlapping contributes to the definition of the industry's boundaries.

Initially there are no firms active in the industry and no consultants offering their services; both exist only in potential. There is neither actual demand for technological services nor is there any supply. In deciding whether to enter the market or not, firms and consultants act on their -mutually dependent- expectations.

The pattern of mutual dependency is determined by the extent in which the unique capabilities of individual consultants meet the particular needs of individual firms, or -from a different standpoint- the extent in which the needs of individual firms support the emergence of specific capabilities. We describe such a pattern by distinguishing between the skills that a firm can and cannot use, representing it by a matrix in which the rows are firms, the columns consultants, and an x in a cell (i,j) indicates that firm i requires the services of consultant j (figure 2). Call this the dependency matrix.

At the same time, the development of a specific consulting capability requires the investment of a fixed amount of resources which can only be recouped if sufficient demand materializes for that particular skill. Thus the mutual dependency described by the matrix raises the possibility of multiple equilibria -and offers scope for market-building policy.

Two related issues arise in this context: the need to achieve a critical mass of consulting services, and the need to coordinate the different requirements of the various firms. We consider each in turn through simplified versions of our general model.

IV.1.1 Critical Mass

To address the issue of critical mass assume that each of the firms can use the services of all consultants (i.e., the dependency matrix is full), and that each firm generates the same amount of demand for services from each of the consultants it uses. This amount is an increasing function of the number of active consultants, as we assume that the effective marginal product of consulting services increases with the scope of services offered. At the same time, firms have different threshold levels that determine when they are viable: some may require the services of three consulting capabilities to survive, some may require four, etc. And different consultants have different break-even points, each requiring a different level of demand to cover the fixed costs of generating the services they provide.

These assumptions imply an increasing functional relation between the number of firms that are active and the number of different capabilities that are available. This is represented by the NN schedule in figure 3; the marginal firm is earning positive economic profits (i.e., a better than competitive return on its investment) to the right of this curve, and negative economic profits (a less than competitive return) to its left. The supply of consultants is regulated by a zero-profit-equilibrium, represented by the MM curve in figure 3. The shaded area above this curve is the area of positive economic profit

Figure 2: Dependency matrices

CONSULTANTS

	1	2	3	4	5	6	7	8	9 = m
1	X	X		X		X		X	X
2	X					X			
3		X			X			X	
4	X			X		X	X	X	X
5 = n	X		X		X		X		

FIRMS (row labels)

(2a)

CONSULTANTS

	a	b	c
1	X	X	
2	X		X

FIRMS (row labels)

(2b)

Figure 3: Equilibria in the market for infrastructure services

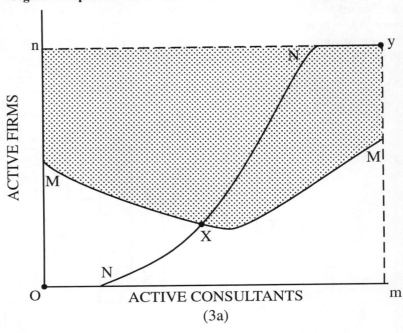

(3a)

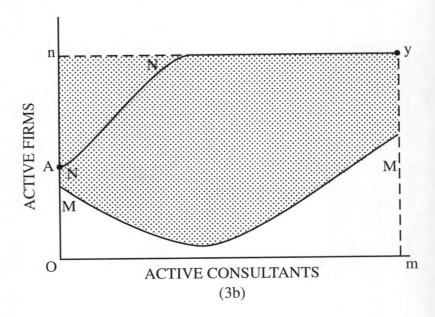

(3b)

These profit levels imply a dynamic system represented by the directional arrows in the figure, indicating that of the three equilibrium points in figure 3a -O, X and Y- only O and Y are stable. The intermediate point X is unstable: a slight increase in the scope of consulting services will trigger convergence to the higher equilibrium, Y; a slight decrease will lead to convergence to the low level equilibrium, O.

Thus X represents a critical mass of consulting skills that must be reached if the market is to settle on its high-level equilibrium, Y, in which all firms and consultants are active. This can be achieved through an initial subsidy of firms' activities in the market, causing the NN schedule to shift to the left, and a concomitant support to emerging infrastructure services, shifting the MM curve down (figure 3b). As a result, some firms can operate profitably without the support of infrastructure services (point A in 3b) and their custom is sufficient to allow the emergence of the first consulting practices. From that point on the market "snowballs" towards its only equilibrium point Y, and subsidies can be gradually removed as the profitability of both firms and consulting services improves.

IV.1.2 Coordination

To address the issue of coordination, consider the industry described in figure 2b. Potentially, it comprises two firms (1 and 2) and three consultants (a, b, and c), where firm 1 requires the services of a and b and firm 2 requires a and c. We further assume that each firm is viable only if it has access to both the capabilities it needs; that consultant a requires the demand generated by both firms to cover its fixed costs; and that b and c each require the demand of one firm to operate profitably. Clearly, this industry has are two possible equilibria, both equally stable: one in which there is no activity in the industry, and another in which all firms and consultants are active. The capacity of the industry to move from the first to the second may depend on the extent in which industry agents can be expected to make "leaps of faith".

When there is no industry activity, no firm can profit -immediately- from unilateral action. But assume firm 1 begins operating at a loss, without the aid of infrastructure services; then consultant b becomes viable and enters the market, reducing firm 1's losses, and perhaps firm 2 realizes that if it begins operating the services it needs will become viable, whereupon firm 1 will also have all the services it needs to be profitable. Alternatively, consultant a could enter an empty market, in the hope that this action would trigger a similar chain of fortuitous events (firms 1 and 2 each realizing that its entry would trigger the emergence of the other services it needs.)

215

Whether such a leap of faith is made depends on the extent of the early losses that are incurred, the lags that are involved, firms' confidence in their knowledge of the dependency matrix and the profitability conditions of the other actors in the market, the possibility of extraneous factors disrupting the desired chain of events, etc. It cannot be taken as a foregone conclusion that such a leap will generally be made, even when it is advantageous for all concerned, and is this that leaves scope for some form of publicaction.

IV.2 Policy Implications

Our simple conceptual model examined the interaction of supply and demand in the creation of infrastructure services, highlighting both the need to coordinate the specific choices of users and suppliers, and to assure the emergence of a critical mass of capabilities. It rested on the underlying assumption that the set of necessary capabilities -and the uses for which they were needed- were known or knowable. This assumption is suitable for mature technologies, in which the problem of assuring a critical mass of capabilities may be viewed as a variant of the potential inefficiency that is generally associated with natural monopoly, when imperfect possibilities of price discrimination in service provision imply that a socially profitable capability will not be created [cf. Justman and Teubal (1986)].

High-tech industry, however, may also require the development of new capabilities at the cutting-edge of technology. In such cases, user involvement may be critical for arriving at the right structure of capabilities and services, one that reflects true "needs"; these are not as obvious at the outset as they are with regard to more mature capabilities. Agency problems that arise in this context imply that entrepreneurs may not be able to operate independently in such a market: users themselves may need to own -or invest in- he capabilities they need.

When this is the case, our focus must shift from coordination of supply and demand to coordination among potential users. This suggests a discussion of the policy implications of our analysis in two parts: policies for promoting user-producer or supply- demand coordination between the "firms" and "consultants" in our model; and policies aimed at promoting user coordination and cooperation, so as to assure that a critical mass of infrastructure materializes.

IV.2.1 Promoting the Coordination of Supply and Demand

The least intrusive course of action in this regard might involve generating information regarding the structure of the dependency matrix and the viability conditions of the various actors, which industry agents could use in making their decisions -in other words, initiating industry studies that focus on the need for technological infrastructure. A second mode of intervention would be for government to serve as a clearing-house cum arbiter cum guarantor of agreements undertaken within the industry to take concerted action. These modes would seem most appropriate when the number of agents involved is relatively small so that the information needs and the challenges of coordination are manageable

When the number of agents is larger and the complexity of the dependency matrix is daunting, a different approach might be warranted. For example, government can subsidize the supply of technological capabilities, effectively lowering the threshold for consultant entry. This can be achieved through a broad-based subsidy that requires little information but may place a substantial burden on the budget; or it can be aimed more narrowly as a catalytic measure aimed at starting a self-fuelling process of capability generation. However, if the "dependency matrix" (figure 2) is sufficiently complex, a finer aim may require intimate knowledge of the capabilities and strategic needs on both sides of the market for technological infrastructure services, which may not be available initially, or may only be obtained through a costly effort.

Alternatively, the government may be in a position to guarantee that it will provide industry firms with a basic level of infrastructure services if the market does not. If this undertaking is made with a thorough understanding of technological needs, and if it is widely believed, it may not need to be acted upon: the confidence it generates may be enough to spur the market to provide the necessary services.

Yet another possible course of action is for the government to act on the demand side of the market for technological services, again either through broad-based subsidies aimed at the users of these services or through a more finely-tuned information-intensive effort.

IV.2.2 Facilitating User Coordination

User cooperation and coordination will generally be facilitated by expectations that other users are willing to invest in the capabilities they need, or by the knowledge that the government has targeted the industry (and indirectly, the infrastructure). In such circumstances a clearly stated government "vision" or at least a credible announcement may be critical. For example, the government may declare its determination to create a significant level of infrastructure no matter what level of industry financing and participation is forthcoming. This will significantly reduce the uncertainty facing any individual user intent on investing in the new capabilities since it reduces its dependence on the actions of other users. A clearly stated vision may have a significant effect in inducing individual user participation even when the level of explicit government commitment is low.

Indeed, in some extreme cases, user coordination may be unthinkable without some confidence that a critical mass of infrastructure will materialize, i.e., a level that will assure participating firms of a minimal level of profits, given the uncertainties associated with the emergence of the necessary capabilities and the possibility of incurring substantial losses. In such cases direct and even massive government subsidization of capability development -or other risk reduction schemes- may have to be adopted, especially for initial projects of this kind

The need for government initiative and significant government involvement in triggering cooperation on an agreed project configuration is more likely to arise when significant levels of cutting-edge technological capabilities are required. Such involvement may be feasible if there is a history of Government support to individual firms which provides it with a vantage point for assessing sectoral needs -as opposed to the needs of individual firms

IV.2.3 Organizational Aspects

The model rests on the assumption that the new infrastructure services will be provided by independent business units each specializing in a different skill, neglecting the possible existence of economies of scope in generating the various elements of infrastructure. When such economies of scope exist, some initial concentration of infrastructure capabilities within a public or semi-public body -a "Technology Center" or an academic "Center of Excellence"- may be warranted. The organizational configuration in this more general case would then involve interaction between this Center and a population of consultants. A more complete model of market creation in high- tech industries would examine the dynamics of this relationship, with capabilities initially housed in a central location and then spun off as independent consultancies as they mature.

V. NOTES ON POLICY DESIGN AND IMPLEMENTION

Technological infrastructure is a relatively new and little-understood notion that should be distinguished from the related and better known concepts of innovation and scientific research. In contrast to these, the approach to successful capabilities development is neither a pure supply-push approach (which characterizes curiosity-oriented research) nor a pure demand-pull one (which characterizes most innovations). It is in fact a hybrid of the two; it must satisfy an industry-relevance criteria, one which cannot, in general, be based on existing demands (since the market for the activities flowing from the capabilities is not yet developed) but on

218

general, un- determined needs. The absence of markets also explains the necessity of generating alternative mechanisms of linking needs to capabilities. These distinctive characteristics separate technological infrastructure policy from traditional support to industrial R&D and from traditional science policy.

Like any new policy area, strategic technology policy requires especially careful consideration of policy formulation and implementation issues, and many aspects of implementation can only be fully examined within a specific institutional context. Nonetheless, a number of broader principles seem to cut across most of the current spectrum of applications.

"Market failure" analysis does not provide a sufficient foundation for implementing strategic technology policy. Nelson's (1987) exegesis of the general limitations of "market failure" as a policy guide is very much applicable to the role of government in developing technological infrastructure. The term market failure carries the presumption that government intervention is a last resort. This is clearly appropriate in some contexts (e.g., in setting the price of dinner jackets) but not in establishing infrastructure where there is no reason to assume that the market works well at all. Where market failure is ubiquitous such an analysis cannot serve as an effective screening device for identifying suitable projects for intervention. At nodes of structural change, projects must be chosen on the basis of comparative effectiveness rather than on the mere fact of being a socially profitable project which the unhindered market will not bring about. Rather than focus our attention on what the market can or cannot accomplish, we must also ask: "(i) How much government infrastructure is needed for profit incentives and markets to work well?... (ii) What are the things that government can do well?... (iii) Can an appropriate government structure be defined and if so be established?" [Nelson (1987)].

Strategic technology policy is still in an experimental stage. Therefore, a linear, "planning" approach to policy, in which formulation clearly precedes implementation is not appropriate here; rather an evolutionary approach that emphasizes sequential experimentation should be adopted. Initial projects should be chosen not only for their intrinsic value, but also for the new information they generate on technological infrastructure, for the capabilities they develop, and for their demonstration effect [cf. Nelson and Winter (1982)]. In this early stage, the boundary between policy formulation and implementation is fuzzy.

The government's role is catalytic. Strategic technology policy has little to do with subsidization of R&D, certainly not long-term subsidization. It is about fostering private-sector intiatives for cooperative infrastructure development. The government's first objective in promoting strategic technology policy is to trigger an endogenous process of generating "reasonable" cooperative technology programs with the active participation and financing of the private sector.

219

Its first goal is to generate a set of "reasonable" projects. Past experience with previous cycles of innovative policy initiatives (e.g., Israel's industrial R&D support program in the early seventies) sugests that, initially, the budget constraint may not be binding, i.e. there will be very few suitable candidates that meet the requirements of a bona fide cooperative technology infrastructure project, and funding needs will be easily met. In this stage government may have to take a more active role in gathering information (surveying local firms, mapping global trends) and in building coalitions in the private sector. This may involve some "mild targeting", directing government initiatives at projects that are likely to contribute most to the general momentum -projects that address key industry bottlenecks and promise strong learning and demonstration effects.

In later stages, greater selectivity implies a need for an explicit technology strategy and a broad vision of the future. Once the endogenous process has been triggered and greater demands are placed on the public sector in supporting private initiatives for infrastructure development, it may be necessary to exercise greater selectivity in allocating government support. This will be an inherently political process; its subversion to the interests of narrow constituencies within government and in the private sector is a possibility that must be addressed. Consequently, adequate safeguards against such misuse are a prerequisite of any such policy. This requires that the government's strategy be explicit [Justman and Teubal (1990)], and grounded in a clearly defined, consensual vision of the country's future path of social and economic development [cf. Johnson (1982)]. Explicitness in strategy formulation and a consensual vision of the future provide a coherent framework for evaluating competing policy measures on the basis of widely held goals.

Successful strategic technology policy will require institutional innovation. The machinery of government must be able to provide an effective interface between political forces and professional analysis. Central banks perform this function in determining interest rates. In Japan, MITI fills this function in formulating and implementing its industrial technology policy. Other countries will find other organizational solutions, for example, an independent Technology Council acting outside the normal channels of government on the strength of the personal standing of its leading members and the professional reputation of its technical staff; a tripartite commission comprising representatives of government, labor and industry acting as a steering committee for a professional staff seconded on a similar basis; or an inter-ministerial committee, within government, with a professional staff that is part of the civil service.

A second organizational\institutional aspect relates to the "housing" of particular infrastructures. Cardinal issues which must be resolved include: the choice between temporary and permanent structures and between a "secretariat" and an operational organization [Ouchi (1989), E. Arnold & K. Guy (1987)]; the relationship between the organization and existing industry associations; and its relationship with universities. While members of the academic community clearly have a prominent role to play in the development of technological infrastructure involving Generic Research performed at universities, this does not imply adoption of the management approach and incentive structures of a university.

Whatever the organizational form adopted, the responsible government agency should serve as a clearing house for information and learning. Whatever the form of institutional innovation, it has an important role to play in infrastructure development in scanning the horizon and suggesting an agenda for future infrastructure projects; as a locus of evaluation capabilities and coordination services; and, not least, as a source of proposals for a broad vision of the nation's future. To meet these objectives it must adopt a systematic approach to information-gathering, learning, and experimentation, requiring both detailed knowledge gained from in-depth case studies, and a broad strategic understanding of the substance of generic capability development. Systematic knowledge implies that information should be codified and related to broader contexts and existing theory and integrated in an educational effort aimed at creating a supply of capable professional staff. Such an effort could be undertaken in cooperation with an inter- disciplinary university graduate program combining business and economics training on the one hand, and science and engineering on the other; there are quite a number of such programs in existence today.

Government itself requires substantial capability in order to trigger action in others. Ultimately, the success of strategic technology policy will depend on the business sector, on its initiative, on its forward-mindedness, on its ability to work together for common goals. But triggering a process that culminates in such success requires an awareness on the part of government, of the role of technological infrastructure in industrial development, and the capability to act on its awareness. Government agencies need some internal R&D capabilities in order to successfully procure innovative equipment embodying private-sector R&D [Nelson, Peck and Kalachek (1967)], and for much the same reasons, triggering strategic technology initiatives in the private sector requires that the same capabilities be present, in some measure, within the government itself.

Recent work on "national systems of innovation" [Lundvall (1988), Nelson and Rosenberg (forthcoming)] should help further our understanding of strategic technology policy, allowing us to compare one country's system of innovation with that of other countries, in the context of "paradigmatic changes" [Freeman and Pérez (1989)] and other global trends that define its techno- economic environment. Such a comparison can help identify possible foci for policy action, revealing the institutional underpinnings that are necessary for bringing about desired changes;

221

e.g., the development of a sectoral technological infrastructure may be furthered through the promotion of effective industry associations.

Israel's national system of innovation, for example, has numerous gaps on which its industrial and technological policies might focus in coming years. Among the missing elements are:

(i) Sector-specific technology centers to house new technological capabilities for conventional industry -including management reorientation of existing institutions (from a simple shift to demand pull to "strategic" capability development) and to promote technology diffusion.

(ii) University-related Centers of Excellence in multi-sector functional areas of science and technology such as biotechnology, microelectronics, opto-electronics, and new materials. Existence of these centers can eliminate fragmentation and duplication of effort in that part of overall national effort which should essentially remain curiosity oriented. An intelligent interface with industry will allow such Centers to play a role in generating new technological infrastructure.

(iii) A Technological Council to assist in the identification of desired technological infrastructure; to coordinate the establishment of the various infrastructures (e.g., CIM for a particular sector with a nationwide, multi-sector CIM program); to develop criteria for evaluating the economic and other contribution of cooperative infrastructure-type R&D programs; and to represent a national locus of capabilities in the technological infrastructu-re policy area.

(iv) Secretariats and temporary joint labs in high-tech areas, such as those involved in the Esprit program of the EEC and the microelectronics and opto-electronics programs of Japan.

(v) Vision exercises and policies, at the branch, sectoral (e.g., high-tech industry) and national levels. These are essential ingredients in the effective development of new technological capabilities.

Most of these institutions are critical nodes in important networks that link firms with other agents. Their absence implies insufficient networking between scientific research, technological development, and industrial activity

VI. SUMMARY

The underlying theme of this paper is that the development of a new industry requires a technological infrastructure of firm-based capabilities and market links, that transcends individual firm training and experience. Our purpose here has been to describe this type of infrastructure, characterize the conditions that enable it to emerge, and explore the potential role for structurally oriented public policy in facilitating this process. There are an increasing number of examples of such policy intervention. We refer to this new area of public policy as "strategic technology policy" to distinguish it from "tactical" intervention at the micro-economic level and from macro-economic policy.

Several key distinctions mark our analysis. First, the distinction between conventional infrastructure, supplying a familiar, homogeneous good or service, and technological infrastructure, supplying a multi-dimensional array of new goods and services in a variety of combinations. Second, the distinction between the capability-creating phase of infrastructure development and its market building phase. Third, a distinction between the static and dynamic roles of technological infrastructure. And fourth, the different roles that public policy must play in periods of routine growth and at nodes of structural transformation, when the critical factor of development may be entrepeneurship in establishing new technological infrastructure.

In such times, government cannot rely on simple market-failure analysis for policy guidance as market failure is ubiquituous. The appropriate questions are: what infrastructure is necessary and what can public policy do to help it come about. Government intervention at this point should be aimed at triggering initiatives for infrastructure generation in the private sector. This indicates an emphasis not on subsidies but on gathering information; encouraging and initially coordinating cooperative efforts in the private sector; disseminating information and the lessons of past experience; and serving as a nurturing locus for the development of new skills and capabilities that play a part in this process. As a new policy area it calls for an evolutionary approach that relies on sequential experimentation with strong feedback mechanisms between policy implementation and formulation. It is essentially a catalytic effort aimed at creating new market agents which if successful should result in its own redundancy.

This paper is a step towards a more complete understanding of the role of capability–creation and market-building in economic development. More work is needed in defining and clarifying the issues raised here through both theoretical analysis and continued evaluation of the experience of a growing number of countries in implementing strategic technology policy. In this regard, there are parallels to be drawn with recent work on national systems of innovation [Nelson and Rosenberg (forthcoming)].

The analytical framework that we have developed in this paper should also be of use in understanding the role of public policy at nodes of structural change other than those on which we have focused here. One possible extension in this vein would be to examine the importance of capability-creation and market-building for the re-industrialization of Eastern Europe and, more generally, for the difficult transition from a centrally-planned to a market-based economy. This could indicate modes of public intervention that might help make for a more efficient process of industrialization and for a shorter and smoother transition.

REFERENCE

ARROW, K. (1962), "Economic Welfare and the Allocation of Resources for Invention" in R. Nelson (ed.), **The Rate and Direction of Inventive Activity**, NBER.

ARNOLD, E.; GUY, K. (1986), **Parallel Convergence: National Strategies in Information Technology**, London, F. Pinter.

ARNOLD, E.; GUY, K. (1987), **Information Technology and Emerging Growth Areas**, Science Policy Research Unit, August.

BARZEL, Y. (1968), "Optimal Timing of Innovations", **Review of Economics and Statistics**, 50 (August): 348-55

CHENERY, H. (1959), "Interdependence of Investment Decisions" in M. Abramovitz (ed.), **The Allocation of Economic Resources**.

DAHLMAN, C.; ROSS-LARSEN, B.; WESTPHAL, L. (1985), "Managing Technological Development: Lessons from the Newly Industrialized Countries", **The World Bank**.

ERGAS, H. (1987), "Does Technology Policy Matter?" in B. Guile and H. Brooks (eds.), **Technology and Global Industry: Companies and Nations in the World Economy**, National Academy Press, 191-97.

FREEMAN, C.; PEREZ, C. (1988), "Structural Crises of Adjustment, Business Cycles and Investment Behaviour" in G. Dosi et al., **Technical Change and Economic Theory**, London. Pinter.

GUY K.; ARNOLD, E. (1991), "IT Diffusion Policies", Typescript, **Science Policy Research Unit**, Univ. of Sussex, September.

HELLER, W.; STARRET, D. (1976), "On the Nature of Externalities" in S. Lin (ed.), **Theory and Measurement of Economic Externalities**, New York, Academic Press.

HIPPEL, E. VON (1976), "The Dominant Role of Users in the Scientific Instruments' Innovation Process", **Research Policy**, 5(3).

HOFFLINGER, B. (1989), "Cooperative Research on Application-Specific Micro-electronics", **Proceedings of the IEEE**, Vol. 77, no. 9, pp. 1390-1395.

JUSTMAN, M.; TEUBAL, M. (1986), "Innovation Policy in an Open Economy: A

Normative Approach to Strategic and Tactical Issues", **Research Policy**, 15, pp. 121-138.

JUSTMAN, M.; TEUBAL, M. (1988), "A Framework for an Explicit Industry and Technological Policy for Israel and some Specific Proposals" in C. Freeman and B.A. Lundvall (eds.), **Small Countries Facing the Technological Revolution**, London, Pinter.

JUSTMAN, M.; TEUBAL, M. (1990), "The Structuralist Approach to Economic Growth and Development: Conceptual Foundations and Policy Implications" in R. Evenson and G. Ranis (eds.), **Science and Technology: Lessons for Development Policy**, Westiview Press.

JUSTMAN, M.; TEUBAL, M. (1991), "The Structuralist Perspective on the role of Technology in Economic Growth and Development", **World Development**, Vol. 19, no. 9, pp. 1167-1183.

JUSTMAN, M.; TEUBAL, M. (1991b), "Economic Growth, Structural Change and Technological Progress", **Industrial Development Policy Group**, The Jerusalem Institue for Israel Studies, Jerusalem.

KLINE, S.; ROSENBERG, N. (1989), "An Overview of Innovation" in R. Landau and N. Rosenberg (eds.), **The Positive Sum Strategy**, Washington D.C., National Academy Press.

LUNDVALL, B.A. (1988), "From Learning by Interaction to National Systems of Innovations" in C. Freeman and B.A. Lundvall (1988), **Small Countries Facing the Technological Revolution**, London, Pinter.

MALERBA, F. (1991), "Italy: The National System of Innovation", to appear in R. Nelson and N. Rosenberg (eds.), **National Systems of Innovation**, Oxford University Press, forthcoming.

MURPHY, K.; SHLEIFER, A.; VISHNY, R. (1989), "Industrialization and the Big Push", **Journal of Political Economy**, 97(5), pp. 1003-26.

NELSON, R.; PECK, M.; KALACHEK, E. (1967), **Technology, Economic Growth and Public Policy**, Brookings Institute, Washington D.C.

NELSON, R.; WINTER, S. (1982), **An Evolutionary Theory of Economic Change**, Harvard University Press.

NELSON, R. (1987), "Roles of Government in a Mixed Economy", **Journal of**

Policy Analysis and Management, Vol. 6, no. 4, pp. 541-557.

NELSON, R.; ROSENBERG, N. (eds.), **National Systems of Innovation**, Oxford University Press, forthcoming.

OUCHI, W. (1989), "The New Joint Labs" in **Proceedings of the IEEE**, Vol. 77, no. 9, September, pp. 1318-1326.

PACK, H.; WESTPHAL, L. (1986), "Industrial Strategy and Technological Change: Theory versus Reality", **Journal of Development Economics**, Vol. 12, June, pp. 87-128.

ROSENBERG, N. (1963), "Technological Change in the U.S. Machine Tool Industry, 1840-1910", **Journal of Economic History**, 23, pp. 414-443.

SENKER, J. (1991), "Evaluating the Funding of Strategic Science: Some Lessons from British Experience", **Research Policy**, Vol. 20, no. 1, pp. 29-44.

SIGURDSON, J. (1986), "Industry and State Partnership in Japan: The VLSI Circuits Program", Discussion Paper no. 168, **Research Policy Institute**, Lund, Sweden.

TEECE, D. (1988), "Profiting From Technological Innovation: Implications for Integration, Collaboration, Licencing and Public Policy" in G. Dosi et al., **Technological Change and Economic Theory**, London, Pinter.

TEUBAL, M. (1979), "On User Needs and Need Determination" in M. Baker (ed.), **Industrial Innovation: Technology, Policy, Diffusion**, London, Macmillan.

TEUBAL, M.; YINNON, T.; ZUSCOVITCH, E. (1991), "Networks and Market Creation", **Research Policy**.

TEUBAL, M.; ZUSCOVITCH, E. (1991), "Networks and Evolutionary Differentiation", **typescript**.

TEUBAL, M. (1991), "The Innovation System of Israel: Description, Analysis and Outstanding Issues" in R. Nelson and N. Rosenberg (eds.), **National Systems of Innovation**, forthcoming.

TASSEY, G. (1991), "The Functions of Technology Infrastructure in a Competitive Economy", **Research Policy**, 20, pp. 345-361.

WESTPHAL, L.E.; KIM, L.; DAHLMAN, C.J. (1984), **Reflections on Korea's Aquisition of Technological Capability**, Washington D.C., The World Bank.

227

10 LOCAL SYNERGY AND INNOVATION IN PERIPHERAL AREAS[72]

Walter B. Stöhr

IIR. U. of Economics and Business Administration, Vienna.

I. INTERNATIONAL DIVISION OF LABOR AND THE REGIONAL COMMUNITY

Since about the middle of the 1970s the changing international division of labor has had an increasing impact upon local and regional communities. This paper deals with the more specific question of how regional policy and action can lead to innovation and put communities in a better position to cope with the impact of changes in the international division of labor. The term "regional policy, in this context, refers mainly to central (for example State) policy measures "from above" [Stöhr and Taylor (1981)], while "regional action" refers mainly to regional mobilization "from below".

Changes in the international division of labor have recently taken place with accelerated speed and have increasingly put local communities into a state of instability. There have been numerous examples, however, which show that such externally induced instability can also promote spurts of local creativity, particularly if combined with the existence of certain local factors such as local competence and synergy.

Regional policy under these changing external circumstances has been forced to become more inventive, and, instead of mainly centrally supported measures (such as regional capital incentives, infrastructure investment, interregional income transfers, and the promotion of interregional factor mobility) has experimented with new measures geared towards the promotion of regional innovation and the integrated mobilization of endogenous regional resources [Stöhr (1984)]. In many cases, successful action to improve the international competitiveness of regional communities also started at the regional level; we shall analyze some relevant test case studies later.

---

[72] This presentation draws on the author's earlier work quoted in the References.

X. Vence-Deza and J. S. Metcalfe (eds.), Wealth from Diversity, 229–240.

II. ON THE DEFINITION OF TECHNOLOGICAL INNOVATION AND HIGH–TECHNOLOGY INDUSTRY

Technology can be defined usefully for our purpose as "a formal and systemic entity of knowledge and skills in order to realize and control complex production techniques/processes" [de Smidt (1981)]. *Innovation* can be defined as the first commercial utilization of new scientific or technical knowledge within one enterprise [Eckert (1985)]. Following these, we can define technological innovation as the (first) commercial utilization of a formal and systemic entity of knowledge and skills within an enterprise to realize and control complex production techniques/processes. This definition can refer to the introduction of new products (product innovation) and of new production processes (process innovation).

High–technology industries, on the other hand, have been defined as those in which firms have (1) above average ratios of R&D expenditure to net sales, (2) above average percentages of the labor force engaged in engineering, scientific, professional, and technical work, and (3) rapid growth in terms of employment and output. These are useful operational variables for finding firms or industries that are in a good position to apply innovation in their corporate strategies to attain competitive advantage over rivals.

As for their spatial distribution, high–technology industries and R&D activities would appear much more mobile and footlose than the more traditional industries.

III. RECENT LOCATIONAL ANALYSES OF HIGH–TECHNOLOGY INDUSTRIES AND R&D ACTIVITIES: THE MONOCAUSAL TRAP?

To date, locational analysis of the distribution of high–technology industries and R&D activities has –very much like that for traditional industries– mainly taken a monocausal approach. Apart from general conditions such as favorable tax climates and regulatory conditions, discrete factors have been analyzed regarding their spatial correlation with the emergence or sustainability of high–technology industries and entrepreneurial innovation. Factors that have been studied include the availability of universities, public research institutes, venture capital, a highly skilled labor force, urban facilities, a diversified urban base, high entrepreneurial density, consulting and information services, and rapid transport facilities. These monocausal analyses typically were done either with a macro approach, generally correlating the spatial distribution of individual factors with that of high–technology industries, or with a micro approach, usually via firm surveys.

The results of these analyses have been rather ambiguous or have shown a high degree of mutual interaction. We shall argue that this ambiguity is mainly due to the analyses of each factor in isolation from all others and to an assumption of additivity of universally discrete factors. In reality, innovation generally seems to be created by the mutual —and occasionally quite unique— interaction (synergy) of various of these and other factors within rather different local or regional environments.

The analyses of these individual factors would indicate that only the major urban centers and the main transport and communication axes would be able to generate innovation and attract high–technology industry; there would be few opportunities left for the remaining areas except to specialize in low–technology activities or to wait until products become mature or obsolete and are displaced from the major urban centers. However, as we shall show later, areas outside of the main centers have also been able to innovate. It seems that an important condition for this has been precisely the interaction (synergy) of important local factors, rather than just the pure existence of (possibly isolated) single factors, as the analyses quoted above have implicitly assumed.

Frequently found examples of where such regional interaction is lacking are the many universities which (at least in European countries) often have very little local innovative effects: either their staff may have an "ivory tower" attitude towards their environment, their contacts may be mainly along disciplinary lines in an international context, or the local community may not be able to use their products. Relevant contrasting situations are frequently found in neighboring universities, such as in northern Italy, where the university of the old town of Pavia is described as "representative superstructure without direct influence on (local) entrepreneurship which in view of its traditional small firm structure has no demand for academics", quite distinct from the role played by the universities in closeby Milano.

IV. REGIONAL SYNERGY AS A MORE POWERFUL EXPLANATION OF INNOVATION

In explaining growth within local and regional economies, an important step was taken from industrial location theory (which basically has been concerned with the relevance of individual factors for specific sectors) to growth pole theory [Perroux (1955)] and its spatial extension growth center theory [Boudeville (1966)]. Growth pole and growth center theory explicitly introduced interindustry relations into the explanation of local growth. Not only the sectoral composition of a local economy was considered important, but also the functional interrelations between sectors, particularly between "leading" and other sectors.

In explaining local and regional innovation, similar explanations based on unrelated, single determinants of innovation have obscured rather than clarified spatial causality. This has led to the demand for a network approach. We now contend that not only the availability of all these factors, but their dynamic regional interaction is essential for sustained regional innovation.

This increasing functional specialization of regions (cutting across and going beyond traditional sectoral specialization) has deprived many regions, particularly the peripheral ones and the old industrial areas, of most of the key functions needed for innovation. What these areas need, therefore, is a regional reintegration of these key functions in a synergetic form.

The concept of synergism was first used in chemistry and pharmaceutics, where it denotes that the "effect obtained from the combined action of two distinct chemical substances is greater than that obtained from their independent action added together" [Encycoplaedia Britannica (1978), p. 740]. In regional development this concept would denote that not only the presence of specific agents and institutions within a region, but also their mutual dynamic interaction, is a prerequisite for optimizing regional creativity and innovation under conditions of structural instability.

V. EXAMPLES OF REGIONAL INNOVATION COMPLEXES IN PERIPHERAL AREAS

In this section we shall briefly analyze three examples of regions that have shown relatively high rates of technological and institutional innovation in recent decades. We have intentionally chosen nonmetropolitan –within their national context peripheral–areas to disavow the frequent assertion that innovation can only emerge in major metropolitan centers. In all three cases local and regional initiatives triggered innovation, in what I have elsewhere [Stöhr and Taylor (1981)] called "development from below." Only in the last case (the Japanese Technopolis policy) has this local initiative been complemented by systematic support on the part of the national government. Furthermore, the three examples chosen are taken from different types of social systems: the first one has a cooperative structure, the second one is essentially based on private enterprise (mainly small– and medium–sized firms), while the last one is of a mixed "third sector" type (combining local government, local university, and private enterprise). This was done to show that regional innovation does not depend on any one specific social system.

We shall focus our analysis mainly on the regional synergetic structures underlying these innovation processes. Therefore, I have attempted to show important patterns of interaction within each regional system as well as important outside interactions. Emphasis is thereby placed on functional and institutional

232

interaction rather than on the usually depicted (physically or financially defined) input–output flows of commodities and production factors.

V.1. Regional Innovation Complex I: A Cooperative Model

The Mondragon Cooperative Group in the Basque Country, Spain, has been widely analyzed and documented [Thomas and Logan (1982), Stöhr (1984)]. The following refers to the situation of the mid–1980s. Begun back in the 1940s, it now comprised about 160 cooperative enterprises geographically dispersed and in a wide variety of manufacturing sectors (ranging from metal working and capital goods, to intermediate products and durable consumer goods), industrial services, training and education, housing, agricultural processing, community services, and a consumer cooperative. It is spatially decentralized in numerous medium–and small–sized towns and villages adjacent to the major old industrial areas traditionally dominated by the steel industry and shipbuilding.

While the traditional, old Basque industries have, for several decades. been losing jobs and closing down plants, the Mondragon Cooperative Federation has, even during the most severe international structural adjustment, been able to increase its number of plants and stabilize, in part even increase, its workforce. To a considerable extent this has taken place with sophisticated technology including process electronics, computer–aided design, and robot development. But advanced technology has also been developed in more traditional sectors, such as household electrics, in which the Mondragon Cooperative plants were among the most technologically advanced and most efficient in the nation, with a large share of outputs targeted to export markets.

The relatively high innovation capacity of the majority of the Mondragon Cooperatives is due, to a considerable extent, to the fact that within the Cooperative Federation there exists intensive interaction among its training, research, and technological development units, its consulting services, and its own financing institution. This regional Cooperative Federation is organized basically like a large (private) multilocational company, except that it is regionally defined and has territorial identification and responsibility; it is not "footloose", as are most other multilocational firms. This endogenous training–research–innovation–financing-production complex [Thomas and Logan (1982)], with its intensive feedback mechanisms, appears to be mainly responsible for the high innovation rate and the competitiveness of most of the Mondragon plants [Stöhr (1984)]. The Consulting and Management Unit identified potential new product lines that, with the aid of the Financing Unit and individual production plants, are developed to the prototype stage by IKERLAN, the Research and Development Unit. The Training Unit participates in this development work so that adequate new job skills are developed at the same time that the new processes and products are created.

An interesting mechanism has been devised for the acquisition and development of new technologies from outside. Because universities in the Basque Country have been oriented insufficiently towards technological innovation, the Mondragon Cooperative Federation sent personnel for medium–term stays "in residence" at various outstanding foreign universities and research centers in order to establish contacts and collect relevant information. This stock of information has been used successively for joint R&D projects and for internal technical development within IKERLAN, the Group's R&D core group, and within individual firms in cooperation with IKERLAN. This is an interesting example of how specific key functions for innovation, still missing within the region, can be successfully internalized from outside. An important condition for this, however, appears to be the existence of an innovation–oriented, regional (synergetic) interaction system combining other key elements.

A second group of reasons for the relatively high organizational and institutional innovative capacity are the participatory structures within individual cooperatives, and between them in the frame of the Cooperative Federation. These are related to the Cooperative's territorial, cultural, and ethnic identification with the Basque Country. These are reinforced by an increasing degree of autonomy granted to the provinces, and to the Basque Region, by the Spanish central government.

The Mondragon Cooperative Federation, in spite of its relatively small size (about 19,000 members, representing only 2.5 percent of the active population of the Basque Country), has been playing an important role in the Basque Country in institutional terms. Upon Franco's death in 1975, the Cooperative's members, due to their previous experience with organizational structures, participated in key roles in the local planning committees that emerged before the formal establishment of democratic institutions at local and regional levels in Spain. Since their establishment, the Mondragon Cooperative Federation has been supplying personnel for key positions in the local, provincial, and regional governments of the Basque Country.

While the Cooperative Group operates in an essentially open market environment, with free commodity and factor flows, its major external inputs are technological innovation. A certain regional "closing off" is only effective in decision–making structures (cooperative decision–making process, regional political and economic autonomy from central government) and in terms of capital flows. Capital outflows are restricted in that the financing institution of the Mondragon Cooperative Federation (Caja Laboral Popular) was able to invest the substantial surplus it makes only within the Basque Country (interestingly including the Basque areas in France). Because Caja Laboral Popular was not able to shop around for the most profitable investment on a worldwide scale (as banks normally would), it was forced to generate profitable projects within the Basque Country and promote facilitating institutional structures, like the synergetic ones described above [Stöhr (1984)]. This regional "locking in" of capital and surplus, embedded in a competitive international market situation, has created –together with the internal synergetic structures described– a self–propelling regional innovation and adjustment mechanism.

V.2 Regional Innovation Complex II: A Private Sector Model

As a second case study we consider the groups of regionally interacting, originally small– and medium–sized private firms that have achieved high innovation rates in what is called "Third Italy", in the north–east of the country, as distinct from the highly developed north–west, and the "underdeveloped" southern part of Italy. Details on this and the following case study are contained in Stöhr (1986/b).

V.3 Regional Innovation Complex III: A "Third Sector" Model

Here the case of the Japanese Technopolis policy is used [see also Stöhr (1986 and 1992)]. A major characteristic is the close interaction at the regional level between local governments, local universities (mainly science and technological departments), and private enterprise.

It is important to note that these Technopolises signify not only technological but also institutional innovation (by the establishment of local Technology Research Centers and Technology Promotion Organizations) via the synergetic interaction between important local and regional factors for technological development. The central government provides, by a well–designed set of instruments [described in Stöhr (1986)], a favorable environment and subsidiary support for local innovation.

VI. SOME CONCLUSIONS

A number of conclusions can be drawn from this paper:

1. Innovation is a synergetic process and a complex phenomenon that requires technological, institutional, and social change. The existence (or provision) of single factors to promote innovation –e.g. public research institutes, knowledge centers, management consulting services, venture capital firms– is usually not a sufficient condition for the actual emergence of innovation.

2. Although regional innovation is often triggered by external pressure (structural instability, sectoral competition, etc.) and inspired by external examples (technological development), for it to become a self–sustaining process it requires specific intraregional synergetic processes and structures. These are similar to what Colombo and Lanzavecchia call a "scientific apparatus" in which "technology is born and develops as a form of scientific knowledge in itself in a close interaction between science, industry, information, education, financing and government" in a synergetic form at the regional level.

3. The emergence of innovation is not restricted to highly developed core regions where interaction and synergy usually are considered to be highest, but, as the three case studies show, can also take place under certain conditions in peripheral areas or structurally weak industrial areas.

4. It has been shown historically that if such synergetic interaction is missing, even core areas with initially high rates of technological inventions will not innovate. Countries or regions possessing such synergetic interaction structures, however, have often been able to innovate even if their initial rate of inventions was comparatively low. The reason seems to be that only with the availability of the forementioned "scientific apparatus" and synergetic interaction is it normally possible to apply inventions effectively and to adapt technology to different (regional) socio–economic and cultural conditions.

5. The recent process of spatial functional specialization has led, particularly in non–metropolitan areas (both of the peripheral rural as well as of the "old" industrial type), to the disruption of such synergetic interaction networks. For non–metropolitan areas and for those with low invention rates the creation of these regional synergetic interaction structures appears to be an important prerequisite for innovation. Three relevant case studies for peripheral areas have been analyzed in this paper, taken from three different socio–political systems.

6. Considering these case studies, important components of regional networks of synergetic interaction appear to be: educational and training institutions, R&D, technological and management consulting, risk financing, production, and locally–rooted decision–making functions. This interaction can take place either within or between specialized regional institutions or –if regional institutional specialization has not proceeded that far– by informal cooperation between (frequently functionally less specialized) small– and medium–sized firms, as in the "Third Italy" case study.

7. If one or a few of the regional functions mentioned above are missing it seems possible to (at least temporarily) substitute external ones, provided that

236

regional interaction between the remaining functions is operating. For example, in the Basque case study the lack of an adequately—oriented, regional university could be bridged temporarily by contacts with foreign research and university centers until an adequate regional one is developed.

VII. CHARACTERISTICS OF PREDOMINANTLY ENDOGENOUS REGIONAL DEVELOPMENT IN PERIPHERAL AREAS

Endogenous development is primarily sustained within one specific region, rather than being mainly externally supported. These are some of its key characteristics:

1. Relatively wide differentiation of development strategies applied: As distinct from more centrally steered regional—development policies, which were essentially based on the relatively uniform success model of early industrialization and urbanization, these endogenous programs appear to be much more differentiated. To a high degree they are based on the specific historical, cultural, institutional, and natural conditions of the respective areas and are aimed at the broadest possible mobilization of local and regional resources (natural, human, capital, and so forth) for the satisfaction of basic needs of the regional population.

2. Societal complement to market mechanism: Many of these initiatives give priority to the production of goods and services considered to be socially valuable in the region as well as to the satisfaction of basic needs of specific target groups, both criteria that it was felt were not sufficiently taken care of by the market mechanism. The Mondragon Cooperative network in the Basque Country, for example, guarantees its members employment within its regionwide federation and within a radius of 50 kilometers from the member's location of residence. It has also in principle excluded arms and nuclear components from its production program (although the latter principle is said to have recently been undermined).

3. Participation as a necessary but not sufficient precondition: Most programs provide for as broad as possible a participation of its members, both in decision making and forthcoming benefits. In some cases this participation is primarily firm or sector related and only in the second instance regionally organized (for example, Mondragon). In other cases their organization is primarily territorially based (for example, the community economic—development cooperatives on the Atlantic coast of Canada or of western Scotland). The Mondragon cooperatives, which specialize along product or sectoral lines, have recently, however, found it useful to also form territorially organized subgroups (Grupos Sociales) in order to increase their collaboration. In many of the cases mentioned, this participation and collaboration takes place parallel or outside (sometimes even as a counterweight to)

237

the constitutionally provided local and regional representative bodies (which, in many cases, have traditionally been dominated by small elites or central government representatives).

4. Trans–sectoral orientation: In contrast to the economic monostructures that have emerged in peripheral areas during the last decades (frequently specializing in the exploitation of natural resources or of cheap labor) –and in part as a reaction to this fact– most of these initiatives aim at a more diversified, multisectoral development ("standing on more than one leg") and also at an increased intraregional mutual interaction between sectors and economic functions.

5. Promotion of regional economic and financial circuits: counterbalance the increasing internationalization of economic and financial circuits, many of these initiatives aim at the strengthening of intraregional economic and financial circuits. This is to facilitate the retention of a higher share of value added within individual regions, to safeguard regional investment requirements, to increase the innovative capacity within the respective regions, and to make them more resilient against the direct impact of worldwide economic shocks. The Mondragon Cooperative Federation, with its more than 160 enterprises, has even during the recent crisis years, been able to increase employment and generate new enterprises, while the greater remaining part of the Basque economy is in serious crisis and burdened with an unemployment rate of more than 20 percent. Depending on the type of region, a higher degree of processing of regional resources, of regional research and training and more intensive regional interaction among economic sectors and/or between producers and consumers (for example, producer–consumer cooperatives) are aimed at.

6. Innovation orientation, multilevel and not restricted to technological innovation: While the concept of innovation–oriented regional policy, which developed in recent years, was mainly related to technological innovation (either by process or product innovation), many of these programs also include innovation in the organizational and institutional spheres, such as in the forms of decision making and cooperation, in organization of work, and so forth. Most of them contain new (or revitalized old) forms of entrepreneurial regional cooperation and broad democratic decision making. In the sphere of technological innovation, they include the promotion of parallel technologies, for example, of potentially decentralized (human–capital intensive) technologies applicable to small and medium–sized plants along with the traditional promotion of (finance–capital intensive) large–scale technologies. This may also require changes in the organization of work and in the institutional sphere. Of particular interest in this context is the Mondragon cooperative federation in the Basque Country, which as mentioned, provides for an endogenous research–training–production–innovation–financing complex with direct feedback loops. This has in various cases permitted its enterprises a much higher

238

innovation rate compared with corresponding enterprises outside its network.

7. Promotion of territorial identity: Most of these programs are either based on or aim at a high degree of regional identity, be it retrospectively in the sense of ethnic or historical communality or prospectively in the consciousness of a common future fate. The presence of regional identity appears as an important prerequisite both for the cooperation among diverse (often for economic and political reasons divergent) interest groups within the region as well as for the retention or recuperation of initiative and creative personalities in the region.

8. Synergy and integration among regional economic functions, regional identity, and decision–making structures: in most of these programs, a direct linkage between economic (production, service) functions and regional decision–making processes (of workers and/or consumers) in various forms of entrepreneurial or territorial self–determination are provided for. This usually provides also for a high degree of identification of the local/regional population with these programs. Many of them take the form of cooperatives. The eastern Canadian community–economic-development programs furthermore are open for participation to all members of the respective municipalities and, at the same time, subjected to various forms of democratic control by them through contact and coordinating committees. In many cases, the economic crisis situation together with a high degree of regional identity appear to have had a strong mobilizing effect upon the local population, particularly if a high share of younger age groups was still present and no rigid social stratification existed within regions.

9. Promotion (or at least permissiveness) on the part of central authorities of linkage to transregional cooperative networks: In almost all cases, such endogenous development was only possible if the respective central government agencies were willing to either tolerate them or promote them in a way guided not primarily by central–agency interests. Furthermore, it proved essential that such development was supported by (frequently informal, but in any case, not politically dominated by established institutions) cooperative networks or "committed link cadres". These link cadres frequently fulfilled important functions in strengthening the bargaining position of regional groups vis–à–vis central authorities or external (multiregional) economic enterprise in training and consulting as well as in increasing the local consciousness of the reasons underlying existing problems and of the required self–organization for overcoming them.

Central government therefore does have a role also in the promotion of endogenous development.

REFERENCES

BOUDEVILLE, J. (1966), **Problems of regional economic planning**. Edinburgh, The University Press.

SMIDT, M. DE (1981), **Innovatie, industriebeleid en regionale ontwikkeling**. Geografisch tijdschrift 15, 3, pp. 228–38.

ECKERT, D. (1985), **Dezentrale Förderung von Forschung und Technologie in Regionen**. Universität Siegen, FRG: Diskussionsbeiträge zur Ökonomie des Technischen Fortschritts 4.

ENCYCLOPAEDIA BRITANNICA, MICROPAEDIA (1978), Vol. 9. Chicago, Hellen Hemingway Benton.

PERROUX, F. (1955), **Note sur la notion pole de croissance**. Economie Appliqué, 7: 307–20.

STÖHR, W. (1984), **La crise économique demand–t–elle de nouvelles strategies de developpement regional?** in Crise et Espace, P. Aydalot (ed.), pp. 183–206. Paris, Economica. English version: Changing external conditions and a paradigm shift in regional development strategies, in Europe at the crossroads (agendas of the crisis), S.A. Musto and C.F. Pinkele (eds.), pp. 283–308. New York, Praeger.

STÖHR, W. (1986a), **Regional technological and institutional innovation: the case of the Japanese Technopolis Policy**, in Technologie nouvelle et ruptures regionales, J. Federwish and H. Zoller (eds.), pp. 123–40. Paris, Economica.

STÖHR, W. (1986b), **Regional Innovation Complexes**. Papers of the Regional Science Association, Vol. 59, pp. 29-44.

STÖHR, W.; PÖNIGHAUS, R. (1992), "Towards a Data–based Evaluation of the Japanese Technopois Policy: The Effect of New Technological and Organizational Infrastructure on Urban and Regional Development", **Regional Studies**, Vol. 26, 7, pp. 605 ff.

STÖHR, W.; TAYLOR, D.R.F. (eds.) (1981), **Development from above or below?. The dialictics of regional planning in developing countries**. Chichester, England, John Wiley and Sons.

THOMAS, H.; LOGAN, C. (1982), **Mondragon: an economic analysis**. London, Allen and Unwin.

11 INTER-FIRM COOPERATION: A MODE FOR THE INTERNATIONALIZATION OF SME

Claire Charbit
Economics Department
Ecole Nationale Supérieure des Télécommunications, Paris

The ongoing globalization of markets has led economists and policy makers to carefully examine the modes of coordination of firms' international activities in the context of global competition and production. The search for greater flexibility and the race to innovate lead naturally to a consideration of the benefits and limits of firms' international cooperative strategies.

From an analytical point of view, the explanation of international inter-firm cooperation faces a two-fold challenge:

- that of establishing the benefits of intermediate coordination modalities

- that of investigating the international dimension of the phenomenon.

In this article, we will examine the relationship between industrial cooperation and a particular economic agent -the SME (small and medium-sized enterprise)- in its internationalization process. This relationship has heretofore been insufficiently studied to the extent that inter-firm cooperation has featured primarily, in both empirical and theoretical analyses, as one of the strategies of large industrial groups in their global competition and production. In order to elucidate the correspondance between the internationalization of SME and inter-firm cooperation, we will successively examine the following questions:

– What is international inter-firm cooperation?.

– Why are SME particularly concerned by this phenomenon?.

– What are the criteria which distinguish, amongst SME, those which will most favour this mode of internationalization?.

– What are the necessary conditions for successful international cooperations between SME?.

X. Vence-Deza and J. S. Metcalfe (eds.), Wealth from Diversity, 241–266.

– How does this strategy concern SME from the peripheral regions of the European Union (EU)?.

We will conclude by emphasizing the interest for European institutions of fostering inter-SME cooperation between the member states of the EU.

I. WHAT IS INTERNATIONAL INDUSTRIAL COOPERATION?

The literature on this topic is undergoing a spectacular proliferation, all the more remarkable because it comes from a variety of disciplines, ranging from international economics to the economics of the firm, and including elements from management science and industrial economics. We will consider below the difficulties of synthesizing the variety of international cooperative modalities within a narrow definition. We will then evaluate the contributions and limitations of the two principal schools on the topic: the first of these schools focuses on the analysis of globalization, asking "why does the internationalization of activities encourage cooperation?"; the second school focuses on the analysis of international transaction costs, asking "how are the international activities of firms organized?".

The international dimension of firms' activities does not radically modify the nature of the modalities of industrial coordination encountered within a given economic system. Indeed, according to G.B. Richardson's definition of the continuum of inter-firm relations (1972), one can observe a strict correspondance between modalities of coordination of industrial activities and the forms of international relations. Thus, the "market" corresponds to international exchanges of a commercial nature (import and export of goods of all types), "integration" corresponds to the multinationalization of firms as a result of foreign direct investments, and finally, "intermediate forms of coordination" are found in international cooperation (alliances, international sub-contracting, etc.).

Nonetheless, the international dimension does increase the complexity of the discourse, and international trade theory, which bases the differences of the international environment on the factoral endowments of each country, cannot be of great use to us here. Thus the broad principles of differentiation between countries have become increasingly inadequate to explain exchanges, and particularly the intra-branch exchanges, which have developed tremendously over the last 20 years. It has today become appropriate to update this very "passive" vision of nations with a concept of states in action, a concept which is enriched by the abundant literature on innovation systems [Freeman (1987), Lundvall (ed.) (1993)]. It has also become necessary to restitute to international exchanges their true nature: not as exchanges between states, but rather as exchanges between firms. This is all the more appropriate when these "exchanges" no longer correspond to market relationships

242

between foreign firms, but rather to industrial relations, particularly in the case of international cooperation between firms.

* International Cooperation Between Firms: A Problem of Definition

The definitions of international industrial cooperation regularly proposed are not particularly satisfactory for our purposes. For example, Contractor proposes a model under which cooperation is represented in terms of the degree of participation of one multinational firm in the capital of its foreign partner, a value which can be 50% in a "fifty/fifty equity joint venture", or 0% in licensing agreements considered as within the domain of cooperation. In this approach, the obvious flaw lies in its treatment of the nature of the cooperative modality in relation to recourse to the market. What distinction can one validly make between a market exchange and a cooperative alliance, if the only criterion is that of the participation in the capital of the partner [Contractor (1990)]?.

Although, methodologically speaking, international industrial cooperation can be examined here as an autonomous concept, there exist in reality a very great variety of concrete forms filling the space between pure international market transactions and the integration of a subsidiary within a multinational firm: "The forms of (these) cooperative relationships are many, including, for example, research agreement, development plus licensing agreement, joint research and manufacturing agreement, contractual and equity joint ventures, and research plus marketing agreements" [Shan (1990)]. The quoted author has in fact identified nearly 40 different types of cooperative agreements in the sole field of American biotechnology firms [Shan (1990)]. The above type of very broad definition can be found quite frequently in the literature devoted to international inter-firm cooperation [Hagedoorn & Schakenraad (1990), Chesnais (1988), Muchielli (1991)].

A clear conceptual structuring of these modalities is necessary in order to see how certain forms are nothing other than the manifestations of stuctures conceptually identical to those of the market or to integration, while others can truly be considered as specific to the concept of cooperation.

The need for such a clear structuring is borne out by the confusion which has resulted from considering as equivalent, for example, all intermediate modalities as "new forms of investment" [Oman (1984)]. Consequently, it is absolutely necessary to recall that in fact a great deal of international sub-contracting agreements or other forms of partnership do not give rise to any international investment whatsoever. This absence of investment is, moreover, one of the clearest ways of identifying these forms with respect to the concept of integration. The lack of recourse to direct foreign investment is particularly important in that it confirms

243

that cooperative agreements possess the characteristic of *economising resources and economising the risks linked to the new markets or new technologies.*

While it appears here to be inappropriate to conceptually link foreign investments and strategies of international industrial cooperation, it is likewise necessary to conceptually distinguish cooperation from simple market exchanges. However, for most of the authors in this field, license agreements are a form of cooperative agreement. Nonetheless, as emphasized by D. Mowery, the definition of an "international collaborative venture" "... excludes other forms of international economic activity, such as export, direct foreign investment (which implies complete intrafirm control of production and product development activities), and the sale of technology through licensing" [Mowery (1988)]. For Mowery, not only does the sale of the license correspond structurally to a market exchange (although the technology market is a particular one), it does not guarantee a veritable collaboration between the partners, as a consequence of the tacitness of certain technical skills [Teece (1986)]. Nevertheless, Mowery considers, as is generally the case, that joint ventures are clearly forms of international cooperation. Although they involve direct foreign investments, these investments are split between the two (or more) partners. Joint ventures are therefore not a form of integration strategy to the extent that control is shared cooperatively.

Ultimately, it appears that the most pertinent criteria is not the total absence of direct investment, but that rather it is a question of degree. Thus, according to Gullander "There are indications that the difference between a contractual and an equity relationship is highly exaggerated; sophisticated 'cooperators' seem to downplay the importance of ownership control as compared to management control or control through other means" [Gullander (1976), cited by Mowery (1988)]. There is a real problem with analyzing cooperation on the basis of thresholds of ownership of the capital of the partner. Let us consider Contractor and Lorange's citation of the U.S. Department of Commerce's definition of a foreign firm: "An affiliate is defined as a foreign corporation in which the U.S. company has 10 percent or more of voting securities. The surveys distinguish between minority affiliates (10 to 50 percent shareholding) and majority affiliates (one in which the U.S. firm has more than 50 percent of shares)" [Contractor & Lorange (1988), appendix]. International cooperation would therefore be based on international relationships between firms not exceeding a capital stake in the other partner equal to or exceeding 10%. We choose not adopt this criterion here for at least two reasons: first, because these thresholds vary from country to country; and second, because this criterion would identify all capital stakes of less than 10% -which may really be no more than financial portfolio investments- as manifestations of international inter-firm cooperation.

In light of the foregoing analysis, it now appears possible to propose a new definition of the group of agreements which can be qualified as involving international inter-firm cooperation:

All international organizational forms linking at least two partners of different nationality, which deal with at least one industrial production activity in the broadest sense of the term (ranging from research to distribution), excluding simple sales between partners (and therefore assignments of licenses), and not including the taking of a capital stake in the partner except in cases of limited investments made for cooperative purposes rather than as financial investments.

* The Transactional Analysis of International Cooperative Agreements

According to this line of analysis, under which the pure market remains the reference point, firms exist as a result of the occasionally less than instantanous (and therefore costly) character of exchanges, as well as the risks of opportunistic behaviour by exchange partners. Thus, Coase defines the frontier of the firm as the point within which it becomes profitable to internalize activities, a point defined with aid of a comparison of marginal cost between a transaction made according to the mechanism of market prices and a transaction coordinated internally [Coase (1937)]. Williamson has further developed this general corpus by introducing the factors determining transaction costs as well as those affecting individual behaviour [Williamson (1985)]. The specificity of assets and the uncertainty and frequency characterizing certain transactions, on the one hand, and the characteristics of limited rationality and opportunism on the other hand, are the elements which can justify the internalization of a firm's activities (Williamson, id.).

Towards the middle of the 1970's, as has been demonstrated by Kogut, economists of international production shifted their attention from oligopolistic competition to the theme of the profits generated by reducing transaction costs through the internalization of international exchanges. Although some of the strategic motivations considered (such as price discrimination in different markets) are not part of the strict Williamsonian definition of transaction costs, these two approaches are, in general, related [B. Kogut (1989)]. Thus, numerous authors have sought to apply a transactional analysis model to the particularities of firms' international coordination behaviour. The principal results of these analyses can be distinguished as set out above: one part of these studies focuses on the particularities of individual behaviour when international borders must be crossed, while the other part focuses on the particular failures of international markets. The reactions to these different limits determine the organisational modalities that firms choose in order to coordinate their international activities.

What, then, are the limits to the circulation of information, or the market failures, that motivate international cooperation between firms?.

According to the first line of transactional analysis, the principal component of international differences is cultural. This is not understood to be a matter of social or cultural systems, but rather one of differences between individual behavioural criteria, and the methodology of these analyses does not focus on national systems but rather on individual agents. This cultural (and geographical) distance entails different perceptions of confidence which may make it difficult to understand if the foreign partner is "cheating" or not [Buckley and Casson (1988)]. The recourse to industrial international cooperation is therefore justified as a means of guaranteeing confidence and "mutual abstention".

According to the second line of transactional analysis, the principal explanation for the proliferation of intermediate coordination modalities for international activities is to be found in certain kinds of international market failure. Such imperfect markets are those of certain raw materials, components, types of know–how, or financial or distribution services. It as a result of these particular failures that certain authors justify the existence of multinational firms whose international transactions are internalized [Teece (1985)]. Others, in contrast, such as Hennart, maintain that the presence of high transaction costs can also, under certain specific conditions, lead to the establishment of international joint ventures [Hennart (1988)]. Such particular circumstances are those where the "international exchanges" require a long term contact between the two partners due to the tacit nature of the knowledge to be transferred, and due also to the difficulties that firms encounter in the internalization of their foreign activities. Cooperation appears under this view to be a sort of residual solution applied when neither the market nor integration can be utilized to coordinate international activities.

The limits of this theoretical approach lie in its static nature, which does not allow for consideration of the strategic behaviour of firms. Thus, the minimization of international transaction costs generally provides only a weak explanatory criterion for firms' international decisions: "Transaction costs minimization alone does not necessarily equal profit maximization in the short run; nor does it equal strategy optimization in the long run" [Contractor (1990), p. 41]. Furthermore, under the transaction costs approach, the organizational analysis of the firm is according to a "project by project" methodology (or more precisely, "asset by asset"). For each project, the firm is supposed to have asked itself about the optimization of the balance between "internalization costs" and "transaction costs". It appears to us inappropriate, however, to consider the firm as a disjointed ensemble of individual projects, or to expect that the analysis of its behaviour can be separated from the perception of the firm as a "whole" organisation. Indeed, these critiques arise simply from the logical extension of the methodology of transaction costs theory, which does not focus on the whole firm as the pertinent unit of analysis, but rather on individual transactions and on the market.

* The Analysis of Globalization

The unit of analysis here is the firm, within the context of the globalization of markets and therefore of strategies. The principal question addressed by this analysis is the following: why do firms adopt globalization strategies, and why, for our purposes, do they decide to select cooperative international organisation modes?.

The particular firm which is the object of this analysis is the large one, multi-national and multi-product [Dunning (1981), Porter (1986)]. The explanations proposed for its internationalization are based on the competitive conditions that each firm encounters, conditions which are to an ever-greater extent based on technology [Chesnais (1988), OCDE (1992)]. Cooperative modes are increasingly appreciated for the coordination of international activities, to the detriment of the traditional choice of direct foreign investment, for the following reasons: because products are evolving more rapidly and their life cycles are becoming shortened in many industries; because R&D costs have a strong tendency to increase; because firms are making greater use of multi-sectoral technologies (generic technolgies which exceed the individual capacities of these firms) [Delapierre (1981)]; and finally, because firms need to be informed as soon as possible about the setting of new standards (and because they generally seek to directly participate in the definition of these standards). From this perspective, a competitive firm is expected to extend its strategy from the local level to the international level and finally to the global level. The differences between the international level and the global level are linked to the nature of the firm's competitors. An international firm which is established in several different countries but is only opposed by local competitors in each national market can not be defined as a global firm (nor can its competition, which remains "inter-national"). Conversely, a firm acting directly on the global level, and choosing the localization of its activities with reference to the strategies of the other members of a global oligopoly, can be qualified as a global firm, as can the market in which it operates [Porter (1986)].

The international dimension of this analysis, therefore, directly issues from the global character of the firms examined and the markets in which they operate. As a result, a sort of paradox can be observed, in that speaking of globalization reduces the international dimension (and its diversity) to a single market: the global market within which global firms are found. It therefore becomes possible for the authors of this theory to abandon the field of international economics and return simply to the principles of oligopoly theory, or, more frequently, to those of strategic management. The international dimension of cooperation is naturally "subsumed" by the dimensions of competition and firm strategy, and there is ultimately no real difference of treatment here between national and international cooperation.

Two limitations would seem to characterize this approach:

- On the one hand, the limits of international cooperation suggested by the authors are those linked to the stages of maturity of the products. Under this theory, when production becomes standardized, the firm no longer has reason to cooperate. Implicitly, this means that the cooperation which concerns global firms can practically only be justified for R&D phases.

- On the other hand, an important economic actor is ignored: the SME. However, it would seem that it is precisely for the SME that the international dimension constitutes a true leap, a leap which cooperation can facilitate, and this cooperation can not be considered as only valid for R&D activities.

II. WHY ARE SME ESPECIALLY CONCERNED BY THIS PHENOMENON?

In the empirical literature concerning international industrial cooperation, SME are often forgotten, for two reasons:

- First, the principal sources of concrete analysis of international cooperation are found in trade and professional periodicals. The sources of information for these articles tend to be agreements between large groups. SME, therefore, are only marginally represented in this data;

- Secondly, the search for data for establishing statistics is limited, amongst the diverse forms of international agreements, to those which are "measurable". Although recently more open to simple agreements, the press and the concrete analyses have for a long period of time only "measured" those cooperations taking the form of joint ventures. However, SME can only rarely choose this mode of international cooperation, because it often requires a significant foreign direct investment.

This empirical gap should not, however, eliminate the analytical interest of associating SME and cooperative forms of internationalization. We will deal with this in two steps: first, by analyzing the role of SME in the context of global industries so as to indicate where the benefits of internationalization are to be found; second, by investigating the various forms that the internationalization of SME can take.

* The Internationalization of SME in Global Industries

The small size of these firms has often led them to be considered as "passive" economic agents with respect to their environment, whereas larger firms (in terms of personnel and market share) are considered "active". More recent analyses have demonstrated that SME, like other firms, should be considered as active or passive essentially as a function of their innovative behaviour [Amendola & Bruno (1990), Picory (1994)]. Within the framework of global market analyses, SME clearly have an "active" role to play within "niche markets". For Ducros, "...the intra-branch specialisation is brought to a sufficiently fine level of differentiation so that firms, even of small size, are able to profit from monopolistic advantages in narrow niches of the international market" [Ducros (1985), p. 261]. This principle was further developed by Mariti, who offered a true analysis of the industrial economics of global niche markets for SME [Mariti (1993)]. As the SME can only have a "passive" role to play here in terms of mastering global competitive conditions, the SME should not seek to position itself on products in such markets. On the other hand, by correctly evaluating its strengths and the market spaces left vacant by the largest competitors, the SME can have an active strategy to deploy in niche markets.

For Mariti [Mariti (1993), pp. 192-193] a global market can thus leave a place open for even firms of the smallest size to the extent that:

- economies of scale can be generated even for output quantities which are only a tiny fraction of the global market;

- rapid growth in demand in certain markets (notably as a result of the integration of zones corresponding to these markets) leaves an opening for new, small-size entrants;

- the strong tendency for production to utilize "decentralized technologies" enables the maintenance of small units of production in close relationships with the other stages of the productive process;

- the producers in the market seek to diversify their sources of supply by maintaining purchases from small producers of substitutable intermediate goods,

- finally, the economics of scope that multi-product firms can exploit are often linked to economics of scale of their marketing and distribution systems. However, even if the simultaneous distribution of different products enables exploitation of economies of scope, this does not mean that these firms can also expoit economies of scale in the production of these different goods. Therefore there is an opening for firms of the smallest size in narrow segments of production, because there is no reason why such products must be completely fabricated within a single large firm.

In addition to the foregoing, SME possess the following advantages over large global firms:

- SME are often well-established in their home geographical territory, and they are therefore better able to understand specificities of tastes as well as changes in local demand;

- in industries where fashion or style are important or, above all, where contacts with customers must be very close, SME have a better ability to produce small, highly–differentiated production runs,

- SME have, a priori, greater flexibility than larger, more integrated firms.

Despite the foregoing, the strategy of exploiting global niche markets comprises significant risks for SME:

- the first risk of product differentiation is a rather classical one, which counter-balances the advantages of highly-specialized and differentiated products with the disadvantages of the associated higher costs. The consequence is higher prices for these products, which can in the long run reduce demand.

- other risks are linked to the high costs of R&D necessary for the elaboration of such products; these costs can not be of short duration. Indeed, in order to maintain the dynamic of the niche market approach, it is necessary that products evolve rapidly. Firms producing in these niches, where technology is of the essence, take the risk of being acquired by global firms as soon as the benefits of this type of production have been demonstrated. If the absorption of the SME does not occur, the firm may then risk the absorption of its demand, either by imitation from other SME, or by integration of the given activity by larger firms in the market. Finally, "technological SME" can falter as a result of their lack of competence in marketing and distribution [Walsh (1993)].

Having discussed the openings for, and the risks of, internationalization of SME in global industries, we move to a discussion of the strategies which allow for minimizing these risks. We will now seek to demonstrate that the organization of internationalization along cooperative lines presents significant advantages.

* The interest of the cooperative form for the internationalization of SME

One of the possible benefits that Contractor and Lorange assign to the various possible forms of international cooperative agreements is that they facilitate initial international expansion by inexperienced firms [Contractor & Lorange (1988)]. In effect, international industrial cooperation presents two basic types of advantages which are each linked to a different theoretical field and each related to the drawbacks of alternative forms of internationalization. For industrial economists, the favour accorded the cooperative form in comparison to integrative strategies of direct foreign investment is justified by the avoidance of investment costs -which are frequently sunk costs. In addition, international partnerships- as opposed to exports (as international economists can observe) -assure greater proximity to foreign demand. The foreign partner is in very close contact with its own local market and can therefore offer the other partner precious knowledge of this import market. These two types of advantages of cooperation over the two extreme forms of coordination of international activities are nonetheless quite general in nature and concern large firms as well as small ones.

One particularity of SME with respect to international cooperation is that often they have no choice! In fact, the decision to internationalize by SME is often based on very few alternatives: small firms can only rarely realize direct foreign investments, and they can only with great difficulty establish foreign local distribution systems for their exports. Their products, as we have seen, tend to be localized in very narrow market niches of global markets, so that they must generally establish direct relations with their foreign clients. The solution of international partnerships can turn out to be the only possibility when there is an absence of the possibility of foreign direct investment, an absence of specialized distribution networks, or an absence of an interest in exporting to broad networks which are not appropriate for non-standard products: "For small-sized firms, cooperative ventures are often the only realistic way to reduce risk to tolerable levels" [Contractor & Lorange (1988), p. 124].

Beyond the arguments of lack of choice or of the drawbacks of other modalities, international industrial cooperation presents specific advantages for SME. As we have just seen, it provides inexpensive access to the skills or demands of the local market. This inexpensiveness, because it allows for smaller initial investments, leads to a a higher degree of reversability in the event of risk. The access to local markets by international partnerships also enables marketing of products in conformity with local norms. Numerous international cooperations have thus been concluded with the objective of entrusting the last stages of "bringing into conformity with the standard" to the local partner, and offering the reciprocal benefit in the home market. The only alternative for obtaining this access would be the acquisition of a local firm or the creation of a local subsidiary, integration strategies which are practically inaccessible for most SME. In addition to the access to local standards, international industrial cooperation enables SME to benefit from the different national capacities possessed by the partner firm (e.g., particular know-how,

251

R&D programmes, or financing from which the local firm has benefited in order to build its competences, etc.). Finally, reciprocal marketing agreements enable SME partners to exploit economies of scale by producing a greater quantity of output while permitting economies of scope as a result of the splitting between the partners of the distribution.

These strategies are not devoid of risk any more than are other forms of cooperation. An alliance can not be viewed as a static entity, and it is always possible that one of the two partners will not respect the conditions contained in the agreement (especially those concerning the diffusion of intellectual property). Moreover, the progressive learning which is implied by the agreement with respect to the production processes and output specificities of the partner opens the risk that the partner will not renew the agreement if it learns how to produce the products itself instead of importing them. There are two types of defense against this eventuality: that of choosing a form of cooperation which requires very little commitment and therefore which is reversible at any moment or where the information exchanged is not strategic for the participating firms, or on the contrary, that of a strategy of intense cooperation which is strictly balanced, and where each of the parties possessing strategic information on the partner will not wish to break the cooperation and take the risk of seeing itself threatened. These two types of behaviour have been clearly analyzed in Shan's study of the international arrangements of small innovative American firms in the biotechnology sector [Shan (1990)]. This author has demonstrated that these firms frequently choose cooperative modalities for two reasons: first, to increase the size of their market and therefore maximize return on their innovations, and second, to gain access to complementary assets which they do not have (especially with respect to marketing). In this respect, the author succeeds in demonstrating empirically "... that firm size is negatively correlated with the use of cooperative arrangements" [Shan (1990)].

Nonetheless, the ensemble of SME is not a homogenous group. We will now seek to develop a classification of SME. This classification does not seek to be exhaustive, but merely seeks to discern the major criteria which differentiate the capacity of SME to benefit from cooperation as a means to internationalization.

III. TYPOLOGY OF SME AND THE CHOICE OF INTERNATIONAL COOPERATION

We have previously discussed how SME can internationalize in global industries by penetrating niche markets. Nonetheless, it is clear that PME do not all have access to this type of strategy, and that they do not all have the same ability to choose international industrial cooperation as the mode of internationalization. It appears to us that at least in a first analysis, five criteria are of primary importance in distinguishing SME with respect to choice of cooperative internationalization.

* The Nature of the Product Produced by the SME

The degree of maturity of the product produced by the SME has an impact on the relationship established with the product's users. Thus, a very young product, which is still in the development phase, is difficult to offer directly on the market. The producer must therefore seek another mode of organization. The solution can be a form of very strict cooperation, or a form of integration between the producer and the user. A recently-developed product which is nonetheless non-standard, and which therefore must be realized on the basis of precise specifications from the user, also requires a relationship of proximity between the two participants to the "exchange". This proximity will not necessarily be a geographical one, and the contacts between the two firms can take place at a distance once the modalities have been fixed. However, once again, this proximity is in contradiction with the market relationship: with this latter type of product, the relations between the user and producer must pre-exist production. It is in this sense that we refer to a relationship of proximity, as opposed to what is known in anglo-saxon terminolgy as an "arm's length relationship". Finally, the realisation of standard products does not require a strict relationship between the producer and the user: a market exchange, and therefore in our context a simple export sale, is sufficient.

It is clear that the international behaviour of SME will therefore be linked to the stage of maturity of the product produced. One can already note that the choice of cooperation as a means for conquering niche markets is a strategy that may often be employed with innovative or specific products, but only rarely with standard products[73].

* The Vertical or Horizontal Character of SME Specialisation

Linked to the preceding criterion, this characteristic distinguishes SME realizing a finished product from those specialized in the realization of a stage in the productive process. Finished-product specialization is qualified as a horizontal specialization, whereas production-process specialization corresponds to vertical specialization. A horizontal specialization can therefore give rise to international distribution agreements (in addition to other types of agreements), each partner performing the distribution in its local market of the other firm's products -in addition to its own. In contrast, vertical specialization, and therefore the mastery of know–how concerning the realization of a particular task (and not of a product), can

---

[73] In the case where the local labour costs of a sub-contractor allows him to execute, for a foreign principal, standard production tasks at a low price, then this is no longer a case of positioning in terms of niche markets.

not give rise to distribution agreements but rather to international sub-contracting agreements, to industrial cooperation in the sense of an exchange of technological information, and indeed to R&D cooperation.

This criterion, without discriminating the access of SME to international industrial cooperation, selects for the range of possible agreements into which firms can enter.

* The Size of the SME

As we have already remarked, the SME category is not homogenous, and these firms are quite differentiated according to size. Indeed, it is quite difficult to envision a greater behavioural identity between firms of 10 and 400 employees than between firms of 499 and 600 employees, for example. Nonetheless, the firms with 10, 400 and 499 employees are generally classed as SME, whereas the 600-employee firm is not. It is thus risky and occasionally absurd to rely only the criterion of size to distinguish between the behaviours of firms. Indeed, between a 10-employee "start-up" specialized in R&D, and a 10-employee craftswork firm specialized in a traditional activity, great behavioural disparities can be expected. The size of the firm remains nonetheless a good indicator of the mono- or multi-activity nature of the firm. Moreover, rarely will a very small SME be able to simultaneously realise R&D, production and marketing activities. According to the nature of its activity, certain forms of international cooperation will be possible and others will not, as we have shown in the preceding two points.

While the larger SME can exploit economies of scale and scope, this is rarely the case for smaller SME. As a result the smaller SME will only rarely be able to rival larger firms in price competition, except in those cases of localization in low-wage zones. Rather, the production niche for SME at the international level is that of quality and strictly-defined performance. Their strategy will therefore depend on the search for a niche market, which international partners can help it find. We must also mention here that numerous SME realize design activities (i.e., of lines, of elements of production), often in direct relations with their clients, and these can be assimilated to R&D activities. Therefore, the strategy of niche markets is not reserved only for "start-ups" in R&D -quality and innovation can also be found in the most traditional small firms.

* The Establishment of the SME in a Localized Network

One can further distinguish SME according to their membership or lack of such in a localized network of industrial relations. If an SME habitually functions in tight relations with other firms in the same territory, this can have various effects on its internationalization:

254

- This local network can favour the sharing of means allocated between different firms in the local network. By exploiting, together, economies of scale in distribution, through cost-sharing, the strategies of SME (especially the smaller ones) can begin to resemble those of much larger firms. Without necessarily acquiring a scale sufficient to enable integrative strategies (by acquiring or creating a foreign subsidiary) these firms can practice direct exporting, as by choosing, for example, to have their products exhibited together at international trade shows. An initial effect of this strategy would therefore be a tendency toward local cooperation leading to coordination by the market at an international level, rather than to international industrial cooperation.

- The opposite effect can result from the learning possessed by these firms in terms of cooperative strategies. In effect, the existence of a local convention of long-term cooperation can facilitate the recourse to partnership modalities whether the partner firm is local or foreign.

Between these two dichotomous effects, the distinguishing point appears to be the speciality of the given firm. If the SME produces a finished good which can be offered in a catalogue, recourse to the market via exportation can be chosen. Conversely, if the SME is specialized in know–how rather than in a product, its familiarity with cooperation will doubtless facilitate an openness to international partnerships.

* The Benefits and the Nature of Public Intervention with Respect to the SME

Without going into detail regarding public intervention (we will explore this further below), which can be regional, national or community-wide, a European SME can, without infringing competition rules, benefit from public support for its internationalization strategies. The nature of this assistance may not be directly linked to the strategy of internationalization: it may deal with innovation, training, assistance in linking up with other local firms or with research centers (for regional public institutions) or with other firms belonging to other member States (for EU institutions), employment assistance enabling the firm to increase its personnel and the quantity and quality of its output, etc. These various forms of public assistance can have a two-stage effect: in the first stage, they can improve the quality of the output of the firms concerned, and therefore their competitiveness at the international level; and in a second stage, they can directly encourage the internationalization of these firms, especially via cooperation.

There is thus a major role to be played by the different forms of public intervention.

255

IV. NECESSARY CONDITIONS FOR SUCCESSFUL INTERNATIONAL COOPERATIONS BETWEEN SME

Beyond these five elements, we would like to mention two conditions which appear important in terms of the aptitude of SME to opt for international cooperation. These conditions are not linked to the external environment of firms or to their specialities, but rather, to the prerequisites for successful realization once the strategy of cooperation has been adopted.

First of all, international industrial cooperation "is not invented". This means that it is necessary to be previously trained for partnership, and therefore for the sharing of information, in order to select this form of international coordination. A survey carried out in 1985 on French firms producing within the framework of international sub-contracting agreements showed that the great majority of firms in the sample were already sub-contractors for French principals before broadening the scope of their partnerships to the international level [Berthomieu et al. (1985)]. A prerequisite therefore seems necessary for international industrial cooperation, one based on the application of similar modalities from the national or local environment of the firm. A more recent survey dealing with Franco-Californian partnerships between SME also established a correlation between the willingness of firms to cooperate internationally with the existence of identical strategies at the local level. This survey (which we realized in November 1993) reveals also that the "local" level is distinct for the two groups of firms. The existence of local partnerships for firms in the Silicon Valley no longer needs to be demonstrated. But for the great majority of French SME queried (all were high-tech SME), the local level was in fact the European level. The French firms of this sample did not participate in cooperative industrial networks of close proximity. The existence of Europe-wide partnerships therefore seemed clearly to increase the aptitude of SME to establish similar partnerships globally.

In addition to this prior familiarity with the cooperative form (derived from organizational learning) a more recent empirical analysis has demonstrated the necessity of establishing a prior knowledge of the partner before concluding accords with it [Kaufman (1993)]. An investigation of the cooperative choice of internationalization of a sample of German SME demonstrated that the existence of relationships of trust between the partners, stemming from their reputation and from prior successful transactions, was considered critical by the SME surveyed. The risks of loss of information and of opportunistic behaviour by the partner are so high with these SME, that only a very high level of prior confidence (in addition to the economic conditions examined above) can lead the SME to choose cooperation.

256

V. INTERNATIONAL INDUSTRIAL COOPERATION AND SME AT THE PERIPHERY OF THE EUROPEAN UNION

The international industrial cooperation of PME can be concluded either between SME, or between SME and large firms. This paper does not have the objective of reviewing the ensemble of works devoted to the relations between small and large firms. To the extent that the work concerning international networks of SME is still in an embryonic phase, it is not our objective here to measure or analyze concrete international relations between SME. Our approach, rather, limits itself in this section to presenting a certain number of observations which, by contrasting relations between SME with relations between SME and large firms, will enable us to suggest the proper place for public intervention with respect to international industrial cooperation.

The point of departure for this analysis lies in the observation that SME, by virtue of their size, are limited in their available strategies by the lack of accessible information (even though in networks they may possess advantages in terms of flexibility). In effect, international links between SME can enable them to compensate for the advantages of integration available to multi-national firms. As we have seen above with respect to these types of alliances, each of the partners can exploit advantages which could never have been developed alone. Thus, international SME networks can to some extent respond more flexibly and rapidly to changes than can the large bureaucratic multinationals, which take longer to identify alternative productive processes and which must find projects of a sufficient size in order to justify their involvement. Nonetheless, the major limitation to these advantages restraining the growth of SME partnerships lies in the difficulties that SME encounter in exchanging existing knowledge on current products and skills, as well as on the potential and opportunities for future knowledge. International SME networks can be hampered by difficulties of access: to new markets, to innovations with a generic character (which are beyond the competence of mono-product firms) and frequently to innovations in the process of becoming future production standards.

This problem appears in a later period with respect to the prior difficulty for these SME of identifying adequate foreign partners for cooperation. Informational insufficiencies therefore concern both the search for the "right partner" and the learning of "new standards and technologies".

The necessary informational capacity can be brought to SME in two different fashions:

- The first of these involves cooperating with a large firm. More up to date on these trends and having from its multi-localization a broader field of vision, the large firm can supply, in addition to other types of complementary assets (as in its distribution network), "complementary informational resources". The international nature of this type of link can then take two forms: a relationship between a local SME and the subsidiary of a multinational group established locally, or a direct relationship between a local SME and larger producers and distributors in the targeted international markets (a common case for "start-ups"). The risks of monopsony, of buy-out by a larger firm, and of the less than true sharing of all information, are nonetheless very high in these cases for SME.

- The second solution consists in relying on public assistance both in research and in match-making with foreign SME partners. Three non-exclusive levels of public intervention can be distinguished. First of all, there is regional assistance fostering contact with local universities and firms but also encouraging the establishment of subsidiaries of foreign groups, from which information can be acquired. Regional institutions can also be entrusted with prospecting missions abroad in support of the ensemble of skills of local SME. National agencies supporting innovation and export assistance can also encourage the internationalization of SME. The government of course also has the mission of establishing regulations for protecting innovations of firms in the territory, which reduces the risk of imitation by partner firms, especially large ones. Finally, for firms established in Europe, a third level of is that of common European Union policies intended to open the SME in certain programmes to multiple international partnerships.

In the case of SME in the peripheral regions of the European Union, these three levels of public assistance are especially vital, because these are only rarely orientated toward high-technology and are therefore often unable to autonomously handle their own international opening [Martínez Sánchez (1992)]. To date, the strategy the most often implemented, especially in the peripheral regions of Great Britain (Scotland, Wales and Ireland), has consisted in the promotion of local industrial mutations through the policy of attracting American and Japanese subsidiaries, which aimed in the 1950's and 1970's at access to the European market. The studies of the impact on local SME of the establishment of these major manufacturers, especially in electronics, have observed difficulties in encouraging the multi-national subsidiaries to establish relationships with local SME, and also in the transformation these very hierarchical relationships of vertical sub-contracting into true partnerships. There are, nonetheless, a certain number of indications that positive returns do flow from these establishments, even if it takes a very long time [Haug (1986), for the case of Scotland]. Even if the debate remains open, the goal to be attained by public institutions for the local SME is a very clear one: "The aim for the subcontractors to achieve a level of quality, specialization and overall excellence which will ensure that the principal will itself become somewhat dependent on the sub-supplier" [O'Doherty (1990), p. 217].

It is therefore clear that for SME on the periphery, access to true international industrial cooperation will entail a stage of improving product quality, which must first be based on local connections (such as district SME networks), before being able to develop true international connections through international industrial cooperation within a strategy of conquering global niche markets.

The involvement of SME in European Union cooperative programmes is not at all the same for large enterprises. If the larger firms accept cooperation on pre-competitive research, although not insofar as concerns their "core-technologies", they do so primarily in order to be part of the networks which initiate production standards and to ensure that they receive information on generic technologies. They only rarely aim in these partnerships to improve their market share. SME, on the other hand, have above all the objective of seeing concrete results to rendering their joint research in the marketing of products. This distinction has been demonstrated by an L. Mytelka's evaluation of the Esprit programme in 1990 [Mytelka (1990)]. Without going back into the history of the Esprit programme, we will here only recall that the programme was devoted to information technology research, that it was begun in 1980, and that it has undergone several stages of development, especially after its renewal in 1987. Although relatively disappointed by the weak concrete results from the research undertaken, the SME participating in these networks nonetheless benefited from the following positive returns:

- Although the proportion of European Community funds in the research budgets of large groups remained quite marginal, the above study revealed that the Esprit projects contributed from 25 to 45% of the R&D budget of the concerned SME. It is clear that the goal of large groups, as contrasted with that of the SME, is not linked to the advancement of their core-technologies with the aid of Community contributions;

- The participation of SME in a network constituted under the aegis of the European Commission was a guarantee of the maintenance of their independance, which so-called "spontaneous cooperations" might not have been able to preserve;

- A notable risk, but which does not seem to have been realized, is that the SME itself is not absorbed, but instead its most competent researchers (having participated in the Esprit project) are poached by larger firms in the network;

- Access to network research, and therefore to complementary know–how,

259

has enabled numerous SME to accelerate their innovation processes;

- Participation in Esprit projects, and the learning which firms derived from this, enhances SME in the eyes of their future clients, even outside of Community programmes;

- Esprit has enabled partners in cooperation to better understand the future technological specifications which will be demanded by their principal customers;

- Since an international opening is often difficult for SME, the benefits of programmes such as Esprit can be decisive, all the more so because even if SME often meet large clients or suppliers which are dominant in the sector, they are only rarely brought into contact with other SME working abroad in related fields. This type of horizontal partnership between SME remains, however, a rare outcome of the Esprit project.

- Finally, the fact of belonging to Esprit-type networks in research bearing on standards is particularly important for all firms working in industries where technological evolution is very rapid, but even more so for SME to the extent that, although unable to realize the totality of a product, they must furnish components of perfect compatibility. Participating in the elaboration of standards is therefore a condition of access to future markets for such product segments. The fact that the developments of standards may not be linked to an international concept but rather to one of integrated zones (which sometimes even approaches non-tariff barriers), can be of strategic value for SME within the zone because they can thus become essential partners of the members of the external global oligopoly as these larger firms seek to penetrate the zone's markets. Even in its most international conception, the perfecting of standards is a key informational phase for SME, enabling them to find a broader range of foreign partners.

These elements have doubtless served to improve the adaptation of public programmes for international partnerships. A recent evaluation of the Eureka programme shows to what extent SME have been dealt with specifically, and to what extent their expectations of realization in a market have found support from public authorities [Eureka (1993)]. In effect, in the programme of Eureka partnerships, principally structured on the basis of existant vertical relations between partners, 48% of the firms were SME. The main points of this evaluation are the following:

- SME have a tendency to participate in projects which are not in relationship to their competitors (as can be the case for common research carried out by the large firms) but rather with their clients. Thus, the sought-after improvements by SME in these partnership are essentially of increased sales (not only with users who are partners of European projects but more broadly, owing to their improvement in meeting customer needs, with a much broader demand) while larger groups seek above all to reduce their production costs or determine future standards. As has been noted in the case of Esprit, SME establish partnerships which bear directly on their core-technology, which is not the case for larger firms. They demonstrate therefore a greater degree of commitment (and of risk-taking) than do other firms.

- An important result must be underlined: "...it appears that projects involving only SMEs are more likely to have achieved concrete outputs" [Eureka (1993), p. 45]. In effect, not only have SME obtained more concrete results in these programmes than have the large firms, but in addition the partnerships which have concerned only SME have had better success than those which associated large firms. Finally, and this confirms the basic propositions of this paper, the Eureka projects which have enjoyed commercial success have most often been carried out in very narrow niches of international markets, and by SME.

Nonetheless, a certain number of limitations to these projects must also be noted:

- One can note first of all a relatively weak participation by firms from the peripheral regions of these networks;

- Not independantly, the SME involved in these projects are more export-orientated and R&D-orientated than the average SME;

- Finally, the problems encountered by these SME are more significant than those of large firms; while the collaboration between SME have often succeeded, the projects linking SME and large firms often faced serious dissension during the phase of transformation of R&D into product sales.

In conclusion, we wish to extend our reflections to a perspective which is not only of peripheral SME, but also of European Union institutions. It appears to us that if the EU objective is indeed to guarantee progress for the ensemble of economic entities from the different member states, this can not occur in seeking an ever greater homogeneity in industrial capabilities. On the contrary, it is in the diversity of European production and innovation processes that one can find the sources of progress. For this reason, this industrial diversity must be articulated more coherently [Charbit et al. (1991), Cohendet and Llerena (1991), Gaffard (1992)], by not excluding certain firms (SME) or regions (the periphery). If European Union programmes are already evaluated according to their benefits for SME, the positive effects remain quite limited given the weak involvement of SME in the major

European programmes. Despite this, 82% of of European firms are SME employing from 1 to 9 employees [O'Doherty (1993)]. It is therefore imperative to make these the dynamic actors of the European economy and to give them the means of facing up to international competition. International industrial cooperation between SME seems to us to need to be truly conceived as one of the instruments for achieving this objective. As a result, rather than merely favouring commercial integration of the member states developing simple exchanges between themselves [Charbit, Ravix & Romani (1991)], European integration will then be able to take the path of true industrial integration.

REFERENCE

AMENDOLA, M.; BRUNO, S. (1990), "The behaviour of the innovative firm : Relations to the environment", **Research Policy**, no. 5.

BERTHOMIEU, C.; CHARBIT, C.; HANAUT, A.; RAVIX, J.T. (1985), **L'insertion de la France dans la segmentation internationale des processus productifs par la sous-traitance internationale**, Rapport pour le Commissariat Général du Plan, Octobre.

BUCKLEY, P.J.; CASSON, M. (1988), "A theory of cooperation in international business" in **Cooperative strategies in international business**, F. Contractor and P. Lorange eds., Lexington Book.

CHARBIT, C.; GAFFARD, J.L.; LONGHI, C.; PERRIN, J.C.; QUERE, M.; RAVIX, J.L. (1991), **The study of local systems of innovation in Europe**, Working paper of the Commission for the European Communities, February.

CHARBIT, C.; RAVIX, J.T.; ROMANI, M. (1991), "Sous-traitance et intégration industrielle européenne", **Revue d'Economie Industrielle**, no. 55, 1er trimestre.

CHESNAIS, F. (1988), "Les accords de coopération technique entre firmes indépendantes", **STI Revue**, no. 4, OCDE, Paris.

COASE, R. (1937), "The Nature of the Firm", **Economica**, vol. 4, no. 3.

COHENDET, P.; LLERENA, P. (1991), "Diversity and Coherence of Systems of Innovation in Europe: an overview" in Cohendet, Llerena and Sorge, **Modes of Usage and Diffusion of New Technologies and New Knowledge: the Must project overall synthesis report**, Must Project CEE DGXII, May.

CONTRACTOR, F.J.; LORANGE, P. (1988), "Why should firms cooperate?. The strategy and Economics Basis for cooperative ventures" in **Cooperative strategies in international business**, F.J. Contrator and P. Lorange (eds.), Lexington Books.

CONTRACTOR, F.J. (1990), "Contractual and Cooperative Forms of International Business: Towards a Unified Theory of Modal Choice", **Management International Review**, vol. 3, p. 41.

DELAPIERRE, M. (1991), "Les accords inter-entreprises, partage ou partenariat?. Les stratégies des groupes européens du traitement de l'information" in **Revue d'Economie Industrielle**, no. 55, 1er trimestre.

DUCROS, B. (1985), "Coûts comparés et concurrence internationale" in **Croissance, échange et monnaie en économie internationale, Mélanges en l'honneur du Président Jean Weiller**, Economica, Paris, p. 261.

DUNNING, J.H. (1981), **International Production and the Multinational Enterprise**, George Allen and Unwin, Londres.

EUREKA (1993), **Evaluation of Eureka Industrial and Economic Effects**.

FREEMAN, C. (1987), **Technolgy policy and economic performance lessons from Japan**, London and New York, Pinter Publishers.

GAFFARD, J.L. (1992), "Territoires en Europe et innovations", Communication à FAST Conférence Consensus, **Science technologie et cohésion de la Communauté**, Bruxelles, décembre.

GULLANDER, S. (1976), "Joint-Ventures in Europe: Determinants of Entry", **International Studies of Management and Organization**, Vol. 6, pp. 85-111.

HAGEDOORN, J.; SCHAKENRAAD, J. (1990), "Strategic partnering and technological co-operation", B. Dankbaar et al. (eds.), **Perspectives in Industrial Organization**, 171-191, Kluwer Academic Publishers, Printed in the Netherlands.

HENNART, J.F. (1988), "A transaction costs theory of equity JV" in **Strategic Management Journal**, vol. 9, 361-374.

KAUFMAN, F. (1993), "Internationalization via Cooperation - Strategies of SME" in **EMOT Workshop–European Science Foundation**, Berlin, september.

KOGUT, B. (1989), "A note on global strategies", **Strategic Management Journal**, Vol. 10, pp. 383-389.

LUNDVALL, B.A. (Ed) (1993), **National Systems of Innovation: Towards a Theory of Innovation and Interactive Learning**, Pinter Publishers, London.

MARITI, P. (1993), "Small and medium-sized firms in markets with substantiel scale and scope economies" in M. Humbert (ed.), **The Impact of Globalisation on Europe's Firms and Industries**, Pinter Publishers, Londres.

MARTINEZ SANCHEZ, A. (1992), "Regional Innovation and Small High Technolgy Firms in Peripheral Regions", **Small Business Economics**, 4.

264

MOWERY, D.C. (1988): "Collaborative Venture between U.S. and Foreign Manufacturing Firms: An Overview" in D.C.. Mowery (ed.), **International Collaborative ventures in U.S. Manufacturing**, Ballinger Publishing Company, Cambridge Mass.

MUCCHIELLI, J.-L. (1991), "Alliances stratégiques et firmes multinationales: une nouvelle théorie pour de nouvelles formes de multinationalisation", **Revue d'Economie Industrielle**, numéro spécial, no. 55, 1er trimestre.

MYTELKA, L.K. (1990), **Technological and economic benefits of the European strategic programme for research and development on information technologies (ESPRIT)**, A report to the department of Communications, Government of Canada, Contract no. 36100-0-5616, Ottawa, July 16.

O'DOHERTY, D. (1990), "Internationnalization and Cooperation - The Irish Policy Response" in D. O'Doherty (ed.), **The Cooperation Phenonmenon - Prospects for Small Firms and the Small Economies**, Graham & Trotman, Dublin.

O'DOHERTY, D. (1993), "Globalisation and performance of small firms within the smaller European economies" in M. Humbert (ed.), **The Impact of Globalisation on Europe's Firms and Industries**, Pinter Publishers, Londres.

OCDE (1992), **Technology and the Economy: The Key Relationships**, Paris.

OMAN, C. (1984), **Nouvelles formes d'investissement dans les pays en développement**, Etudes du Centre de Développement, OCDE, Paris.

PICORY, Ch. (1994), "PME, incertitude et organisation industrielle : une mise en perspective théorique", **Revue d'Economie Industrielle**, no. 67, 1er Trimestre.

PORTER, M.E. (1986), "Competition in Global Industries: a Conceptual Framework" in M.E. Porter (ed.), **Competition in Global Industries**, Harvard Business School Press, Boston.

RICHARDSON, G.B. (1972), "The Organization of Industry" in **Economic Journal**, Vol. 82, no. 327.

SHAN, W. (1990), "An empirical analysis of organizational strategies by entrepreneurial high-technology firms", **Strategic Management Journal**, Vol. 11.

TEECE, D.J. (1986), "Profiting from Technological Innovation", **Research Policy**, Vol. 15, no. 6.

WILLIAMSON (1985), **The Economic Institutions of Capitalism**, The Free Press, New York.

12 EXTERNAL CONTROL AND REGIONAL DEVELOPMENT IN AN INTEGRATED EUROPE

Brian Ashcroft
Fraser of Allander Institute
University of Strathclyde

I. INTRODUCTION

The process of European integration is generally perceived as an opening up to competition of both product and factor markets. Indeed, for many, the promotion of competition is the *raison d'être* of the European ideal. However, while competition and market integration are important dimensions of the integration process they do not completely define it. Alongside market integration there exists a process of company integration and functional specialisation which is having, and will continue to have, profound implications for economic development in Europe's peripheral regions. This process is taking the form of cross-border takeovers, movements of company functions between locations, inward investments, management buy-outs and so on. The rationale for these changes may be to take advantage of increased market opportunities, but will just as likely be motivated by concerns to limit and restrict the effect of increased competition. Whatever the motivation behind such change, the outcome is likely to be further increases in the degree to which companies and economic activities in the peripheral regions are controlled by firms outside the region typically in the 'core' regions of the EU. This is because 'core' regions attract a disproportionate number of the headquarters of companies, a process which continues, largely through takeovers and re-organisations, within member states of the EU such as Germany [Schackmann-Fallis (1989)] and the UK [Ashcroft, Coppins, and Raeside (1994)]. Such trends would, I suspect, be found in other EU countries if the data were available. Further European integration is likely to accelerate and adjust this process in favour of the 'core' regions of the wider European economy.

Against this background, it is clearly important that policy makers are aware of the phenomenon and that regional economists are able to advise on the likely consequences of increasing external control for economic development in Europe's peripheral regions. This paper seeks to assist in the development of that understanding. Part 1 of the paper briefly discusses the concept of external control

X. Vence-Deza and J. S. Metcalfe (eds.), Wealth from Diversity, 267–291.
© 1996 *Kluwer Academic Publishers. Printed in the Netherlands.*

and considers some evidence on the scale of takeovers and inward investment in recent years. Part 2 considers the consequences for the regional economy. There is no attempt to review the literature here which is covered fully in our recent book [Ashcroft and Love (1993)]. What this paper does suggest, however, is that much of the previous literature in this area has tended to focus on the company effects of external control to the exclusion of a consideration of the effects on the regional economy. Such regional effects are usually implied rather than considered formally. Specifically, there has been little or no attempt to conceptualise the role of external control in the regional growth process. A preliminary outline of such a conceptualisation is offered which suggests that it is the effect of external control on the supply potential of the economy, particularly entrepreneurial capacity, that is the key to the role of external control in the regional growth process. Part 3 concludes with a summary of the paper and a consideration of the policy implications of the growth in the external control of peripheral region economies.

II. THE CONCEPT OF EXTERNAL CONTROL

The term *external control* can be applied to all those activities that are ultimately controlled by other activities outside the regional economy. The question of where ultimate control lies is usually answered by reference to the location of the headquarters of the parent firm.

Given these definitions, it follows that external control of a region's activities can increase through several routes. *At the level of the individual company*, these may include:

– takeover of a regional company by a firm headquartered elsewhere;

– public acquisition of the assets of a regional company where the headquarters of the nationalised industry is located elsewhere;

– establishment of a new 'greenfield' activity in the region by a company headquartered elsewhere;

– relocation of headquarters functions by independent regional companies to take advantage of perceived localisation economies elsewhere, usually in core regions;

– management buy-outs, where regional -and extra regional- activities are separated from the larger organisation of which they are part by a sale to existing management. Buy-outs will serve to increase the level of external control if management outside the region purchases the regional activity from an organisation which is headquartered in the region.

Additionally, the degree of external control may increase *at the regional level* due to:

— a faster rate of immigration of new plants (allowing for size) from elsewhere, compared with the rate of indigenous formation of new firms in the region;

— a faster rate of growth of output or employment in the stock of existing externally-owned firms than that achieved by the stock of indigenous regional companies.

From the available evidence it is clear that the most important routes to external control at the company level within regions are 'greenfield' investments, and takeovers [Ashcroft and Love (1993)]. Moreover, throughout the 1970s and 1980s takeovers appear to have increased in importance as a method of entry, while the overall level of inward investments of all forms appears to have risen. At the international level, the ratio of 'greenfield' to acquisition investments undertaken by US multinationals in OECD countries fell from 3 to 0.2 in just three years between 1976 and 1979 [Hood and Young (1982)]. The same process appears to have been at work at the regional level. In Scotland, the employment associated with new openings and acquisitions made by *foreign* firms amounted to 11,700 and 2,400, respectively in 1970-74 -a ratio of nearly 5 to 1- but by 1985-89 employment in new openings had fallen to 4,700 while the jobs associated with takeovers had risen to 10,200 -a ratio of just under 0.5 [Industry Department for Scotland (1990)]. At the EU level, inward investment into member states rose during the 1980s, while the number of industrial mergers increased at a 25% annual rate, with the proportion of takeovers involving EU member states rising steadily after 1982-83 [Amin (1992)].

So, we can suggest that external control, within, between, and from outside, EU member states, rose during the 1980s, and might be expected to continue to rise -at least until a steady state is reached- as the single economic market (SEM) develops. Moreover, takeovers and mergers appear increasingly to be the driving force behind the process, with international mergers particularly evident in the fast growth sectors of food, chemicals and electronics [Jacquemin *et al* (1989)]. And, much of this investment has been undertaken by large multinational corporations (MNCs) which, due to the increasing importance of takeovers predominantly of the horizontal kind, has resulted in a rise in seller concentration right across the EU market [Amin (1992)].

III. EXTERNAL CONTROL AND THE REGIONAL ECONOMY

If growth of external control appears to be an inevitable consequence of the process of integration at the European and global levels, what are the implications for the regional economy, particularly in Europe's peripheral regions? Is the location of control of any consequence for regional development?

The belief that the location of control does matter rests on the proposition that in externally-controlled establishments where there is a lack of local control, decisions taken outside the region may not, in the long run, be beneficial to the regional economy. Conversely, a regional economy might benefit from external ownership, in the long run, if there is the opportunity to draw on the resources of a large organisation which would not otherwise be available. The essential issue would appear to be whether the network of intra-company links that has developed, and as we have seen is continuing to develop, between regions and nations, provides an effective mechanism for distributing the benefits of integration, or whether external control must be viewed as a mechanism which fosters polarised and unbalanced regional growth.

The literature that deals with the consequences of external control for the regional economy essentially divides the outcome into two groups: company effects and/or regional effects, or alternatively, internal and/or external effects [Ashcroft and Love (1993)].

Company Effects

The range of potential effects hypothesised in the literature is sketched in Figure 1. While beneficial effects are posited, most researchers in this field are of the opinion, supported by limited evidence, that the long-run company effects are likely, on balance, to be harmful. Top management functions are likely to be eroded, income will be transferred to the parent, technological asset stripping may occur, key operational functions such as Marketing and R&D are likely to be transferred out, linkages to the local economy will be cut, product lines lost, and jobs either cut back or subject to a de-skilling process. These costs are held to be more probable and to more than outweigh the possible gain in new management practices, increased availability of finance, technology transfer, gain of functions, jobs and product lines.

The empirical evidence on the company effects of change in the location of corporate control is limited. The study by Ashcroft, Love and Scouller (1987) of the effects of the external takeover of Scottish companies between 1965 and 1980, which probably is the most comprehensive study of the matter to date, confirmed some but not all of the hypothesised company effects.

The following hypotheses suggesting *harmful* effects were *not* confirmed:

– greater probability of company or plant closure;
– technological asset stripping, or indeed any asset stripping;
– excessive transfer of income to parent;
– reduced labour skills as skilled labour is replaced;
– market-power effects. There was no evidence that increased concentration led to a restriction of supply.

The following hypotheses suggesting *favourable* effects were *not* confirmed:

– improved quality of workforce due to improved training;
– access to new technology;
– new products and increased product development.

A few hypotheses suggesting *favourable* effects *were* confirmed:

– increased availability of finance;
– new management practices to some degree, and improved financial techniques more generally;
– improved growth and expansion of acquired companies;
– access to wider markets (but not a strong effect).

Several hypotheses suggesting *harmful* effects *were* confirmed:

– reduction of autonomy in acquired companies, although this often appeared to be a requirement of improved performance;

– rationalisation of product lines. This did occur, but was not a strong or general effect;

– reductions in key management functions such as marketing and R&D. The most frequent result here was no change; but, where change did occur, functions tended to be reduced. Moreover, while a majority of the companies studied in detail experienced no change in the number of senior posts and promotion opportunities, when change did occur there was a net reduction;

– reduced local linkages. The majority of case-study companies did not change their material linkages to the Scottish economy, but where change did occur there was a net reduction. There was a general reduction in service linkages, particularly those involving professional services;

271

– acquired firms experienced an increase in their capital intensity which was probably a reflection of increased investment in acquired companies which occurred as a result of takeover. Employment opportunities did not worsen, however, because of the expansion of output which followed.

This research identified the acquired company effects of external takeover after making great efforts -both statistical and via interviews- to control for what would have happened in the absence of takeover. The same can not be said for several other empirical studies in this area.

Overall, the research concluded that, on balance, the effects of external takeover were favourable to the performance of the acquired companies. The 'harmful' outcomes were more relevant to the consequences for the wider regional economy.

Figure 1: Direct Company Effects of External Control

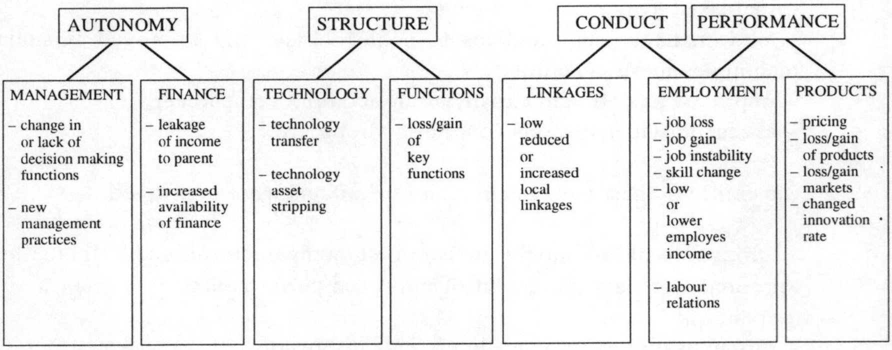

AUTONOMY		STRUCTURE		CONDUCT	PERFORMANCE	
MANAGEMENT	FINANCE	TECHNOLOGY	FUNCTIONS	LINKAGES	EMPLOYMENT	PRODUCTS
– change in or lack of decision making functions – new management practices	– leakage of income to parent – increased availability of finance	– technology transfer – technology asset stripping	– loss/gain of key functions	– low reduced or increased local linkages	– job loss – job gain – job instability – skill change – low or lower employes income – labour relations	– pricing – loss/gain of products – loss/gain markets – changed innovation rate

Regional Effects

One of the difficulties with the literature on external control is the failure to conceptualise the implications for regional development. It can be suggested that there are three requirements:

    – a model (or models) of corporate structure and behaviour;
    – conceptualisation of the effects on companies subject to external control; and
    – conceptualisation of the direct and indirect links between company effects and regional performance.

Much of the research on external control has focused on conceptualising the company effects on the basis of an implicit model of corporate structure and strategy and then, following the collection of evidence on company effects, making deductions about the implications for regional development. So, for example, Schackmann-Fallis (1989) in his important study of external control in the Trier region of Germany writes that:

> "... externally controlled plants will have a functional specialization which influences their regional effects. (p.248)..... (which)... lead(s) to several hypotheses regarding the regional effects of the different plant types."

And after study of the Trier region he concludes that:

> "... the results indicate that the process of decentralization in manufacturing has helped to improve quantitative employment figures in the peripheral regions, but the decentralization of multiplant enterprises has increased the qualitative and functional disparities among regions...... the quantitative deconcentration was accompanied by a functional concentration that was reinforced by the multiplant enterprises. *It disfavours the peripheral regions in the long-run and will affect their indigenous growth potential.*" (p.259 - Italics added).

Schackmann-Fallis offers 5 hypotheses on the company effects of external control that he deems relevant to their regional effects:

    – employment quality will be lower;

    – average income per employee will be lower;

– job stability will be lower as branch plants are used as a buffer during the business cycle;

– intra-regional input and output linkages for materials and business services will be lower; and

– indirect effects on regional technology and innovation diffusion will be relatively low.

However, he offers no indication of their relative importance to regional development, nor how such largely company effects will affect the long-run growth performance of the regional economy and no mention is made of the potentially favourable effects of external control. Moreover, it is clear that the implicit model of corporate structure and behaviour is, to use Amin's phrase, that of the 'mass production economy' which in Amin's view is symbolised by

".... the large, vertically-integrated and centrally-controlled corporation, with its strategic functions and growth-inducing external linkages in core metropolitan regions, and its 'lower-order' functions with negligible external multiplier effects located in less favoured regions." [Amin (1992), p. 142]

Amin and others argue that corporate behaviour is very different today from that in the mass production era. It is suggested that corporations today are more loosely structured, less hierarchical, more likely to decentralise a full range of functions along a value-added chain in each major market, and more likely to encourage local supply networks [Amin (1992)]. The growing sophistication of consumer demands with increased desires for product specificity and differentiation, increased competitive pressure in product delivery and development, are, it is argued, leading corporations to seek to get closer to the consumer and focus on their 'core' business as specialist functions are contracted out to local suppliers. These developments, which still require general empirical confirmation at the regional level, clearly have implications for the regional effects of external control, although I doubt whether they vitiate the insights of the literature. What they imply is that the effects of external ownership and control on the regional economy are likely to be mediated more through dominant customer and supplier relationships via the market rather than within firms' internal hierarchies. Moreover, while these new developments may be occurring on a global scale as companies decentralise to major continental and national markets, it seems to this author somewhat improbable that there will be decentralisation to the same degree at the sub-national and regional scale. In addition, if these new developments are to be associated with sectoral specialisation rather than intra-company functional specialisation across space, then many peripheral regions are likely to lose out absolutely. There have been too many 'false dawns' in regional economics and policy and this looks to be one of them.

These comments should not be taken to imply a negative view of Schackmann-Fallis' work; the implicit assumption of corporate structure and behaviour as vertically-integrated and centralising is supported by his findings and was confirmed in our work on the effects on Scottish acquired firms of external takeovers. Nevertheless, it does perhaps highlight the need to think formally about the corporate processes underlying industrial change, at the very least to ensure that the full range of hypotheses are available to inform empirical work on external control.

Of greater significance in Schackmann-Fallis' work is the effective assertion that his evidence for Trier: low local linkages, narrow and unsophisticated function structure, and a bias towards unskilled and less skilled occupations in externally-controlled companies, will be damaging to regional development. There are strong reasons for believing this could be true but the impact on the regional economy may be more complex than the above findings imply. Without a thorough attempt to conceptualise the possible effects, we cannot be certain *a priori* which effects are likely to be of most significance for regional development and the process through which development effects occur. This latter is especially important because what is being posited here is a relationship between a micro phenomenon: external control, and a macro economic outcome: regional development. It is unlikely that such a relationship can be detected simply by analysing the behaviour of some appropriate macro-economic aggregate such as regional GDP. At this macro level it is exceptionally difficult to demonstrate in any feasible time period that changes in the aggregate are the result of a diffuse process of change at the micro level.

Conceptualising the Effect of External Control on Regional Development

In the present paper no attempt is made to offer a full conceptualisation of the relationship between regional development and external control. What is done here is to suggest that it is principally the supply potential of the regional economy, particularly its entrepreneurial capacity, that is the key to the concern about the effect of external control on regional economic development.

There are several approaches to the explanation of regional growth but it is perhaps not too much of an oversimplification to suggest that all stress, in one way or another, the importance of the availability and quality of the resources present in a region. Demand driven growth models such as the cumulative causation approach highlight the key role of productivity changes, while supply-based models focus on factor supply, real factor prices, and technical change. Understanding the role of external control in regional development requires an appreciation of first, the

275

size and nature of any demand stimulus or contraction, second, how the quantity and quality of resources available to a region are affected, and then thirdly, the microeconomic routes through which resource change affects regional growth.

Figure 2 provides a simple schematic diagram of the potential links between external control and regional development.

Figure 2: External Control and Regional Development

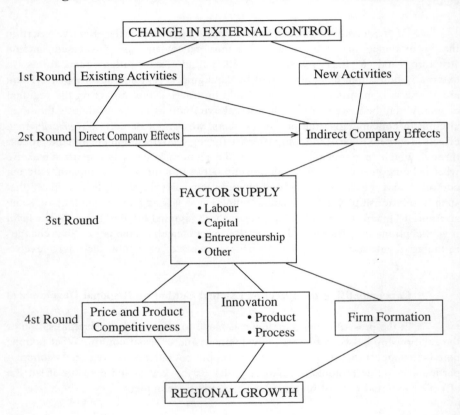

(i) The First-Round Effect

The first point to note is that in considering a change in external control, an important distinction should be drawn between external control that occurs through the change in the location of ownership of existing activities and that which occurs as a result of the introduction of new activities. In the latter case, inward investment unambiguously introduces new products, new capital, creates additional jobs and perhaps transfers in new technology. The *first round* effect is clearly positive. Moreover, while an inward investment, other things remaining equal, produces a once-and for-all effect on the *level* of regional output and jobs, there will be a direct, positive and sustained effect on regional growth if the new activity grows more quickly than the average performance of existing regional activities. Inward investments are frequently in the fast growth sectors such as electronics, chemicals, and food. Moreover, evidence on the generally stronger growth performance of incoming, and particularly foreign, firms, even after controlling for industrial structure, is provided by Schackmann-Fallis (1989) for Trier and by McNie (1983) for Scotland. However, jobs growth in the foreign owned sector in Scotland deteriorated sharply in the 1980s [Young, Peters and Hood (1993)].

Change in the location of ownership and control of existing activities, on the other hand, is neutral in the *first round*: only a change of title occurs.

(ii) Second-Round Effects

In the *second round*, the direct company effect of inward investment is neutral since it represents an activity new to the region. At this stage the effect of inward investment on the regional economy will be indirect through other companies. The effect will be positive if linkages are created to other firms in the region and negative if other regional company sales are displaced through increased competition in domestic and extra-regional markets. To the extent that increased competition to domestic firms occurs, this could be beneficial in the long-run, if not the short term, as domestic firms are forced to become more efficient.

Conversely, in the *second round,* takeovers and other changes in the location of ownership of existing activities may produce direct changes in the autonomy, structure, conduct and performance of the regional company as listed in Figure 1 and discussed above. Competitive effects on other regional companies may be present in much the same way as might occur following an inward investment, if the performance of the acquired firm is improved as a result of takeover. Alternatively, the acquired firm's link to the acquirer, as with the inward investors links to its parent, may be damaging to other regional companies through anti-competitive actions such as cross-subsidisation, tie-in sales and reciprocal buying. Such restrictions will be of little or no long-term benefit to regional

277

development and are more than likely to be damaging. Further effects occur in other regional companies if, as the evidence suggests, domestic material and service linkages are broken and switched outside the region. However, what Schackmann-Fallis and others have failed to note is that there are likely to be favourable demand effects on other regional companies following say a takeover of a regional company even when linkages are broken.

Ashcroft, Love and Scouller (1987) found that the sales performance of acquired firms in Scotland improved as a result of takeover leading to increased purchases on this account from other regional companies. Love (1990) took this finding a stage further and tried to estimate the *net* effect on the demand for regional material and service inputs resulting from takeovers in the Scotch whisky industry. Love found using an input-output model of the Scottish economy that the economy-wide effects of changes in the level of sales in the acquired company following takeover were greater, and possibly very much greater, than the effects from reduced Scottish linkages. This was the case regardless of the direction of the sales effect itself. So, where the sales effect is positive, and this was generally the case for acquired Scottish firms, positive multiplier effects tend overall to outweigh negative linkage effects on other regional firms. This finding should not to be taken to suggest that the reduced linkages effects of takeover are unimportant, rather that the net demand effects on other regional companies can still be, and in Scotland appear to be, positive even with linkage reduction. What appears to be important about linkage change is not so much the demand effects, although these cannot be discounted, but the sectoral distributional effects of linkage reduction. The sectors affected appear to be concentrated in the high-growth, high-entrepreneurship parts of the local economy.

Demand effects are clearly an important consequence of changes in external control. This is particularly the case with inward 'greenfield' investments where new products often in fast-growth sectors are brought into the economy. Takeovers are found to stimulate sales growth directly perhaps by offering access to wider markets. Growth is likely to be stimulated as a result. However, it is our contention that the direct and indirect company effects of changes in external control will be of greater significance for the future performance of the regional economy if they also affect the availability and quality of the regional economy's resources. The effect on the region's factor supply conditions -defined broadly- is identified in Figure 2 as the *third round* of the link between external control and regional development.

(iii) Third-Round Effects: Impact on Factor Supply

Again a distinction must be made between inward investment and increased external control of existing activities. On the face of it, inward investment adds to the factor supply position in a region. Additional capital will be brought in, possibly new technology, and new managerial, organisational and production practices. The potential negative effects of inward direct investment on a region's factor supply position are more subtle. The possibility that incoming plants may exhibit low local input and output linkages, low rates of technology transfer and innovation, job instability, and poorer job quality and lower employee-income levels than domestic regional activities, are not sufficient conditions for damage to regional development. It can be suggested that such investments will only diminish the supply potential of the regional economy and damage future regional development prospects if at some future stage the claim made by the incomer on the region's resources is less beneficial to regional development than otherwise would be the case. The necessary condition for damage is that resources are bid away from other regional activities which would have used them more 'productively' for regional development.

In a peripheral regional economy experiencing high unemployment and general under utilisation of resources, the possibility of resource diversion to the external sector would appear less likely than in fully-employed economies. However, there are several routes through which resource diversion can occur even in economies with significant spare capacity:

– through competition -fair or unfair- and product displacement in the domestic regional market for goods and services;

– through competition in the regional labour market as the offer of higher wages and salaries diverts workers of varying skill profiles to jobs in the external sector;

– through competition in the regional capital market as the possibility of a higher return in the external sector diverts funds away from domestic activities;

– through competition for regional policy and local development agency funds leading to a diversion away from the domestic sector;

– through faster growth of the external sector leading to a greater proportionate claim on the region's resources as the above diversion mechanisms are brought more readily into play.

It can be suggested that it is only when the above mechanisms operate and resources are diverted to the external sector that the characteristics of incoming plants have the potential for affecting the quantity and quality of resource supply in the region and future development. Once this occurs then the potential for an effect on the resource base of the regional economy is formally equivalent to an increase in external control via takeover and the other routes noted earlier.

Our evidence from study of the external takeover of Scottish companies identified function loss and linkage severance as a key outcome of the process. Takeovers which lead to a changed requirement for certain functions to be performed within the acquired company and/or produce a change in the demand for local material inputs and services may precipitate significant supply-side responses within the regional economy. This appears especially likely when, as in most regional economies, the prices of factors of production are relatively insensitive to variations in local supply and demand conditions.

The evidence suggests that external takeovers of regional companies will tend to be associated with a net loss of functions to an acquirer headquartered in a core region of the national economy such as the South-East of England. Similarly, some of the acquired firm's local purchases, particularly of professional services, are likely to be switched to the acquirer's existing suppliers in the core region. Both the transfer of functions and the switch in suppliers will be motivated by the acquirer's desire to reap economies of scope and scale in production and in the purchase of inputs. These are more likely to be realised at the location of the acquirer if its activities are considerably larger than those of the acquired firm. In these circumstances, following the reduction in functions performed by the acquired company and the severance of local linkages, staff will be displaced in both acquired and supplying companies and the senior management remaining in the acquired firm will be subject to restrictions of varying severity on the scope of their decision making.

Displaced staff, including displaced senior management from the acquired firm, might either migrate to positions elsewhere in the merged group or take up employment in other companies. Those who regain employment in the region of the acquired company may in turn displace other staff of similar skills and so on, so that, as with a decrease in the demand for local labour, there will eventually be an impact on regional unemployment. Generally, in these circumstances one should expect a supply response in the form of increased net emigration and unemployment. Moreover, the quality of labour supply and skill lost to the regional economy is likely to be greater the more takeovers are concentrated in, and linkages diverted from, the high growth, high-entrepreneurship sectors of the economy [Love (1990)], as appears to have been the case in Scotland.

A distinction should be made between managerial and other key staff displaced following a reduction in the functions performed locally in the acquired company, and displacement due to the replacement of staff by more efficient personnel who may be brought in from outside the regional economy. In the latter case, there should at worst be a temporary increase in unemployment and/or net emigration without the sustained increase that might follow if the number of posts and functions had been reduced. Future entrants into the local labour market would not be faced with diminished job opportunities, so any supply response would probably be zero or minimal. Moreover, the acquired company would have gained through existing functions being performed more efficiently.

Senior management remaining in the acquired firm will, in varying degrees, face restrictions on their autonomy. These restrictions can be properly seen as a threat to the supply potential of the economy since they may limit the scope for entrepreneurship and managerial discretion compared with the independent firm. Changes in the skill levels of certain employees in the acquired company might also occur following external acquisition. A reduction in functions could require staff remaining in place to work at a lower skill level, with the result that previously acquired skills atrophy and disappear. In our study of the external takeover of Scottish firms we found no evidence of de-skilling [Ashcroft and Love (1993)]. However, it is a point worth bearing in mind, especially since incoming plants, particularly branch plants, are found on average to contain a higher proportion of 'blue-collar' workers (Schackmann-Fallis, 1989). If the external sector eventually claims an increased share of the regions resources so that there is effective diversion away from the domestic sector then, other things equal, this will produce a de-skilling effect at the level of the regional economy.

There is also a set of non-market and extra-company effects following increased external control of existing regional activities which can affect both the quality of the region's labour supply, business leadership and entrepreneurship. These are difficult to verify empirically. One typical benefit often cited relates to the transfer of superior technical and managerial skills from an acquirer through the acquired firm to other firms in the region. The transfer of information might occur through face-to-face contacts with management and staff of the acquired company at local conferences, Chamber of Commerce meetings, educational institutions, trade associations, technical councils, local enterprise networks etc. A less direct transfer might be effected through the labour market as staff eventually move from the acquired company to other local firms. In contrast, it is frequently alleged that the status of key figures in the local business community might decline following takeover due to an expected reduction in their independence of action and thought. Leadership in the whole range of public and community affairs might be weakened with damaging consequences for regional development.

(iv) The Fourth Round: Impact on the Micro-Mechanisms of Growth

The above-mentioned effects on a region's supply potential of change in external control appear likely to affect regional growth and development principally through the following three micro-mechanisms or routes:

 – price and product competitiveness;
 – product and process innovation;
 – firm formation.

Price and Product Competitiveness

External control will *improve* a region's price and product competitiveness through the change in supply potential brought about by the direct introduction of new resources such as investment finance, physical capital, new technology, access to wider and more efficient distribution networks, importation of key workers and skilled personnel, new management practices, new production practices, and a greater commitment to employee training and welfare. These changes appear more likely to be associated with incoming direct investments rather than takeover or other forms of change in external control. Favourable indirect effects on the price and product competitiveness of other regional companies may also occur through the mechanisms noted above such as labour turnover, face-to-face contacts, and information transfer along the local supplier/customer chain.

External control will *harm* a region's price and product competitiveness through the change in supply potential brought about directly by de-skilling, rationalisation of product lines, and technological asset stripping. These changes appear more likely to be associated with takeover rather than incoming investment. Unfavourable indirect effects on the price and product competitiveness of other regional companies may result from anti competitive behaviour by the external sector, transmission of higher wage and salary costs, any diminution in the status of local business leadership, and reduction in the availability and quality of managerial and labour skills via the mechanisms discussed in the previous section.

We are not aware of any research on the effect of external control on the price and product competitiveness of peripheral regions.

Product and Process Innovation

Product and process innovation are increasingly recognised as key 'micro' ingredients in the regional growth process [Malecki (1991)]. They also can be viewed as a key aspect of price and product competitiveness but they are worth treating separately because of their perceived importance to the overall competitive advantage of regions and nations [Porter (1990)].

External control can *benefit* the regional innovation process because the regional firm may be able to draw on the resources of a large group much of which is located outside the region in question. It can be argued that externally-controlled plants are likely to be less isolated technologically and more linked to extensive information networks than independent regional firms. One variant on this view is that access to the greater resources of a larger group will facilitate the *transfer* and *imitation* of new products and *adaptation* of existing products much more readily than in independent firms, but not the *independent generation* of innovations.

External control may, therefore, *harm* the regional innovation process because the lack of autonomy in externally-owned plants inhibits entrepreneurship and managerial discretion resulting in a lower *generation* of innovations than in independent regional firms. This may also be reinforced if incoming externally-controlled plants are producing products in the later phases of their life cycle where innovation rates tend not to be strong [Markusen (1985)] and lack key operating functions relevant to innovation such as R&D and marketing. However, the presence of such functions may be of little relevance to on-site innovation if local managerial discretion is constrained by the policy of the group on the generation and first production of innovations.

Recent research by the present author and colleagues Ashcroft, Dunlop and Love (1993)] using survey data on self-assigned innovation, collected from 417 manufacturing firms producing in Scotland and employing 80,000, or 21.5% of Scottish manufacturing employment, sheds some interesting light on the role of external control in the regional innovation process.

Univariate statistical analysis indicated that:

– foreign-owned companies were more than twice as likely to be product innovators as rest of UK-owned firms and more than four times as likely as independent Scottish companies;

– for process innovations the pattern was broadly similar; while

– for organisational innovations, the differences in the pattern of ownership and innovation just failed to be statistically significant.

Application of multivariate statistical analysis (logit) to control for the simultaneous influence of several factors such as size, sector and presence of certain functions produced the following key results:

– there was a systematically positive effect of foreign ownership on the likelihood of a firm being a product innovator compared with independent Scottish firms;

– no systematic effect on product innovation was evident for rest-of-UK owned firms compared with independent Scottish firms.

However, when the statistical analysis was confined to the set of innovating firms and the rate of innovation per employee considered the multivariate analysis produced the following key findings:

– foreign and rest-of-UK ownership were associated with a lower rate of innovation than in independent Scottish innovators; and

– the negative effect of foreign ownership on the innovation rate was greater than that for rest-of-the UK owned firms.

These findings therefore offer a paradox. Scottish-owned plants are less likely to be innovators than externally-owned firms, but for those that do innovate their rate of innovation appears to be superior to firms in the external sector. However, other evidence from the survey and subsequent interviews with companies revealed that identification and development of an innovation in the external sector offered no guarantee of production in Scotland. Moreover, externally-controlled *branch plants* which constituted about 50% of the external sector cited an 'inappropriate corporate culture' as a barrier to innovation at the Scottish plant, reflecting the lack of local autonomy and corporate centralisation in the sector. Hence, once domestic Scottish firms achieve an innovation 'breakthrough' they may be able to keep on innovating, whereas plants in the external sector may successfully introduce one innovation but have no guarantee of being able to innovate in the future due to the absence of local autonomy.

A further point of relevance here is that while the probability of innovation was higher in the external sector many of the 'innovations' so defined by the companies were actually developed elsewhere in the group. So what we were observing in the external sector was frequently the production of new products rather than innovation *per se*. There is clearly scope for further research on the role of external control in the regional innovation process both in Scotland and other peripheral regions of the EU.

Firm Formation

During the 1970s and 1980s disillusionment with the role of the large externally-owned corporation in regional development led to increasing policy emphasis being placed on the need for regions to pursue endogenous growth. Policy attention therefore turned to local firms and particularly the small firm. The view that regional regeneration might lie with small firms was significantly supported by the research of Birch who in 1979 published findings from a vast survey of job creation in the USA which showed that 60% of *net* job creation could be attributed to small firms. However, the shift in policy attention towards new firm formation and small firm growth perhaps failed to recognise the link between the two and the activities of the external sector producing in the region.

External control might be expected to be *beneficial* to indigenous firm formation and small firm growth if spin-offs or spin-outs from large firms are not discouraged, and if links to local suppliers are extensive and reinforced by strong customer/supplier networking. However, there are substantial reasons for believing that external control will be *harmful* to indigenous new firm formation. First, links to local small firms have tended historically not to be strong, although this could change if the predictions of the proponents of the localisation view of corporate behaviour are fulfilled. Secondly, and perhaps most importantly, if external control affects the regional supply of managerial talent and skilled labour, following function loss, or low functional sophistication, and weak local linkages and/or linkage severance, then the firm formation rate in the region could be damaged.

Research on the link between external control and firm formation is lacking. Casual empiricism does suggest that regions with high levels of external control such as Scotland and the North of England do have low rates of new firm formation. However, there has been no attempt to explain directly the link to external control, largely because of data deficiencies in the measurement of external control at the UK regional level. Nevertheless, in recent research undertaken by the present author and colleagues, we sought to explain econometrically firm formation at the UK county level in terms of a range of structural variables, some of which were relevant to external control [Ashcroft, Love and Malloy (1991)]. The research indicated that Scotland's poor firm formation rate could not be accounted for by an adverse industrial structure and that differences in the rate across UK counties reflected differences in the socio-economic structure of each area. Of key relevance to the role of external control was the finding that the location of managerial skills had a positive effect on the birth rate of new firms. This finding has been subsequently confirmed by research in Germany [Fritsch (1992)]. While the link between external control and firm formation is far from proven and awaits further research, this finding is instructive because it is precisely displaced individuals with managerial skills who may be tempted to emigrate as a result of function loss and linkage severance following an increase in external control.

IV. CONCLUSIONS

This paper has sought to explore the consequences for the economic development of Europe's peripheral regions as increasing integration leads to an extension of the external control of regional activities largely through takeovers. Key points made in the paper were as follows:

– previous research has largely focused on the company effects of external control, especially takeovers, with the key findings from research by the present author and colleagues that acquired firms benefited from increased availability of finance, new management practices, access to wider markets and improved growth. Harmful effects were identified: reduced autonomy, a net loss of key functions such as R&D and marketing, and severance of linkages to the local economy, particularly in professional services. However, these latter effects were judged harmful more to the wider regional economy than to the acquired companies themselves which, on balance, were judged to have benefited from takeover;

– the literature on external control has failed to conceptualise the implications for regional development. In one key representative work they are simply deduced from a partial understanding of company effects on the basis of an implicit model of corporate structure and behaviour. The potentially favourable effects of external control also tend to have been ignored;

– a simple and partial conceptualisation of the effect of external control on regional development was offered. The link between the two could be decomposed into four rounds of effects. It was suggested that a key distinction should be drawn between the effect of external control via inward investment and that via other forms, principally takeover;

– the first-round effect of inward investment is clearly positive as new products, capital, jobs and perhaps technology are brought into the region. For takeovers the effect is neutral in the first round: only a change of title occurs;

– the second-round effects, which relate to the direct and indirect company effects of a change in external control, are for inward investment, neutral in terms of the direct effects, while the indirect company effects can be either positive or negative. For takeovers and other changes in the location of ownership of existing activities, the second-round direct effects can be positive or negative according to the nature of the changes in the autonomy, structure, conduct and performance of the firm. The evidence of Ashcroft, Love and Scouller (1987) suggests that, on balance, they will be positive. Indirect second round effects are likely to be negative on account of severance of linkages but may be positive if the sales performance of the acquired firm increases due to takeover. Many researchers have failed to note that favourable demand effects on suppliers might occur even when linkage severance occurs. However, even in these circumstances the evidence suggests that there are likely to be sectoral distributional implications, with linkage severance affecting the high-growth, high entrepreneurship parts of the economy;

286

– the third-round effects refer to the impact of changes in external control on the factor supply potential of the regional economy. For incoming investments, the necessary condition for damage to a region's supply base is that resources are bid away from other regional activities which would have used them more 'productively' for regional development. This condition does not appear to have been recognised in the literature. For takeovers *et al*, there would appear to be a high probability of damage due to function loss and linkage severance leading to a diminution of entrepreneurship and managerial discretion in acquired firms, and the potential emigration of individuals with managerial skills from companies directly and indirectly affected by the change. A set of non-market, extra-company effects was also identified which could affect the quality of a region's labour supply, business leadership and entrepreneurship;

– fourth-round effects concerned the impact of changes in the factor supply position on the 'micro' mechanisms of growth: price and product competitiveness, product and process innovation and new firm formation;

– incoming investments were seen as likely to improve a region's price and product competitiveness due to the introduction of new resources. Takeovers were considered likely to harm it. No research appears to have been undertaken on the link;

– external control appears unfavourable to the generation of innovations, but favourable to the transfer, imitation and adaptation of new products. Recent research by the author and colleagues does not fully support these hypotheses. Externally-controlled firms were found more likely to be innovators than independent Scottish firms. However, Scottish innovators had a higher innovation rate than firms in the external sector;

– it is expected that, on balance, external control will be harmful to firm formation. Recent research on new firm formation provides evidence of the importance of the location of managerial skills. This is suggestive of a link to external control and takeovers in particular because of the expected emigration of displaced individuals with such skills, following function loss and the severance of linkages to the wider regional economy.

It should be clear that the relationship between external control and regional economic development is complex, indeed more complex than suggested here. Further increases in the external control of economic activity in Europe's peripheral regions seem inevitable as European integration proceeds. While the effect of external control on Europe's regions is by no means unambiguous, there is sufficient in the theory and evidence to expect *prima facie* that the process will serve to dilute the supply potential of the recipient regions and retard the 'micro' mechanisms of growth. Takeovers appear to offer the greatest potential for harm at the wider regional level, although not necessarily for the acquired firms. Inward investment, on the other hand, brings new resources to the region and will only damage regional development if such investments bid resources away from other regional activities and use them less 'productively' for regional development. This is indeed a possibility because of the characteristics of such investments and because of competitive effects on domestic regional firms via product, labour and capital markets, particularly as inward investments tend to be in sectors which are growing at faster than average rates.

There is a message in all this for EU policy makers. First, the use of regional policy to achieve cohesion by the stimulation of the supply potential of peripheral regions through the promotion of entrepreneurship, small firms and technology transfers, is commendable, but unlikely to be of sufficient scale to offset the centralising tendencies of the integration process. Secondly, the fortunes of the small-firm domestic sector cannot be separated from the structure, corporate strategies and performance of the external sector in peripheral regions. A policy is required for both. Thirdly, policy must recognise the inevitability of increasing external control of peripheral regions and seek to maximise the benefit to the domestic sector from the process. Encouragement of mobile inward investments still appears realistically to offer one of the best ways forward to secure movements in the direction of cohesion. However, policy must find ways of encouraging mobile investments which ensure that their characteristics are 'productive' for regional development. This requires encouragement of a full range of functions, decentralised decision making, and strong local supplier links. In short, EU regional policy must find ways of influencing the corporate strategies of internationally mobile firms. Finally, takeovers appear to be increasingly in the vanguard of the integration process. Against this background it is absurd that EU mergers policy takes no account of the potential regional effects of prospective takeovers and mergers. It is therefore a matter of some importance that the EU accept the recommendation of Love (1993) that:

> "... genuine concerns relating to the effects of merger on the economic cohesion of the Community should explicitly be regarded as grounds for investigation by the European Commission." (p. 19)

This is not to suggest that companies in the peripheral regions should be 'ring fenced' and takeovers prohibited, because to do so would ignore the potential benefits from external ownership that might be secured by the acquired firms. Rather the Commission should be prepared to trade off the potential for benefit to the acquired firm against the potential for harm to the wider regional economy. And, in operational terms, the potential for harm largely reduces to the issue of headquarters

288

location and function loss. Without such a change, the Commission will be continuing to address the symptoms of regional imbalance while steadfastly ignoring a principal cause.

REFERENCES

AMIN, A. (1992), "Big firms versus the regions in the Single European Market" Ch. 6 in Dunford and Kafkalis (eds), **Cities and Regions in the New Europe,** Belhaven Press, London.

ASHCROFT, B.; COPPINS. B.; RAESIDE, R. (1994), "The Regional Dimension of Takeover Activity in the UK", **Scottish Journal of Political Economy**, Vol. 41, No. 2.

ASHCROFT, B.; DUNLOP, S.; LOVE, J.H. (1993), **Innovation in Scottish Manufacturing Firms**, Scottish Foundation for Economic Research and Glasgow Caledonian University, Glasgow.

ASHCROFT, B.; LOVE, J.H. (1993), **Takeovers, Mergers and the Regional Economy**, Edinburgh University Press, Edinburgh.

ASHCROFT. B.; LOVE, J.H.; MALLOY, E. (1991), "New firm formation in the British Counties with special reference to Scotland", **Regional Studies**, Vol. 25, pp 395-409.

ASHCROFT, B.; LOVE J.H.; SCOULLER, J. (1987), **The Economic Effects of the Inward Acquisition of Scottish Manufacturing Companies 1965 to 1980**, ESU Research Paper No 11, Industry Department for Scotland, Edinburgh.

FRITSCH, M. (1992), "Regional differences in new firm formation: evidence from West Germany", **Regional Studies**, Vol. 26, pp 233-241.

HOOD, N.; YOUNG, S. (1982), **Multinationals in Retreat**, Edinburgh University Press, Edinburgh.

JACQUEMIN, A. *ET AL* (1989), "Horizontal mergers and competition policy in the European Community", **European Economy**, No 40, (May), CEC Directorate-General for Economic and Financial Affairs, Brussels.

LOVE, J.H. (1990), "External takeover and regional linkage adjustment: the case of Scotch whisky", **Environment & Planning A**, Vol. 22, pp101-118.

LOVE, J.H. (1993), "EC mergers regulation, economic cohesion and the public interest", **Strathclyde Papers in Economics**, 93/1, University of Strathclyde, Glasgow.

MCNIE, W.M. (1983), "Employment performance of overseas-owned manufacturing units operating in Scotland 1954-77" Statistical Bulletin No A1.1, Industry Department for Scotland, Edinburgh.

MARKUSEN, A. (1985), **Profit Cycles, Oligopoly and Regional Development**, MIT Press, Cambridge, Mass.

PORTER, M. (1990), **The Competitive Advantage of Nations**, Macmillan, London.

SCHACKMANN-FALLIS, K.P. (1989), "External control and regional development within the Federal Republic of Germany" **International Regional Science Review**, Vol. 12, No 3, pp 245-262.

YOUNG, S.; PETERS, E.; HOOD, N. (1993), "Performance and employment change in overseas-owned manufacturing industry in Scotland 1980-90", **Scottish Economic Bulletin**, No 7, pp 29-38.

13 EUROPEAN MONETARY INTEGRATION, ENDOGENOUS CREDIT CREATION AND REGIONAL ECONOMIC DEVELOPMENT

Sheila C. Dow
Department of Economics
University of Stirling

I. INTRODUCTION

The purpose of this paper is to consider the implications of European monetary integration for peripheral economies within Europe. The view represented in the European Commission's own research is summarised as follows:

'As regards the regional distribution of the impact [of EMU], which is relevant to the objective of longer-term convergence of economic performance, *there are no a priori grounds for predicting the pattern of relative gains and losses.* There are risks and opportunities of different types affecting both categories of regions. Policies are already at work to reduce locational disadvantages of the least favoured and geographically peripheral regions. However, the key to the catching-up process lies in obtaining synergies between Community and national efforts to upgrade the least favoured regional eocnomies. The fixing of clear policy objectives, such as for the single market and EMU, are also highly relevant here for mobilizing such efforts' [Commission for the European Communities (1990), p. 12, emphasis added].

This paper will focus on the implications of EMU for the financial sector, and for its influence on peripheral regions. The focus will thus be on the single market for financial services and the coordination of national monetary policy, and bank supervision, within a new European institutional structure. The currency unification and fiscal coordination features of EMU will not be addressed directly.

The premiss of the argument to be developed is that credit creation is relevant to regional economic development. Then EMU will affect regional economic development in this sense insofar as it affects credit creation. Consideration will be given not only to the supply side, and the possibilities of credit-rationing, but also to the demand-side, and firms' and households' preferred

X. Vence-Deza and J. S. Metcalfe (eds.), Wealth from Diversity, 293–305.
© 1996 *Kluwer Academic Publishers. Printed in the Netherlands.*

financial structure, with its implications for firms' propensity to invest and consumers' propensity to spend. The argument starts by examining the structure of the financial sector in Europe, and how it is likely to be affected by the aspects of EMU on which we are focussing here: increased competition, unified regulation, and a move to a unified institutional structure. The second stage of the argument is to analyse the implications for credit creation and its distribution. Here we will draw on Chick's (1993) stages of banking framework, which provides a stylised account of the evolution of banking systems and the macroeconomic implications of that evolution. Finally we consider the particular implications for peripheral regions. Risk is the key concept here; given the inevitable limitations on objective risk assessment, the way in which risk is assessed for peripheral regions is crucial. Given too that risk may be assessed differently for different classes of borrowers in peripheral regions, the resulting pattern of credit creation may in turn determine the industrial structure of each region. When we put these supply side considerations together with an analysis of the characterisitics of credit demand in peripheral economies, we see this conclusion reinforced.

It is concluded that, while banks have increasingly gained control over the volume of credit creation, they still choose to ration credit in aggregate, and to particular classes of borrowers. The process of European monetary integration may set in train forces which exacerbate the credit availability problems of peripheral regions. Particularly when consideration is given to the associated discouragement to credit demand, EMU is likely to impinge on these regions' economic development efforts.

II. EUROPEAN FINANCIAL STRUCTURE

The European Commission (1990) research report suggests that peripheral regions will benefit from improved cost and availability of credit as a result of the single European market:

> 'With full monetary union local banks will loose(sic) [local] monopoly power assuming that borrowers will have direct access to foreign banks, either locally established or not, after the opening of the domestic financial market. Besides this credit availability effect, the borrowers in peripheral countries will also benefit from the level of interest rates which very likely will stay below those prevailing in the region before monetary unification. These two effects - availability and lower price -will represent a clear benefit for the borrowers of lagging regions.

However, local banks may be disrupted because foreign banks will rapidly seize the best segment of the market leaving the local banks with the less performing borrowers. Besides, the new and open financial market will develop in a context of tighter financial discipline and stricter rules for credit granting which may crowd out the (marginal) borrowers of the lagging regions. The net effect is ambiguous. ... Naturally, the situation will disappear once the local bank[s] ha[ve] recovered and adapted to the new conditions.'

[Commission for the European Communities (1990), p. 225]

This quotation involves several presumptions about the financial structure which is likely to pertain in an integrated European financial system:

1. increased competition in which small local banks will initially be disadvantaged;

2. an eventual resumption of balance between large national and multi-national banks and regenerated local banks;

the eventual consequences being:

3. increased efficiency resulting in lower interest rates in peripheral regions;

4. reduced credit rationing, but lesser availability for high-risk borrowers in remoter regions.

The likely outcome for banking structure is most usefully discussed with reference to historical experience with increased competition in banking, either due to changing market conditions or changing regulatory environment. What that experience suggests is that increased competition initially encourages a proliferation of institutions, but eventually a concentration among relatively few banks. This was the experience for example of the Scottish and Canadian banking systems [see Dow and Smithin (1993), Dow (1990), respectively].

A review of the literature on Europe in the 1980s and 1990s by Bisignano (1992) suggests that concentration is the likely eventual outcome of European financial integration, although he predicts that fragmentation will persist for some time. Gentle and Marshall (1992) and Gentle (1993) identify a significant spatial element in the concentration in the financial services sector in the UK in the 1980s, which is anticipated to continue with further integration in the European financial sector. The EC (1990) prediction of a competitive reassertion by small local banks draws support from Branson's predictions of the emergence in Europe of a two-tier banking structure whereby local markets will still be served by local banks operating alongside large multinational banks [a predicton shared by Goodhart (1987)]. But what they seem to have in mind is local in the sense of national, which would still be compatible with concentration away from peripheral regions within nations. Further, the type of two-tier system they have in mind has much in common with that traditionally identified in the US [as analysed for example by Moore and Hill (1982), Harrigan and McGregor (1987), Dow (1987)]. Weisbrod (1991) explicitly uses the US case to argue that concentration will not occur in Europe. Yet financial liberalisation in the US has significantly increased the degree of concentration in banking, eroding the importance of the second tier of banking. But the concentration has not been spatial in the same way as the UK. Indeed, Santomero (1993) argues that the political and cost pressures which have led head offices to disperse operations spatially may well operate similarly in Europe. This could be to the advantage of peripheral regions in terms of employment and output although not necessarily in terms of credit availability.

Overall, then, the prediction that local banks will eventually regain their competitive position relative to the large national and multinational banks does not find much support from past experience or the balance of the relevant literature. But there is the suggestion of advantage to these regions in terms of more favourable terms for financial services. Certainly, ceteris paribus, increased competition and economies of scale in a multinational market would be expected to increase efficiency [see Bisignano (1992)]. But Eckbo (1991) argues that efficiency gains will not necessarily be passed on, due to increased concentration in European banking.

In addition, the experience of Britain as an already integrated financial market suggests that regional differentials may persist. McKillop and Hutchison (1990) identify two forms of segmentation present in the British banking sector: institutional and market segmentation. The first refers to the existence of local banks institutionally distinct from the national banks; the second refers to national banks having regionally distinct markets. As far as Europe is concerned the first would be the case anticipated by the EC, with the persistence of local banks. The second would be the case of concentration in large multinational banks.

In both cases, segmentation creates the opportunity for different interest rates and/or charges. These latter are becoming an ever-more significant element in bank profits and may be a means more appealing to the banks to differentiate between markets than interest rates as such. In the case of institutional segmentation, small local banks have less ready access to funds than large banks, as well as some local monopoly power in terms of credit availability, both of which would explain higher interest rates in remote regions. The observation that this is the case is consistent with the theoretical literature with respect to a two-tier regional banking market in the US. In the case of market segmentation, Hutchison and McKillop identify the higher costs to national institutions of operating in remote regions: costs of acquiring information about borrowers in the region, transaction cost premiums for remote regions on trading in extra-regional assets, and differential attitudes towards risk. Again, the outcome for remote regions was identified to be higher interest and/or other charges. The implication for Europe is that, even if a two-tier banking system emerges, institutional segmentation can be expected to lead to higher charges in remote regions, just as would the alternative of a single-tier system with market segmentation. If market segmentation can persist within a nation state like Britain, then it is even more likely to persist in Europe. Montgomery's (1991) analysis, based on the experience within Italy, supports this view. Regional differences in interest rates and/or bank charges are therefore likely to persist even within a financially integrated Europe.

At the same time as the financial sector is undergoing spatial concentration, its overall structure is also undergoing evolution. Banks have over the last decade faced increased competition from non-banks, while the policy of financial liberalisation has progressively broken down many of the regulatory distinctions between banks and non-banks. This process has been termed market structural diffusion [see Gardner (1988)]. Insofar as the large multinational banks have developed competitive strategies more quickly and successfully than small local banks in order to address the increased competition from non-banks, the tendency towards concentration in banking will be reinforced. The influence of financial liberalisation in Europe on credit creation in peripheral regions will thus be the outcome of two forces:

1. the spatial concentration of creators of credit

2. the effects of non-bank competition on the preferences of banks with respect to the volume and distribution of credit.

We examine these forces more closely in the next section.

III. CREDIT CREATION

The process of market diffusion is simply one stage in the evolution of banking systems. Chick (1993) has developed a framework which sets out a series of stages of evolution of banking systems. The particular significance of this framework for our purposes is its implications for credit creation. As banking systems evolve, inspiring more confidence in depositors, their capacity for credit creation increases. The introduction of a lender-of-last-resort facility by central banks, designed to ensure confidence in the banking system as its credit creation expands, further fuels that capacity. The struggle over market share among banks, and then with non-banks as financial liberalisation was introduced, promoted an ever-greater expansion of credit whose impetus came more from the financial sector than the rest of the economy. The consequence was a level of bank debt which could not be sustained by the rest of the economy, such that the banks have been curtailing their credit creation activity in order to improve the risk profile of their portfolios. The perception of the market of the scale of bad debts still held by banks has limited banks' capacity to raise capital, which in turn has curtailed their capacity to create credit.

The financial instability arising from these latest stages in banking evolution has also been identified by Davis (1992). Davis offers a persuasive analysis of the effects of financial liberalisation on the structure of banking in a wide range of countries. He identifies the direct effect of liberalisation as being financial instability due to an excessive extension of credit, which in turn was a consequence of increased market entry:

> 'the transmission mechanism between entry and instability includes features such as lower levels of information, heightened uncertainty over market responses, and herd-like behaviour among lenders, as well as the more general consequences of heightened competition in terms of prices and quantities' [Davis (1992), p. 239]

Banks (broadly defined) are still unique in the sense that only their liabilities are means of payment; this allows a capacity for credit creation not open to non-banks. But non-banks have increased their share of the credit market. First, the scope for funding in wholesale markets allows non-banks considerable latitude to extend credit, competing directly with the banks for business; financial liberalisation policy which has reinforced this trend may indeed be seen as a market-led policy. Second, the increased competition in credit markets has directed profit-seeking more towards financial services for which fees may be charged. This has reinforced the trend for banks increasingly to securitise debt in order to meet capital adequacy requirements. There is no particular niche for banks as opposed to non-banks in providing financial services as a result of regulatory differences. Nevertheless the long experience of banks of their customers and monitoring economic conditions provides them with a comparative advantage in information which allows them to compete successfully in providing financial services. Thus for example banks' experience with risk assessment gives them a comparative advantage in administering, underwriting etc securitised debt which may never appear on their books as credit. Here curiously the German universal banks may be at a competitive disadvantage. The close involvement of German banks with a narrow range of customers may not provide as broad an informational base as that enjoyed by British-style banks with more short-term involvement with a wider range of customers. More particularly, the market may favour capitalisation of banks whose profit base relies more on provision of services than returns on illiquid debt. The character of European banking is thus likely to shift more towards the British than the German style of banking. [See Eliasson (1991)].

Increased liberalisation in Europe will exacerbate this shift of focus with respect to credit creation among the banks. Central to the question of the actual structural outcome and its consequences for credit is the concept of risk: how systematic risk will be affected by increased competition given the supervisory structure likely to pertain, and how risk in peripheral regions will be assessed by large national and multinational banks. In spite of the process already underway of financial integration in Europe, differences persist in the regulation and supervision of banks. These differences have marked consequences for the structure of banks' portfolios and also for the market's evaluation of banks, and thus the banks capacity to capitalise their credit creation. Even in EMU, bank supervision is to be administered by national authorities, each with their own conventions and history. At the same time, the risk of moral hazard is likely to increase with a more concentrated European banking sector, threatening the stability of the sector as a whole [see Bisignano (1992)]. The need for regulatory and supervisory reform in Europe as part of the financial integration policy has been emphasised for example by Folkerts-Landau et al (1991) and Santomero (1991). Without such reform, greater scope is likely to be created for financial instability, with large swings in the volume of credit banks are willing and able to create.

This instability has its origins in the instability in banks' own perceptions of risk. Recent experience indicates the presence of 'disaster myopia' among banks [see Guttentag and Herring (1986)]. Not only was there an unconscionable delay in the 1980s between bank debts becoming bad and banks recognising the situation, but there was a similarly long delay before banks made that recognition public by writing off the bad debts. The source of the problem, which is insoluble, is that default risk cannot be fully quantified. Yet banks expand credit when default risk is perceived to be low, and contract when it is perceived to be high [see Minsky (1976, 1982)]. Further, given the limitations on objective risk assessment, banks use conventional methods to estimate risk for different classes of borrowers, so that there is systematic credit rationing for particular classes of borrowers [see Dow (1993)].

In the next section we consider the implications of the analysis for peripheral regions and their prospects for economic development.

IV. ECONOMIC DEVELOPMENT IN THE EUROPEAN PERIPHERY

The foregoing analysis predicts increased competitive pressures within European banking, adding to the increasing competitive pressures from the non-bank financial sector. Small local banks in peripheral economies are likely to find themselves increasingly at a competitive disadvantage. Because of small scale and previous experience of some degree of local monopoly, these banks are unlikely to have innovated at the same pace as the large national and multinational banks. Further, the pattern of innovation has been designed to protect banks from bad debt, so that small local peripheral banks will find themselves unduly exposed to default risk. These banks will accordingly find it difficult to raise capital. Even if they make progress towards holding onto their market, therefore, they will be vulnerable to take-over. As a result, peripheral economies are likely to find the banking sector dominated by large national and international banks.

Since these banks have a stronger base than the local banks, they are not so constrained in credit creation, so that opportunities which are perceived as profitable and low-risk will find greater availability of finance. Thus for example a natural resource boom in a peripheral region will find plentiful access to credit; yet the collapse of the boom will see an equally strong withdrawal of finance; more liquid alternative investments will be found in the financial centre, not in the periphery. Thus the financial instability predicted for the system as a whole is likely to be exaggerated for peripheral regions. This argument is developed more fully in Chick and Dow (1988).

More generally, credit availability to peripheral regions depends on banks' assessment of risk. Small local banks have the advantage of superior information; large national and international banks are likely to have limited information on peripheral economies. Given that risk must be estimated partly on a qualitative basis, personal contact is the most effective means for gathering information. Yet the cost of doing so for the large banks may be prohibitive relative to the expected gains from the resulting lending. Thus for example, while the banks have improved their capacity for assessing sovereign risk, they do not have detailed country-by-country knowledge, and have only solved the problem of debt exposure to countries of which they have inadequate knowledge by relying on the IMF which does not face the same limitations.

Of potential borrowers in peripheral regions, the large banks will have most information on large national and multinational corporations; indeed these are the corporations which already use the large banks rather than the small local banks. Indeed many large corporations have internalised their provision of financial services to limit their own risk exposure, thus acting as a competitive threat to the large banks themselves. For both reasons, large corporations do not face the same credit limitations as small firms, and thus increase their competitive advantage. The financial constraints on small firms have been explored in terms of informational problems by Fazzari et al (1988). To the extent that small firms are crucial to peripheral regions, the arguments can be applied directly to these regions, with more force on account of spatial peripherality. Further, as Smith and Walter (1991) point out, a concentration of financial resources further fuels merger and acquisition activity in the industrial sector. For both reasons therefore, the process of financial concentration is thus likely to encourage industrial concentration. This too has a spatial dimension which disadvantages peripheral regions [see Gentle and Marshall (1992)]. The existence of financial constraints on small firms has already been recognised by the EC, which has established programs to alleviate them [see Henderson (1993), p. 277].

So far we have concentrated on the supply of credit. But demand for credit also has a regional dimension. One of the statistical difficulties in identifying regional credit constraints is that there is an interdependence between supply and demand. Thus peripheral regions with relatively poor economic prospects will be characterised by defensive financial behaviour [see Dow (1992)]. Relative low wealth levels and past experience of economic vulnerability (including incidence of credit rationing) encourage a tendency to hold relatively high proportions of liquid assets (ie assets issued in the financial centre) and to be reluctant to incur debt. [This pattern has been identified for the regions of Canada in Dow (1990), and for Scotland relative to the rest of the UK in Dow (1992)]. This behaviour is rational at the micro level but irrational at the macro level. Defensive financial behaviour fails to achieve its object if the result is less local consumer and investment expenditure, which further weakens the domestic economy. This further encourages a concentration of activity in production by large national and multinational corporations to meet demand in external markets; this activity already has more ready access to finance.

V. CONCLUSIONS AND POLICY RECOMMENDATIONS

The prospects painted here for peripheral economies are rather bleak. To the extent that the retrenchment of the banks following their overenthusiastic expansion slows down the process of financial integration, the processes described here will work more slowly. What has been suggested here is that the European Commission is right to qualify the argument that financial integration will increase credit availability to peripheral regions. But it has also been suggested here that the qualification is more complex than the EC (1990) report implies. First, the notion that high risk borrowers in peripheral regions will have reduced access to credit implies that risk is an objective concept. But the riskiness of a project may be perceived quite differently by the potential borrower in the peripheral region on the one hand and the large external bank with limited information on the other. The high risk perceived by the bank may arise directly from the lack of information, not from the nature of the project as such. Second, even if the peripheral region receives on average as much credit as before, this credit may be much more volatile than before, due to the instability of risk assessment based on limited information. Third, that credit may be extended more to large national and multinational corporations than when credit was allocated more by local banks. As a result the very character of the peripheral economy may change, with increased dependency on outside corporations and all that goes with it.

What can be done to counteract these tendencies is somewhat limited by the other features of European integration, such as the fiscal limitations on member governments and the single market policy. Nevertheless there is a growing awareness that the economic convergence necessary for EMU will require a system of fiscal transfers among members. The reasoning parallels that which has long been applied to the case for regional policies within nations.

As far as the banking sector itself is concerned, measures to encourage small local banks to innovate in order to find a market niche, together with measures to protect against take-over, are necessary to halt the initial stage of concentration; once the banking sector has concentrated, reliance must be placed on the much more problematic possibilities for regional direction of credit. At the same time, the future of a local banking sector depends on the level of confidence in the local economy, which in turn relies on an effective industrial policy. If, and it is a big if, the weakness in confidence can be reversed, then there is hope for overcoming the defensiveness of local financial behaviour. Without financial resources staying in the local economy, and without demand for credit to finance local investment coming forward, local banks are relatively powerless to encourage economic development.

REFERENCES

BISIGNANO, J. (1992), "Banking in the European Economic Commmunity: Structure, Competition, and Public Policy" in Kaufman, G.G. (ed.), **Banking Structures in Major Countries**. Boston, Kluwer.

BRANSON, W.H. (1990), "Financial Market Integration, Macroeconomic Policy and the EMS" in Brago de Macedo, J.; Bliss, C. (eds.), **Unity with Diversity within the European Economy: the Community's Southern Frontier**. Cambridge, Cambridge University Press.

CHICK, V. (1993), "The Evolution of the Banking System and the Theory of Monetary Policy" in Frowen, S.F. (ed.), **Monetary Theory and Monetary Policy: New Tracks for the 1990s**, London, Macmillan.

CHICK, V.; DOW, S.C. (1988), "A Post Keynesian Perspective on Banking and Regional Development" in Arestis, P. (ed.), **Post Keynesian Monetary Economics**. Aldershot, Elgar.

COMMISSION FOR THE EUROPEAN COMMUNITIES (1990), "One Market, One Money: An Evaluation of the Potential Benefits and Costs of Forming an Economic and Monetary Union", **European Economy**, 44 (October).

DAVIS, E.P. (1992), **Debt, Financial Fragility and Systemic Risk**. Oxford, Clarendon.

DOW, S.C. (1987), "The Treatment of Money in Regional Economics", **Journal of Regional Science**, 27(1), pp. 13-24.

DOW, S.C. (1990), **Financial Markets and Regional Economic Development**. Aldershot, Avebury.

DOW, S.C. (1992), "The Regional Financial Sector: A Scottish Case Study", **Regional Studies**, 26(7), pp. 619-31.

DOW, S.C. (1993), "Horizontalism: A Critique", University of Stirling mimeo.

DOW, S.C.; SMITHIN, J.K. (1992), "Free Banking in Scotland, 1695-1845", **Scottish Journal of Political Economy**, 39(4), pp. 374-90.

ECKBO, B.E. (1991), "Mergers, Concentration, and Antitrust" in Wihlborg, C. et al (1991).

ELIASSON, G. (1991), "Financial Institutions in a European Market for Executive Competence" in Wihlborg, C. et al (1991).

FAZZARI, S.M. ET AL (1988), "Financing Constraints and Corporate Investment", **Brookings Papers on Economic Activity**, Vol. 2.

FOLKERTS-LANDAU, D. ET AL (1991), "Supervision and Regulation of Financial Markets in a New Finanacial Environment" in Wihlborg et al. (1991).

GARDNER, E.P.M. (1988), "Innovation and New Structural Frontiers in Banking" in Arestis, P. (ed.), **Contemporary Issues in Money and Banking**. London, Macmillan.

GENTLE, C.J.S. (1993), **The Financial Services Industry: The Impact of Corporate Reorganisation on Regional Economic Development**. Aldershot, Avebury.

GENTLE, C.J.S.; MARSHALL, N. (1992), "The Deregulation of the Financial Services Industry and the Polarization of Regional Economic Prosperity", **Regional Studies**, 26(6), pp. 581-5.

GOODHART, C.A.E. (1987), "Structural Changes in the British Capital Market" in **The Operation and Regulation of Financial Markets**. London, Macmillan.

GUTTENTAG, J.M.; HERRING, R.J. (1986), "Disaster Myopia in International Banking", **Essays in International Finance**, no. 164, Princeton University.

HARRIGAN, F.J.; MCGREGOR, P.G. (1987), "Interregional Arbitrage and the Supply of Loanable Funds: A Model of Intermediate Financial Capital Mobility", **Journal of Regional Science**, 27(3).

HENDERSON, R. (1993), **European Finance**. London, McGraw-Hill.

MCKILLOP, D.G.; HUTCHINSON, R.W. (1990), **Regional Financial Sectors in the British Isles**. Aldershot, Avebury.

MINSKY, H.P. (1976), **John Maynard Keynes**. London, Macmillan.

MINSKY, H.P. (1982), **Inflation, Recession and Economic Policy**. Brighton, Wheatsheaf.

MONTGOMERY, J.O. (1991), "Market Segmentation and 1992: Toward a Theory of Trade in Financial Services" in Wihlborg, C. et al (1991).

MOORE, C.L.; HILL, J.M. (1982), "Interregional Arbitrage and the Supply of Loanable Funds, **Journal of Regional Science**, 22(4), pp. 499-512.

SANTOMERO, A.M. (1991), "The Bank Capital Issue" in Wihlborg, C. et al (1991).

SANTOMERO, A.M. (1993), "European Banking Post - 1992: Lessons from the United States" in Dermine, J (ed.), **European Banking in the 1990s**, (second edition). Oxford, Blackwell.

SMITH, R.C.; WALTER, I. (1991), "The European Market for Mergers and Acquisitions" in Wihlborg, C. et al (1991).

WEISBROD, S.R. (1991), "Comment" on Smith and Walter (1991) and Eckbo (1991) in Wihlborg, C. et al (1991).

WIHLBORG, C.; FRATIANNI, M.; WILLETT, T.D. (eds.) (1991), **Financial Regulation and Monetary Arrangements After 1992**. Amsterdam, North Holland.

14 BANKING AND FINANCE IN THE ITALIAN MEZZOGIORNO: ISSUES AND PROBLEMS[74]

Marcello Messori
University of Cassino
(Italy)

I. INTRODUCTION

In this paper I analyze the specificity of the workings of capital markets, and in particular the behavior of the banking system in the Italian *Mezzogiorno*. As is well known, the South is the least developed part of the Italian economy: it represents a marginal area of an industrialized economic system, and can therefore be considered an industrialized area in itself. I maintain that, beyond some negative institutional features (mainly, the heavy economic role played by illegal organizations such as the *Mafia*), the *Mezzogiorno*'s mix of marginality and industrialization represents an extreme case of the growth paths followed –or to be followed– by other areas in the European pheripheries. On the other hand, the development of finance and banking institutions and their (active or passive) role are crucial in the development of the economic systems[75]. Hence, my analysis aims to highlight the financial constraints which hinder the economic evolution of Southern Italy and, possibly, of other European marginal areas. To this end, I stress some of the more general and theoretical aspects.

The distinguishing characteristics of the workings of capital markets in Southern Italy can be summarized in three elements: (i) the level of average interest rates on credit supply is higher in the *Mezzogiorno* than in the other Italian economic regions; (ii) the impact of credit rationing is stronger in the South than in the North of Italy; (iii) financial markets, which are poorly organized in Italy (with the exception of Treasury bond markets), play an even more negligible role in the *Mezzogiorno* than in the other Italian economic regions.

There have been a number of empirical analyses of elements (i)–(iii), in

---

[74] I gratefully acknowledge Dr. Damiano Silipo's collaboration in compiling the tables and figures of the paper. I am also grateful to MURST for financial support (grant 40%).

[75] The links between financial and economic development have been emphasized for some time by Gerschenkron (1962), Goldsmith (1969) and McKinnon (1973); a more recent empirical analysis of these links is offered by King–Levine (1993) (see also the related bibliography). The (active or passive) role of financial institutions has been investigated, among others, by Cameron (1967). Moreover, both these problems are analyzed by Chick-Dow (1988).

X. Vence-Deza and J. S. Metcalfe (eds.), Wealth from Diversity, 307–342.
© 1996 *Kluwer Academic Publishers. Printed in the Netherlands.*

particular of the first two [e.g. Banca d'Italia (1990), Messori–Silipo (1991), Faini-Galli–Giannini (1992)]. They have proposed the following arguments. In Southern Italy, the default risk of borrowers is very high and banks' balance sheets ratios are somewhat unsatisfactory (in particular, there is a too low ratio of bank's own capital to total liabilities and a too high ratio of deposits to total liabilities). The last situation persisted over time because of the extreme segmentation of the credit market in the *Mezzogiorno*. This segmentation has not been overcome by the competitive pressure which could have been exerted in recent years by the financial markets (see iii above) and by the Special credit institutions (*Istituti di credito speciale*). A negligible part of the Southern firms finance their activities by means of the stock or bond markets, and the financial portfolio of Southern wealth owners includes a share of bonds and equities which is largely lower than the Italian mean[76]. The result is that the exit of many inefficient local banks from the market has been blocked.

According to this interpretation, the determining factors of Southern Italian banks' behavior are: (a) the high risks due to insolvent borrowers; (b) the inefficient liabilities management; and (c) the extreme segmentation of the credit market. It is difficult to weigh the relative importance of these three factors. Faini et al. (1992) offer one of the most sophisticated attempts to solve the problem. They point out that (c) (and, possibly, b) ease the establishment of long–term relationships between local banks and firms, and that (a) is strengthened by a significant lack of information on the part of the banks (asymmetry of information). The segmentation of the Southern credit market and the related long–term relationships create the possibility for borrowers to be "informationally captured" by the local banks [cf. Sharpe (1990), also Stiglitz (1987b)]; and, because of the distribution of default risks due to (a), information asymmetry can lead to Southern borrowers over–investing and therefore can imply an inefficient credit allocation in the Southern credit market [cf. De Meza–Webb (1988)].

I could criticize Faini et al. (1992) for some theoretical inconsistencies of their paper[77], but instead let me confine the discussion here to the results of their econometric tests. According to the evidence reached by these tests, the long–term relationships between local banks and firms and the asymmetry of information in the Southern credit market are linked to the crucial role the selection and default risk of borrowers play in determining the behavior of the banking system. I do not follow this econometric approach here since I maintain that the clue to assess the relative

---

[76] I do not concern myself here with the difficult task of explaining the inefficient relationships between the behavior of Special credit institutions and commercial banks in Southern Italy. See on this, Messori–Silipo (1991), Faini et al. (1992).

[77] It is evident that the concept of "informationally captured" does not easily fit with the "signaling models" which justify overinvestment. Moreover, the origins of information asymmetries which lead to overinvestment [see also De Meza–Webb (1987)], are incompatible with those which explain credit rationing [Stiglitz–Weiss (1981), see also: Milde–Riley (1988)]. Nevertheless, Faini et al. (1992) refer to the results of De Meza–Webb as well as to those of Stiglitz–Weiss.

importance of factors (a)–(c) is the interdependence of these factors and, above all, the influence which the behavior of the Southern banking system itself exerts on factor (a).

The paper takes up this interdependence and, in particular, the links between the behavior of the Southern banking system and the default risk of Southern borrowers in three steps. First, I stress that, in order to analyze the interdependence of factors (a)–(c), we must attempt to identify the critical drawbacks for the industrial development of the Southern economy. Secondly, I specify a possible role for the banking system in the *Mezzogiorno* which should overcome these drawbacks. It is worth noting that I analyze both these problems by turning to the theoretical suggestions provided by some strands of "New Keynesian economics" and "New institutional economics". Thirdly, I emphasize the reasons why the potential role the banking system could play to overcome the current drawbacks for development of the *Mezzogiorno*, is far from being realized.

In the first three sections I offer an empirical framework which analyzes factors (a)–(c) above through a number of proxies. In particular, section 1 quantifies the differentials of interest rates on credit supply between the South and the other Italian economic areas, and it presents some data which could explain the geographic differences in the borrowers' default risk; section 2 presents data on Southern banks' balance sheet ratios and make a comparison with the equivalent ratios of the banks in the rest of the Italian economy; and section 3 presents the data delineating the extreme segmentation of the Southern credit market.

Given this empirical framework, the remaining three sections of the paper stress and explain the interdependencies of factors (a)–(c). In particular, section 4 examines the crucial drawbacks which bind the industrial development of the Southern Italian economy, and section 5 investigates the potential role of the banking system in the *Mezzogiorno* which could help to overcome these drawbacks. This analysis, which defines the theoretically optimal role of the banks, suggests a new interpretation of the workings of Southern Italian capital markets. The crux of this re–interpretation, which is examined in section 6, is to point out the gap between the optimal and the actual behavior of the Southern banking system. The *Conclusions* stress some unresolved problems which require policy intervention and which cannot be tackled in this paper.

II. THE EMPIRICAL FRAMEWORK: THE BORROWERS' DEFAULT RISK

I begin with a reference to the average interest rates of short–term credit supplied by the banking system in three different areas of the Italian economic system: the North–West, which was industrialized early on, the Central–North–East, whose economy is centered on industrial districts of small firms, and the South, which is a marginal sub–system[78]. Table 1 shows that, in the last period (1975-1993), interest rates of short–term credit have been systematically higher in the *Mezzogiorno* than in the other two areas. In comparison with the North–West, the Southern differentials have swung from a minimum of 0.95 points (1981) to a maximum of 3.05 points (1985); and, in the last year examined, this gap has reached 2.18 points (2.01 points in the first year of our time series). In comparison with the Central–North–East, the Southern differentials have swung from a minimum of 0.39 points (1982) to a maximum of 2.83 points (1985); and, in 1993, it has reached 1.62 points.

Table 1.– Short–Term Interest Rates in Three Different Areas (North-West:NW; Central–North–East: NEC; South: S)

Years	NW	NEC	S	Italy	S–NW	S–NEC
7506	15.91	15.45	17.92	15.92	2.01	2.47
7606	17.51	17.33	19.45	17.63	1.94	2.12
7706	19.13	19.26	20.63	19.32	1.50	1.37
7806	15.80	16.27	17.95	16.19	2.15	1.68
7906	15.43	15.57	17.15	15.67	1.72	1.58
8006	19.80	19.84	20.96	19.95	1.16	1.12
8106	22.88	23.36	23.83	23.19	0.95	0.47
8206	22.90	23.61	24.00	23.34	1.10	0.39
8306	21.08	21.92	23.07	21.69	1.99	1.15
8406	18.95	19.23	20.81	19.29	1.86	1.58
8506	17.92	18.14	20.97	18.32	3.05	2.83
8606	16.11	16.63	18.63	16.71	2.52	2.00
8706	12.62	12.67	15.02	13.13	2.40	2.35
8806	12.81	13.32	15.11	13.37	2.30	1.79
8906	14.06	14.44	16.07	15.09	2.01	1.63
9006	14.12	14.67	16.15	14.66	2.03	1.48
9106	13.80	14.31	15.77	14.32	1.97	1.46
9206	14.24	14.63	16.06	14.73	1.82	1.43
9306	14.27	14.83	16.45	14.85	2.18	1.62

Source: Own calculations based on Bank of Italy data (Centrale dei Rischi).

My first question is: is the higher level of interest rates on credit supply in the *Mezzogiorno* partly due, besides other possible factors (see sec. 2 and 3, below),

---

[78] It is worth noting that the geographical distribution depends on the location of the declarant's bank branch.

to the higher default risk of Southern borrowers[79]?. It is difficult to assess directly the borrowers' default risk, therefore it is convenient to have recourse to a set of proxies. One of the most standard proxies is given by an indicator of the "quality" of the realized credit demand: the ratio of bad loans to total loans in the different areas of the Italian economy[80]. The reference to this ratio is particularly significant because, as shown by Pittaluga (1987), its fluctuations have greatly influenced the level of interest rates in Italy.

Table 2.– Bad Loans to Total Loans in Two Different Areas (Central–North: CN; South: S)

Years	S	CN	Italy	(Bad loans/Total loans)S- (Bad loans/Total loans)CN
1981	7.9	3.7	4.3	4.3
1982	8.1	4.3	4.8	3.8
1983	8.6	4.9	5.4	3.7
1984	9.0	5.2	5.7	3.8
1985	11.3	7.5	8.1	3.8
1986	13.3	7.5	8.3	5.8
1987	14.8	7.7	8.8	7.1
1988	13.2	6.4	7.4	6.8
1989	11.7	5.1	6.1	6.6
1990	11.0	4.6	5.5	6.4
1991	11.4	4.9	5.9	6.5
1992	12.5	5.0	6.1	7.5
1993	15.9	6.9	8.2	9.0

Source: Own calculations based on Bank of Italy data (Centrale dei Rischi)

Table 2 shows that, during the Eighties and the Nineties, the ratio of bad loans to total loans was constantly higher in Southern than in Northern Italy: in 1981 as well as in the last period (1988–'93), the value of this ratio in the *Mezzogiorno* was more than double the rest of Italy. This means that, during the past thirteen years, the geographic differential in this ratio followed an increasing trend correlated to the escalating increase in loans granted by the Southern banking system [see also: Fazio (1985), pp. 34 and 42; and D'Amico–Parigi–Trifilidis (1990), pp. 329–31]. As a result, in the first half of the Nineties, the worrying rise in the weight of bad loans

[79] As mentioned above, the impact of credit rationing is stronger in the South than in the North of Italy. However, unlike the interest rates on credit supply, it is almost impossible to estimate precisely the amount of credit rationed in equilibrium. This is the reason why I do not include credit rationing in the empirical framework. I should recall that high interest rates and strong credit rationing are probably due to the same cause.

[80] In Italy, the definition of "bad loans" includes not only loans under judicial proceedings but also loans granted to defaulting borrowers or to borrowers in structural economic difficulties. It is worth noting that, besides some sampling problems, the economic interpretation of Italian data on bad loans can be distorted by banks' different tendency to enter under this item only those loans under judicial proceedings. From the point of view of geographic location, these data refer to the borrowers' legal residence.

in the whole of Italian banking system, has particularly hit the Southern banks: in the last year examined, the geographic differential in the ratio of bad loans to total loans reached its maximum (9.00 points).

My previous observations above suggest that the borrowers' default risk could crucially determine the specificity of the behavior of the *Mezzogiorno*'s banking system. In order to verify this further, we need to compare the credit allocation with the scale of borrowers' activity in the different Italian economic areas. It is a common belief that, other things being equal, the smaller the borrower the riskier the bank loan. Moreover, according to several empirical analyses [see Siracusano–Tresoldi (1990), Giannola–Marani (1991)], the main indicators of economic profitability emphasize that the negative gap between South and the other Italian areas is more accentuated in the case of small firms than in the case of medium or big firms. Finally, if it is true that the ratio of bad loans to total loans is inversely correlated to the size of loans in Southern as well as in Northern Italy, it will also be true that the positive differentials of this ratio between the South and the North are particularly large for loans under 5 billion Italian lire [cf. Banca d' Italia (1989), pp. 196 and 293–4].

To assess the importance of the borrowers' default risk in explaining the specificity of the Southern banking system, it is thus significant that the economic structure of the *Mezzogiorno* has a large concentration of firms ranging from 1 to 9 employees; the number of medium and big firms is considerably lower than in the North–West and in the Central–North–East. It is even more noteworthy that the allocation of different size loans supplied by Southern banks, follows the dimensional structure of the Southern economy[81].

As shown in tables 3, in the second half of the Eighties and in the first half of the Nineties the branches of Southern banks recorded the highest proportion of loans below 5 billion Italian lire. These branches held around the 55%–60% of loans in that range (around the 26%–28% of loans below 500 million Italian lire) during the period 1987–1993. In contrast, the corresponding figures for the branches of North–Western banks were around the 30% (around the 13% below 500 million Italian lire). On the other hand, in the range over 50 billion Italian lire, the branches of Southern banks underwent a negative gap of around 26–30 points in comparison with the branches of North–Western banks. These data show that the allocation of the bank loans of different amount has a role in determining the different default risk of borrowers in the three Italian economic areas. This obvious result is underlined by two more interesting observations: (a) the differentials in the interest rates of short–term credit supplied by banks in these areas are inversely correlated with the unit amount of loans, and (b) the correlation between variations in the weight of bad

---

[81] The range of loans is determined by the amount of credit granted. The geographical distribution of borrowers is determined by the location of the bank branch where the operation takes place.

loans and changes in the interest rates is particularly high for the smaller loans

Table 3a.– Loans Allocation by Size in Three Different Areas (87–06)

Loan size (million of lire)	NW	NEC	S	S–NW	S–NEC
0-80	4.23%	6.83%	10.86%	6.63%	4.03%
80-249	4.97%	7.41%	9.39%	4.42%	1.98%
250-499	3.83%	5.31%	7.34%	3.51%	2.03%
500-999	4.70%	6.33%	8.53%	3.83%	2.20%
1000-4999	13.51%	14.78%	21.81%	8.29%	7.03%
5000-99999	6.31%	6.95%	7.23%	0.92%	0.28%
10000-49999	14.09%	13.37%	15.14%	1.05%	1.77%
50000 and over	48.35%	39.03%	19.70%	-28.65%	-19.33%
Total	**100**	**100**	**100**		

Table 3b.– Loans Allocation by Size in Three Different Areas (90–06)

Loan size (million of lire)	NW	NEC	S	S–NW	S–NEC
0-80	2.92	4.65	9.76	6.84	5.1
80-249	6.16	6.88	9.66	3.49	2.78
250-499	3.64	4.77	6.62	2.97	1.85
500-999	4.66	5.83	8.08	3.42	2.24
1000-4999	13.89	17.38	21.04	7.14	3.66
5000-99999	6.77	6.92	7.89	1.13	0.97
10000-49999	15.58	13.85	16.51	0.93	2.66
50000 and over	46.37	39.71	20.45	-25.92	-19.26
Total	**100**	**100**	**100**		

(a) deserves further analysis. Let me refer to some data on the trend of interest rates for short–term loans of different size (see tab. 4). The higher unit cost for smaller loans and the stronger competition for larger loans help to explain the inverse correlation between the size of bank loans and the level of interest rates. The data, from the second half of the Eighties to the first half of the Nineties, confirm such a correlation in each of the three Italian economic areas. However, the most interesting point is that, during this period (1987–'93), the differentials between the Southern and the North–Western interest rates decreased as the size of bank loans increased without systematic exceptions. This same result holds true if we compare the South and the Central–North–East.

Table 3c.– Loans Allocation by Size in Three Different Areas (93–06)

Loan size (million of lire)	NW	NEC	S	S–NW	S–NEC
0-80	3.07	4.43	11.29	8.22	6.86
80-249	5.56	6.58	9.63	4.07	3.05
250-499	3.46	4.19	5.87	2.41	1.68
500-999	4.39	5.14	7.13	2.74	2.00
1000-4999	12.97	14.05	19.62	6.66	5.57
5000-99999	6.32	6.24	7.86	1.53	1.61
10000-49999	14.95	13.28	15.69	0.75	2.42
50000 and over	49.29	46.09	22.91	-26.38	-23.19
Total	**100**	**100**	**100**		

Source: Own calculations based on Bank of Italy data (Centrale dei Rischi)

Table 4a.– Short–Term Interest Rates by Loan Size in Three Different Areas (87–06)

Loan size	NW	NEC	S	S–NW	S–NEC
0-99	14.78%	16.02%	17.44%	2.66%	1.42%
100-249	14.90%	15.60%	16.92%	2.02%	1.32%
250-499	14.66%	15.18%	16.60%	1.94%	1.42%
500-999	14.30%	14.63%	16.21%	1.91%	1.58%
1000-4999	13.44%	13.71%	15.19%	1.75%	1.48%
5000-99999	12.64%	12.83%	13.99%	1.35%	1.16%
10000-49999	11.97%	12.14%	13.65%	1.68%	1.51%
50000 and over	10.83%	10.99%	11.76%	0.93%	0.77%

Table 4b.– Short–Term Interest Rates by Loan Size in Three Different Areas (90–06)

Loan size	NW	NEC	S	S–NW	S–NEC
0-99	16.51	16.91	18.69	2.18	1.78
100-249	16.12	16.69	18.02	1.90	1.33
250-499	15.70	16.26	17.69	1.99	1.43
500-999	15.32	15.83	17.17	1.85	1.34
1000-4999	14.59	15.02	16.30	1.71	1.28
5000-99999	13.97	14.29	15.34	1.37	1.05
10000-49999	13.42	13.67	14.58	1.16	0.91
50000 and over	12.53	12.99	13.77	1.24	0.78

Table 4c.– Short–Term Interest Rates by Loan Size in Three Different Areas (93–06)

Loan size	NW	NEC	S	S–NW	S–NEC
0-99	17.04	17.48	19.17	2.13	1.69
100-249	16.47	16.90	17.96	1.49	1.06
250-499	16.04	16.42	17.83	1.79	1.41
500-999	15.58	15.91	17.34	1.76	1.43
1000-4999	14.84	15.26	16.65	1.81	1.39
5000-99999	14.20	14.61	15.85	1.65	1.24
10000-49999	13.68	14.01	15.11	1.43	1.10
50000 and over	12.70	13.20	14.52	1.82	1.32

Source: Own calculations based on Bank of Italy data (Centrale dei Rischi)

III. THE SPECIFICITY OF SOUTHERN BANKS' BALANCE SHEET

The higher interest rates of Southern banks on their loans, are matched by the difference between the unit return of the loans and the unit cost of deposits, both calculated in Italian lire[82]. For, as shown in table 5[83], during the period 1979–88

---

[82] Leaving out the bank's transactions in foreign currency allows us to analyze the different geographic profits of banks in the domestic market. Moreover, it is worth noting that the geographic differentials in the average interest rate on bank deposits have been, at least in the recent years, negligible.

[83] This table, as well as the following table 7, is not updated since it is taken from Marullo Reedtz (1990); however, related indicators are updated in table 6 below. Marullo Reedtz divides the Central–North–Eastern area into two sub–areas (North–East and Center) in order to separate the data of the large banks located in Lazio, from those of the

Southern banks achieved greater level of this difference in comparison to banks located in the other Italian economic areas. The relative benefits of Southern banks swung from a minimum of 0.5 points to a maximum of 2.3 points.

The difference between the unit return of loans (r) and the unit cost of deposits (i) in Italian lire indicates one measure of a bank's profitability, which is independent from the composition of the bank's balance sheet and which relates to more traditional and important banking transactions [see for instance: Banca d'Italia (1989), appendix, p. 165]. The margin of interest (Mr) with reference to the interest–bearing funds in Italian lire (F_r) is instead an indicator of a bank's profitability, which takes into account the returns of other assets (the bank deposits at the central bank, DEP_C; bonds owned by the bank itself, BO; net balances towards other banks, DEP_B) as well as the costs of other liabilities (mainly, financing from the central bank, FIN_C) in the bank's balance sheet. Hence, the margin of interest of any given bank depends on the composition of its balance sheet.

Let me continue to refer to a given bank where all transactions are in Italian lire. The margin of interest of this bank with reference to its interest–bearing funds can be expressed as:

$$(2.1.) \quad Mr/F_r = (r\,L + r_C\,DEP_C + r_F\,BO + r_B\,DEP_B)\,/\,F_r - (i\,DEP + i_C\,FIN_C)$$

$$/\,F_r \qquad\qquad \text{with } DEP_B \;\lessgtr\; 0;$$

where: L denotes the loans made by the bank under examination, and the different interest rates correspond to the various items.

Let me denote with r_a the unit return of interest–bearing assets, and with r_p the unit cost of interest–paying liabilities (PL) of the given bank. Hence (2.1.) can be simplified as:

$$(2.1.\text{ bis}) \quad Mr/F_r = r_a - r_p\,PL/F_r.$$

(2.1. bis) can be changed, in its turn, in:

$$(2.2.) \quad Mr/F_r = r_a\,NWC/F_r + (r_a - r_p);$$

where NWC denotes the net working capital of the bank considered, and hence it coincides with the bank's own liquid capital (i.e. with the difference between the bank's own total capital and the bank's technical and financial locks–up), which is equal to the difference between its interest–bearing funds and liabilities.

---

small banks located in the North–East. Moreover, Marullo Reedtz divides commercial banks from savings banks. Here, I disregard the latter distinction and I aggregate the different types of banks under the general heading of commercial banks.

(2.1.) and (2.2.) emphasize that, the difference between the unit return of loans and the unit cost of deposits in Italian lire being equal, the margin of interest with reference to the interest–bearing funds of a given bank is the higher the larger the weight of its interest–bearing assets with higher unit return [in decreasing order: loans, bonds, net balances towards other banks, bank deposits at the central bank: see Banca d'Italia (1989), appendix, p. 165] and the larger its net working capital.

In this respect, Southern banks suffer a drawback in comparison to banks located in the other Italian economic areas. During the period 1979–88, the average proportion of loans (net of bad loans) on the total assets has been very close in the balance sheets of the banking system located in the different Italian areas. However, during this same period, Southern banks have realized the lowest ratios of bonds owned to total assets in Italian lire; moreover, because of the lower ratios of bank's own capital and the higher ratios of deposits to the total amount of liabilities [cf. Banca d'Italia (1989), p. 297; Marullo Reedtz (1990), pp. 269–70], Southern banks have been compelled to maintain high mandatory reserves and, hence, have recorded the highest ratios of deposits at the central bank to total assets. Finally, Southern banks have undergone the fastest decrease in ratio of net balance towards other banks to total assets [cf. Marullo Reedtz (1990), pp. 256–7; see also: Onado–Salvo-Villani (1990), pp. 92–3]. These data can be specified and updated by comparing the shares of various interests and proceeds on banks' balance sheets in the three different areas. The higher interest rates of Southern banks on their loans bring about that the share of these interests on bank total assets is higher than the corresponding share on the total assets of Northern banks. As shown in table 6, the *Mezzogiorno* is the only area in which this share stayed always above the Italian mean during the period 1983–'92. The weight of loan interests in the Southern banks' balance sheet is so large that it countervails by itself the low value in the shares of many other items.

These data explain why the greater difference between the unit return of loans and the unit cost of deposits, enjoyed by the Southern banking system in comparison to the other Italian banks, did not translate into corresponding differences as far as the margins of interest were concerned. During the period 1979–82, the Southern banking system realized margins of interest on Italian lire transactions, which were above the national mean; nevertheless, during the period 1983–86 and –in particular– 1987–88, its margins drew very close (within a positive or negative range of few fractions of a point) to those of the banking system in the other Italian areas.

317

Table 5.– Bank Returns and Unit Costs (*) in Four Different Areas (North-West: NW; North–East: NE; Center: C; South: S)

	Years	Commercial Banks				Savings Banks			
		NW	NE	C	S	NW	NE	C	S
Loans Returnin	1979	17.9	19.4	17.4	19.1	17.4	17.5	18.4	17.7
Italian lire	1980-82	25.8	26.8	23.8	25.5	24.2	23.7	25.1	23.6
	1983-86	21.4	22.2	20.1	23.0	20.9	20.8	22.8	23.1
	1987-88	15.2	15.9	15.2	17.2	14.9	15.3	16.7	17.2
Deposits Cost in	1979	9.2	9.2	9.4	8.5	9.2	8.9	8.8	8.6
Italian lire	1980-82	12.1	12.4	12.1	11.0	11.9	12.1	11.5	10.7
	1983-86	11.0	11.1	11.0	10.8	11.1	11.3	11.0	11.7
	1987-88	6.8	6.8	6.9	7.0	6.8	6.9	7.1	7.3
Assets Return in	1979	13.1	13.8	12.8	13.2	12.1	12.5	13.0	13.0
Italian lire	1980-82	17.8	19.1	16.7	17.1	16.2	16.6	17.1	17.4
	1983-86	15.8	16.3	15.5	16.2	15.6	15.7	16.6	16.8
	1987-88	12.1	12.3	11.7	12.7	11.9	12.0	12.4	13.1
Assets Return in	1979	3.9	4.6	3.3	4.6	2.9	3.5	4.3	4.4
Italian lire -	1980-82	5.6	6.7	4.0	5.7	4.5	4.6	5.7	6.3
Liabilities Cost	1983-86	5.0	5.2	3.9	5.2	4.6	4.4	5.4	5.1
in Italian lire	1987-88	5.3	5.5	4.5	5.3	5.2	5.0	5.3	5.3

(*) Interest of loans and deposits are weighted by their respective balance items.

Source: Marullo–Reedtz (1990), p. 258.

The relative position of the Southern banks does not improve if profitability is measured by the so–called "margin of intermediation" on the Italian lire transactions. The margin of intermediation is determined by the sum total of the margin of interest and the net proceeds on non traditional banking services (i.e. the intermediation in bond or equity transactions). It is a commonly understood notion that, during the Eighties, the Central–Northern banks responded to decreasing growth rates in deposits and loans and to the related re–organization of financial markets by expanding the bond and equity transactions. The Southern banks, possibly prevented by the lacking of interest in the financial markets shown by Southern firms and wealth owners (see above), were out of step. Especially after 1983, the contribution to the margin of intermediation by bond and equity transactions has been systematically lower for Southern banks than for the other Italian banks (cf. tab. 7). This negative gap was not compensated for by the proceeds raised from other banking services. In particular, as shown in table 6, the *Mezzogiorno* is the only area in which the shares of the net proceeds from bond and equity transactions and from other banking services stayed always below the Italian mean during the period 1983–'92.

Table 6.– Bank Interests and Proceeds in Three Different Areas

	NW				NEC			
	1983	1986	1989	1992	1983	1986	1989	1992
Interest on loans	49.7	43.0	50.8	62.1	46.6	41.7	50.1	60.0
Interests on bonds and equities	27.7	25.6	19.8	15.6	31.3	26.4	21.6	17.0
Net proceeds from bond and equity transactions	4.5	7.3	5.8	4.9	4.3	6.9	4.9	3.6
Net proceeds from other banking services	8.6	11.4	11.0	8.2	8.3	11.7	10.7	9.1
Proceeds from tax collections	0.6	0.7	0.7	0.3	1.0	1.2	1.5	0.8
Other net proceeds	8.9	12.0	11.9	8.9	8.5	12.1	11.2	9.5
Total	**100**	**100**	**100**	**100**	**100**	**100**	**100**	**100**

	S				Italy			
	1983	1986	1989	1992	1983	1986	1989	1992
Interest on loans	53.8	52.9	58.2	66.7	48.8	43.8	51.5	61.8
Interests on bonds and equities	26.8	23.6	20.2	16.3	29.2	25.7	20.7	16.3
Net proceeds from bond and equity transactions	2.9	4.0	3.2	2.7	4.2	6.6	5.0	4.1
Net proceeds from other banking services	8.0	9.5	8.8	7.1	8.4	11.3	10.6	8.4
Proceeds from tax collections	0.3	0.2	0.3	---	0.8	0.8	1.0	0.5
Other net proceeds	8.2	9.8	9.3	7.2	8.6	11.8	11.2	8.9
Total	**100**	**100**	**100**	**100**	**100**	**100**	**100**	**100**

Source: Own calculations based on Bank of Italy data.

Table 7.– Contribution of the Margin of Intermediation.

	Years	Commercial Banks				Savings Banks			
		NW	NE	C	S	NW	NE	C	S
Proceeds	1979	12.7	7.4	8.3	16.2	28.6	6.9	8.7	5.8
from bond	1980-82	11.3	7.5	11.1	15.0	25.7	8.5	9.9	4.2
and equity	1983-86	15.1	12.2	22.1	9.6	14.8	12.5	10.0	7.5
transactions	1987-88	14.6	11.3	14.6	6.6	16.1	11.1	9.1	6.7
Proceeds	1979	13.0	10.0	16.5	9.1	16.5	18.0	11.3	10.0
from other	1980-82	13.0	9.4	16.5	10.3	13.9	17.0	11.1	10.4
banking	1983-86	12.5	9.3	16.4	14.6	13.0	14.5	11.2	9.7
services	1987-88	12.2	9.5	18.3	14.8	11.3	13.7	10.4	10.7

Source: Marullo–Reedtz (1990), p. 262.

This picture is worsened further because Southern banks bore the highest operating costs, largely due to their policy of high salaries and wages [see Marullo Reedtz (1990), pp. 263–8; Banca d'Italia (1989), pp. 295–6]. During the period 1979–88, Southern banks faced the greatest increase in the unit cost per employee; and in 1988, they in fact bore the highest level of this cost. Given that these same banks were the least productive, it follows that during this period the margin of intermediation of the Southern banks was hit by operating costs to a greater (and increasing) measure than other banks. I cannot enter here into explanations for this trend (for instance: the higher proportion of managers to employees in Southern banks; and the small size of the banks). It is sufficient to stress that, together with the dynamics of the geographic values in the margins of interest and in the margins of intermediation, during the period 1979–88 the high operating costs led to operating profits of Southern banks which were low and largely below those realized by the other Italian banks.

It is possible to conclude that the workings of the banking system in the *Mezzogiorno* has experienced more binding constraints and distortions than the banking system in the other areas. Nevertheless, it is difficult to assess the influence that these constraints and distortions exerted on the high interest rates on the credit supplied in the *Mezzogiorno*. To answer this issue, I must analyze the structure of the credit markets in the different areas of the Italian economy.

IV. THE SEGMENTATION OF THE SOUTHERN CREDIT MARKET

A precise definition of market segmentation could be based either on the degree of differentiation of the products supplied by competing activities, or on the number and size distribution of sellers –i.e. on the market degree of concentration. As to the segmentation of the credit market, I will single out the latter alternative; however, I will also refer to recent analyses of market structure such as the theory of contestable markets, rather than confining myself to the traditional concept of market concentration (see below). This choice makes it difficult to select a set of data which can assess directly the segmentation of the actual credit markets in the different Italian areas. It is therefore convenient to have recourse, as in the case of borrowers' default risk (see above, section 1), to a number of proxies in order to build up a reasonable picture by degrees.

I can already infer some provisional elements from the dynamics of the unit interest rates on short–term credit supplied by the banking system. It is important to underline that in the period examined (1975–1993), the differentials between the average interest rates in the *Mezzogiorno* and in the other two Italian areas have not followed a diminishing trend. Such differentials had instead pro cyclical oscillations in relation to the conditions of the national credit market. This is quite clear with reference to Southern and North–Western regions: as shown in Table 1, the differential between these two areas has widely increased in expansionary phases in which the level of average interest rates decreased (1977–79 and 1982–87), and it has declined in the recessive phases in which the level of average interest rates increased (1975–77, 1979–81, and 1988–'90). Analogous observations apply to a comparison between the Southern and Central–North–East regions.

This implies a greater stickiness in the average interest rates in the *Mezzogiorno* than at the national level and, in turn, such greater stickiness exposes the greater imperfections of the Southern credit market when compared to the Central–Northern areas. Based on the theoretical approach known as "structure-behavior–performance", D'Amico–Parigi–Trifilidis (1990, pp. 332–34) have attributed the larger imperfections of the Southern credit market to its high degree of concentration; for, according to this approach, the degree of monopoly depends precisely on the degree of concentration of the supply[84]. As mentioned above, my analysis also applies to other approaches such as the theory of contestable markets. The degree of contestability of a given market does not depend on the number of suppliers but, given some restrictive assumptions as to the incumbent's price changes [cf. Baumol–Panzar–Willig (1983)], on the potential freedom of entry and on the possibility to exit without costs. As a consequence, the imperfections of a given credit market could imply an excess number of banks with respect to the number

---

[84] The degree of concentration of the supply in a given credit market is traditionally determined by the number (exogenously given) of the banks present and by the distribution of their market shares with respect to the market's extension.

which is compatible with an efficient equilibrium[85]. It follows that, rather than measuring here the imperfections through the controversial link between the degree of concentration and the degree of monopoly[86], I maintain that such stickiness of interest rates in the *Mezzogiorno* is a first indicator of the strong segmentation of the Southern credit market with respect to the other two areas of the Italian economy.

A second indicator of the stronger segmentation of the Southern credit market can be deduced from the comparison between the morphology of the Southern banking system and the rest of the Italian banking system. If the imperfections of a given credit market can also depend on an excess number of banks, it will be important to assess the comparative numerousness and size of Southern banks. As shown in table 8 and figure 1, in the *Mezzogiorno* there is a scarce presence of small, medium and large banks and there is a poor impact of the loans granted by small and medium banks. On the other hand, in this same area there is probably an excess number of minor banks and an excess impact of their loans.

In particular, table 8 shows that, in the period 1978–'91 the percentage decrease of minor banks and the percentage growth of small and medium banks was higher in the Southern area than in the Central–North. Notwithstanding this, in 1991 the incidence of minor local banks in the *Mezzogiorno* was equal to 85.7% (95.6% in 1978) with respect to the total number of Southern banks; the corresponding percentage of minor banks in the Central–Northern region was 64.4% (71.3% in 1978); moreover, again in 1991 the small Southern banks were only 7 (57 in the Central–North), the medium Southern banks 3 (10 in the Central–North), and the major and large banks were 2 in the South (13 in the Central–North). This dimensional distribution was such that in 1993 the credit supply in the *Mezzogiorno* polarized because: more than half the bank loans in the area was supplied by the few major and large banks and around 20% was supplied by the minor banks (cf. fig.1). Instead, in the Central–North in 1993, the share of credit supplied by the major and large banks was around 40% and the share of credit supplied by minor banks was around 4%.

The polarization of the Southern banking system between two large banks (the Banco di Napoli and the Banco di Sicilia) and three medium banks on the one hand, and a plethora of minor banks on the other, is confirmed by an indicator even more synthetic which refers to the end of Eighties [cf. also Galli–Onado (1990), pp. 35–6]. In 1988 the Southern banking system held less than 15% of the national credit market. In the same year the first five Southern banks covered 8% of the national credit market, that is to say almost 60% of the market share held by the

---

[85] Cf. in general: Fudenberg–Tirole (1986); and for an extension to the credit market Di Battista–Grillo (1988).

[86] It is worth noting that one of the main criticisms addressed by the theory of contestable markets to the approach "structure–behavior–performance", is directed at this link [cf. Baumol–Panzar–Willig (1982)].

total Southern banking system, while the first five Central–Northern banks covered 26% of the national credit market, that is to say around 30% of that held by the total Central–Northern banking system. Considering the first 20 banks, the share increased by less than three percentage points in the *Mezzogiorno* (10.7%) but more than doubled in the Central–North (55.2%, equal to a little less than 65% of the share held by the total Central–Northern banking system).

Table 8.– Banks Allocation by Size in Two Different Areas

		1978		1988		1991	
South	Large	2	1.5%	2	1.9%	2	2.4%
	Medium	1	0.7%	3	2.8%	3	3.6%
	Small	3	2.2%	7	6.6%	7	8.3%
	Minor	130	95.6%	94	88.7%	72	85.7%
	Total	**136**	**100%**	**106**	**100%**	**84**	**100%**
Central-North	Large	10	3.7%	14	6.2%	13	5.8%
	Medium	15	5.6%	12	5.3%	10	4.5%
	Small	52	19.4%	58	25.8%	57	25.3%
	Minor	191	71.3%	141	62.7%	145	64.4%
	Total	**268**	**100%**	**225**	**100%**	**225**	**100%**

Source: D'Onofrio–Pepe (1990), and own calculations based on Bank of Italy data.

According to my previous definitions, I can maintain that these data converge to underline the extreme segmentation of the Southern credit market with respect to the rest of Italy. A further confirmation is the data from the calculations of the Herfindahl index which is carried out by the Banca d'Italia in relation to the deposit markets of around 1,500 sub–regions. This index ranges between 0–1 with the former value indicating the maximum of the traditionally defined competition and the latter value indicating the maximum of the traditionally defined monopolistic degree.

Figure 1

Loans by size in two different areas (Central_North: CN; South: S) (1993)

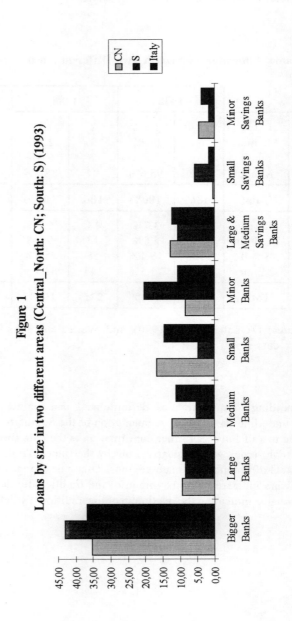

The data in table 9 refer to the distribution of the credit (deposit) markets in the *Mezzogiorno* and the Central–North among four classes of value in the Herfindahl index for the years 1979 and 1987[87]. In this last year 23% of the Southern sub–regions were still served by only one bank (in contrast to 8.2% in the Central–Northern sub–regions). Furthermore, between 1979 and 1987, the increase in the number of Southern credit markets with strong (traditionally defined) competition (0.00–0.30) was not greater than the corresponding Central–Northern markets; in 1987 the differential of the relative share remained equal to 7 percentage points (27% in the *Mezzogiorno* compared to 34.1% in the Central–North). Finally, in 1987 the analogous differential in the next class of value of the Herfindahl index (0.30–0.50) went beyond 10 percentage points.

Table 9.– Herfindahl Index of Concentration in Two Different Areas (1979-1987)

concentration index	CN		S		Italy	
	1979	1987	1979	1987	1979	1987
0.00-0.30	282	315	112	133	394	448
0.30-0.50	302	314	108	115	410	429
0.50-0.99	252	218	138	131	390	349
1-	34	76	115	113	199	139
Total	**920**	**923**	**473**	**492**	**1393**	**1415**

concentration index	CN		S		Italy	
	1979	1987	1979	1987	1979	1987
per cent allocation						
0.00-0.30	30.7	34.1	23.7	27.0	28.3	31.7
0.30-0.50	32.8	34.1	22.8	23.4	29.2	30.4
0.50-0.99	27.4	23.6	29.2	26.6	28.0	24.5
1-	9.1	8.2	24.3	23.0	14.5	13.4
Total	**100.0**	**100.0**	**100.0**	**100.0**	**100.0**	**100.0**

Source: D'Onofrio–Pepe (1990), p. 231.

It is worth stressing that the Herfindahl index is based on the "structure-behavior–performance" approach rather than on the theory of contestable markets. It is therefore meaningful, on the empirical ground, that the data from the

---

[87] This table is not updated since it is taken from D'Onofrio–Pepe (1990). See also: Banca d'Italia (1989), p. 287.

calculations of this index are in agreement with the previous data based on a different theoretical approach. Such data taken together prove that the credit market is more segmented in the *Mezzogiorno* than in the Central–North. Given the dimensional composition of the Southern banking system (see section 1), this means that the many minor Southern banks operate often in local protected sub–markets. The principal element of protection is secured by the limited economic dimensions of these sub–markets, which make the entry of large banks inconvenient and which are due to both the low unit amount of the operations and to the elementary character of the demands for services. The minor banks acquire, besides, a widespread knowledge of their local sub–market and can, therefore, reinforce the barriers to entry by setting up long–term relationships with their borrowers and building–up a "reputation" [cf. Sharpe (1990)]. The result is that these Southern credit sub–markets guarantee the permanence of inefficient local banks.

The setting up of exclusive long–term relationships between a lender and a number of local borrowers in the *Mezzogiorno*'s credit markets can receive an indirect but interesting confirmation by comparing the distribution of the number of loans contracted by borrowers in the diverse Italian economic areas. As shown in table 10, in 1993 82% of the Southern borrowers (around 79% in 1990) had debt relationships contracted with less than ten bank branches; and, again in 1993, a little less than half (49%) of these same borrowers had debt relations with no more than three bank branches (around 47% in 1990). On the other hand, in 1993 only half of the borrowers in the North–West (around 47% in 1990) and about 56% of the borrowers in the Central–North–East (about 54% in 1991) had debt relationships with less than ten bank branches; and, again in 1993, around 27% of the Central-Northern borrowers (less than 25% in 1990) and less than 35% of the Central-Northern borrowers (less than 30% in 1990) had debt relations with no more than three bank branches (cf. tab.10).

Table 10.– Number of Loans Contracted by Borrowers in Three Different Areas (1990–06; 1993–06)

Number of loans	NW		NEC		S	
	1990	1993	1990	1993	1990	1993
1-3	24.4	27.5	29.7	33.1	47.2	49.0
4-6	11.8	12.3	14.3	13.3	19.2	20.0
7-10	10.6	10.2	10.4	9.5	12.3	13.0
> 10	53.2	50.0	45.6	44.1	21.3	18.0
Total	**100**	**100**	**100**	**100**	**100**	**100**

Source: Own calculations based on Bank of Italy data.

It could be thought that the deregulation in the opening of new bank branches, as undertaken by the Banca d'Italia in the last years, succeeded in activating the opening of new branches above all in the *Mezzogiorno* and in such a way that it has mitigated, even if not overcome, the negative situation just described. In fact, at the end of the Eighties the rate of increase of the number of bank branches opened in the Southern area was grater than in the rest of the Italian economy. As shown in table 11, this increase was however, fed by the banks with headquarters in the Central–North: in the three years 1988–90 the local banks had in fact seen the fall of their share of branches in the whole of the *Mezzogiorno* of around 4 percentage points. These latter data are confirmed by the trend in the share of local bank loans in the total loans of the Southern Italian economy: this share decreased from around 68% to less than 59% in five years (1988–'93). Given the scarce propensity of the banks outside the *Mezzogiorno* to enter the protected segments of local banks [see above; Messori–Silipo (1991), pp. 78–80; and in general: Chick–Dow (1988)], this means that the increasing entry of the Central-Northern banks in the *Mezzogiorno* could have had little effect in reducing the degree of segmentation in the Southern credit market.

Table 11.– Allocation of Branches and Loans in Two Different Areas

	Years	S			CN	
		S. Banks	CN Banks		S. Banks	CN Banks
Branches	1978	85.8	14.2	100	1.7	98.3
	1988	78.7	21.3	100	1.7	98.3
	1990	75.0	25.0	100	1.7	98.3
Loans	1980	69.3	30.7	100	3.4	96.6
	1988	67.8	32.2	100	3.0	97.0
	1993	58.8	41.2	100	2.9	97.1

Source: Own calculations based on Bank of Italy data.

V. A POSSIBLE GROWTH PATH FOR THE MEZZOGIORNO

The empirical framework, sketched in the previous three sections, makes it plausible that the high default risk of Southern borrowers, the weakness of some items in the balance sheet of Southern banks, and the strong segmentation of the Southern credit market have determined the specificity in the behavior of the *Mezzogiorno* banking system. However, as I also emphasized in the *Introduction*, these empirical data neither enable us to assess the relative importance of each factor, nor to stress their possible interdependency, nor –most importantly– to point to the behavior of the Southern banking system as having a negative impact on these factors and, in particular, on the high default risk of Southern borrowers.

My thesis is that, in order to arrive at some conclusions from these observations, we need stronger theoretical tools. In order to take possession of such tools, we need to investigate the main drawbacks in the *Mezzogiorno*'s economic development and to identify the consequent contribution with which the banking system could overcome them. This section analyzes the main impediments to the development of the Southern economy in order to prepare, for the following section's discussion, the decisive role the banking system could be called upon to perform.

I maintain that the main difficulties which hinder the economic development of the *Mezzogiorno* are due to the area's inefficient relationship of the market with other economic institutions, and in particular with firms. This is mainly due to the heavy interference exerted on economic performance by various (legal and illegal) mechanisms of political and social regulation [e.g. Messori (1989)]. These mechanisms, combined with the economic weakness of the workings of the market, make the role of "market failures" crucial. In order to overcome these failures in market co–ordination, the development of the Italian Southern economy would need to be centered on productive and institutional innovations. However, these very failures impede further the realization of innovative behaviors in the market and in other economic institutions of the *Mezzogiorno*.

In order to suggest a way out of this vicious circle, we need to define precisely the determinants of these "market failures" from a theoretical point of view. In this respect, it is worth noting that the inefficient relations between mechanisms of political and social regulation, on the one side, and economic institutions, on the other side, imply that market failures in the *Mezzogiorno* include but cannot be reduced to the elements traditionally emphasized by standard economic theory (public goods, externalities, etc.). The failures of the Southern market refer rather to: economic situations characterized by uncertainty and private information; the consequent impossibility to govern economic relationships through contract designs which can account for a limited but important set of the possible future contingencies and for the relative binding behavior; and the consequent widespread practice of short–term and routine choices[88].

It is interesting to note that these last factors, though disregarded by traditional economic theory, are at the core of the new microeconomic approaches. In particular, I turn here to the strand in "New Keynesian economics" which investigates co–ordination failures due to market imperfections as explained by the concept of asymmetric information [Stiglitz (1987a), Greenwald–Stiglitz (1987)]; this approach emphasizes the role played by uncertainty with private information. I also

---

[88] In order to be more concrete, it would be useful to offer some examples from the *Mezzogiorno*. To this end, let me simply refer to Gambetta (1988).

look at the strand in "New institutional economics" which is concerned with the allocation of property rights [Grossman–Hart (1986), Hart–Moore (1990)]; this approach emphasizes the concept of "incomplete contracts", and hence the impossibility to solve an important subset of economic relationships by means of contract designs [Hart–Holmstrom (1987)]. In the New institutional economics another useful strand is"transaction costs economics" [Williamson (1975, 1985, 1989)], which emphasizes opportunistic behavior and the related difficulties to invest in specific assets.

Within the confines of this paper I am not able to analyze the major theoretical elements of these approaches nor can I seek to show their potential convergence towards a unitary scheme of work [see for instance: Kreps (1990), Currie–Messori (1993)]. All the more reason for my beeing unable to show that this possible unitary scheme could give an important contribution to the analysis of the Southern area. Hence let me look only at a case of market failure which could be particularly important in the *Mezzogiorno*: information asymmetry and incomplete contracts allow for opportunistic behavior, and thus hinder the realization of firms' investment in specific assets.

Incomplete contracts due to, for example, information asymmetries[89], mean either that the contracts cannot specify the actions to be undertaken and payments to be made in a large subset of the possible future contingencies, or that certain decisions and actions cannot be stipulated in terms of *verifiable* variables. In both cases, incomplete contracts allow the parties to a transaction to try and exploit their market position in order to fill –also "with guile"– the contractual "gaps" to their own benefit. This kind of behavior, which Williamson (1985) calls of "opportunistic" behavior, becomes particularly important in the case of investment in specific assets. Specific investment involves high "sunk" costs and, once the investment is made, its profitability depends on both parties staying in a bilateral long–term exchange relationship. The possibility of one party threatening to leave this relationship, due to opportunistic behavior, implies that the other party would not find it expedient to invest in specific assets [see Williamson (1985 and 1989), Hart–Moore (1988)]. On the other hand, specific investments do play a crucial role in the implementation of productive and institutional innovations able to overcome a number of failures in market co–ordination.

These theoretical observations apply to a large share of market transactions. However, they lock particularly up the working of Southern markets due to the inefficient relations between the mechanisms of economic and social regulation in this area (see above). Hence, they help to explain the vicious circle which binds the development of the Southern Italian economy. A possible question is, therefore, how

---

[89] It is worth noting that incomplete contracts can also be based on incomplete *but* symmetric information [cf. for instance: Hart–Holmstrom (1987), Hart–Moore (1988)]. Here I leave this problem to one side, and I assume that incomplete contracts are due to asymmetry of information.

to break out of the vicious circle caused by the lack of specific investments. It is well known that transaction costs economics proposes recourse to the co–ordination of firms instead of the co–ordination of the market [e.g. Williamson (1975, 1985)]. Several commentators, however, have recently pointed out that the boundary between the firm and the market is not so clearly determined; specifically, the typical costs of the workings of the market (*transaction costs*) may find a corresponding cost in the workings of the firm (i.e., *influence costs*) [see Milgrom-Roberts (1990)], both these costs may weigh on both organizations, and the efficiency of vertical integration may depend mainly on the optimal allocation of property rights [see Hart–Moore (1990)].

Besides such theoretical ambiguities, the proposals of new institutional economics do not fit the empirical situation of the *Mezzogiorno*. During the Sixties and the Seventies, government intervention in Southern Italy centered on financial incentives to supplement the investment of State–owned firms or private large firms located in the North of Italy. This policy led to the industrialization of Southern Italy but did not overcome market failures. The inefficient relations between the mechanisms of political and social regulation and economic performance, mentioned above, are also due to the government policy in support of the industrial firms. My conclusion is that we need to find potential growth path for the Southern Italian economy through institutions which function in between policy institutions, firms and the market. As pointed out in the contributions of the economic historians quoted at the beginning of the paper (see no. 1), the importance of the banking system becomes thus immediately apparent.

VI. THE POTENTIAL ROLE OF THE BANKING SYSTEM IN SOUTHERN ITALY

It is true that inefficiencies in market co–ordination and incomplete contracts are binding constraints also for the microeconomic behavior of the banking system. I argue, however, that these constraints underline that banks' role in the *Mezzogiorno* cannot be limited to a search for short–term financial and microeconomic efficiency. Southern Italian banks would have, in particular, to perform the Schumpeterian role of taking part in the economic development or, at least, a role which is not limited to a financial microeconomic function. Let me specify the point at a theoretical level before appraising its consequences for the *Mezzogiorno*.

Schumpeter [(1912), ch.3; (1939), chs. XI–XIII] maintains that bank credit is a "claim", issued on behalf of the community, which enables the borrowers to purchase before they sell. Hence, in the Schumpeter's model of economic development centered on entrepreneurial innovation, banks play the role of financing (some of) the innovative decisions undertaken by the entrepreneurs. Banks can fulfil this role by performing two essential functions: (1) building up an accounting system

330

[cf. in particular Schumpeter (1970)], and (2) screening the groups of prospective borrowers and stimulating some of their initiatives. These two functions entitle the Schumpeterian bank to act as an intermediary institution between firms and markets, and in particular to mitigate (some of) the more negative consequences of market failures due to risky innovations and to speculation. This becomes all the more true in a world of asymmetries of information and incomplete contracts. It is worth noting that Schumpeter did not sketch, even embryonically, the latter two concepts. Nevertheless, we can ascertain that, in this kind of world, (1) and (2) become pivotal activities in the sense that lending to an innovator acts as a positive signal for the other economic agents and stimulates imitations and/or other innovative projects.

This intuition is refined by Stiglitz and Weiss (1988). They elaborate on the Schumpeterian credit model in order to apply some of its results to the relationship between a number of informed borrowers and an uninformed bank. As is well known, at the very moment in which a standard debt contract is drawn up, the bank involved must issue an effective loan whereas the borrowers must engage themselves in a simple promise to pay the principal and the interests fixed by contract at a known but future date. Given this temporal lag, the presence of asymmetric information and related credit market imperfections imply not only that the borrowers may default at the date of the repayment but also that their rates of default are influenced by the credit contracts offered by the bank. This means that, because of "adverse selection effect" and "negative incentive effect"[90], the bank's expected profits do not grow as a monotonic function of the interest rates on its lending and, hence, credit can be rationed [see Stiglitz–Weiss (1981, 1992)].

Here I should specify the possible implications of asymmetries of information and credit rationing for the general behavior of the banking system. The obvious and important point is that each bank faces the following trade–off: (1) to search for all available but costly information on its borrowers in order to minimize the impact of adverse selection and negative incentive effects as well as the amount of credit rationing; (2) to use credit rationing and/or the terms of the (standard debt) contract as an imperfect screening device of borrowers in order to minimize the search costs.

Stiglitz and Weiss (1988) solve the trade–off between (1) and (2) by simplifying the problem. Each bank cannot mitigate the asymmetry of information by a costly search. Hence, it must optimize the information at its disposal and offer the best standard debt contracts to its borrowers. That is to say, in deciding on credit supply, each bank has to (i) gather the available information on its prospective

---

[90] By "adverse selection effect" I refer to a situation in which the terms of the contract (in this case, the interest rates) turn the "principal"'s (the bank's) preferred "agents" (the borrowers) out of the market, instead of the "agents" with the least desired characteristics. In the "negative incentive effect" (or moral hazard with hidden actions) the terms of the contract lead the "agents" to put aside the actions preferred by the "principal", and to take up instead the actions which carry higher risk. Moreover, it is worth noting that I refer here only to *ex ante* asymmetries of information; hence, I neglect the moral hazard with hidden information typical of the "costly verification" models [e.g. Townsend (1979) and Gale-Hellwig (1985)].

borrowers, (ii) group them in informationally homogeneous classes of risk, (iii) determine the set of interest rates which maximize its expected utility subject to the incentive constraints and the participation constraint[91], (iv) ration (some of) the prospective borrowers who carry highest default risk, (v) monitor the implementation of the financed projects. In so doing, a bank builds up an accounting system in order to determine the optimal allocation of its loans in the short and in the long period. The banks thus centralize information on the solvency of their prospective borrowers, which enables them to choose the activities to be financed, and pass on (some) information to non–bank agents through credit allocation. Moreover, because of the implied costs in (i)–(v), the banks find it expedient to establish long–term relationships with their non–defaulting borrowers.

In relation to the dissemination of information gathered and centralized by each bank, we should stress that such information is "credible" for at least three reasons: (a) as the banking system maintains relationships with a large number of borrowers, each bank is in a privileged position to assess the absolute and relative degree of reliability of its prospective borrowers; (b) when a given bank draws up specific debt contracts with a sub–set of prospective borrowers, the bank undertakes the risk to vouch for its own assessments; (c) since it is expedient for banks to establish long–term relationships with their non–defaulting borrowers, the loans granted by a given bank to a subset of its prospective borrowers signal that this bank is also betting on the long–term solvency and competitiveness of these borrowers.

Building on Schumpeter (1970), Stiglitz and Weiss (1988) synthetize this behavior of the banking system by stating that banks play the role of *social accountants* for credit allocation. Banks can choose to support the most reliable activities among those available in order to reduce the average default risk of the borrowers, but they can also stimulate innovations by screening a subset of new and promising activities. In both cases the credit allocated for the selected borrowers, by lengthening the expected temporal horizon of economic relationships, discourages opportunistic behavior and makes it expedient to invest in specific assets for the activities of these borrowers, and/or for related sectors.

The banks behavior as conceived by Stiglitz and Weiss is important as to the empirical problem raised at the end of the previous section: the determination of institutions which function in between policy institutions, firms and the market, and which can thus stimulate the economic development of the Southern Italian economy. If the main difficulties which hinder the development of the *Mezzogiorno* are due to the area's inefficient relationships of a poorly organized market with other economic institutions such as the firms (see section 4), the banking role of *social accountants* will be useful since it can introduce positive externalities through its

---

[91] These two constraints mean that the "agents" must find it expedient to sign the contract instead of choosing any other transaction ("participation constraint"), and that they must be stimulated to adopt those choices which are preferred by the "principal" ("incentive constraint").

credit allocation and the relevant inflow of credible informations to non–bank agents. As shown above, these externalities could ease firms' relations and the implementation of their specific investments and innovative projects even in the presence of lacks in market co–ordination.

This banks behavior cannot obviously be the panacea for the theoretical and empirical problems which hinder the efficient working of the Southern Italian economy. It is sufficient to note, in this regard, that imperfections in the credit market, due to asymmetries of information, and the consequent possibility of credit rationing, imply that the behavior of the banking system as *social accountants* does not ensure efficient credit allocation from the static point of view: the interest rate does not act as an equilibrating mechanism between credit demand and supply. It holds that each bank as *social accountant*, however, pursues efficiency in allocation as a necessary (even if not as a sufficient) condition for the realization of its operative goals.

My conclusion is therefore that, from an empirical perspective, to assign to banks the role of *social accountants* is equivalent to maintaining that credit allocation by banks is crucial to the definition of and evolution of prospective borrowers' economic decisions and actions. With respect to the banking system in a given geographic area, this implies that bank behavior heavily influences the formation and evolution of the area's economic specialization. The terms of standard debt contracts and possible credit rationing, set by the Southern banking system, therefore play an important role in determining borrowers' default risk, and the macroeconomic competitiveness and evolution of the *Mezzogiorno*'s economy[92].

VII. THE PRESENT ROLE OF THE BANKING SYSTEM IN SOUTHERN ITALY

The above analysis raises the following question: did Italian banks and, in particular, banks located in Southern Italy, perform in the *Mezzogiorno* the active and important role proposed by the analysis of Stiglitz and Weiss (1988)?. The empirical data in sections 1–3 suggest not [see also: Messori–Silipo (1991)]. Credit allocation follows and enhance the specialization of the Southern Italian economy, and hence it is concentrated in the financing of traditional projects and in that of small and local activities. Moreover the borrowers who were granted an important amount of credit in the *Mezzogiorno* in the recent past, realized the lower values in the indexes of profitability. This means that the banks, which are the lenders in the Southern credit markets, are not even exploiting their limited informations in order to optimally screen and stimulate their prospective borrowers in the short and in the

---

[92] These observations do not imply that the banks' role is exhaustive and monocausal. The interaction between monetary and real aspects is also influenced by the workings of the real economic system on bank behavior. Hence, the competitiveness and development of a given geographic area depend on an efficient credit policy as well as on a suitable industrial policy. Here, I neglect the latter element because it is outside the scope of this paper.

long period.

As indicated by the strong impact of credit rationing in Southern Italy and, at least partially, by the geographic differentials in the average interest rates on credit supply (see section 1), the lack in the screening and promoting procedure cannot be explained just by a peculiar loose policy of loans on the side of the banks in the *Mezzogiorno*. Various banks seem, instead, to have adopted defensive and routine policies in the Southern credit market. They utilized their screening devices not in order to support and stimulate those borrowers who could have contribute to the development of the Southern productive system, but in order to follow passively the inefficient market demand. There are numerous theoretical explanations for such passive behavior. Let me examine just three which can be applied to the empirical case of the *Mezzogiorno*.

A first explanation assumes that bank behavior in marginal areas is characterized by a high degree of risk aversion[93]. This assumption leads us to the conclusion that banks aim to minimize the short–term costs due to the economic weakness of the area and to the related high default risk of the local borrowers. Unfortunately, this implication does not fit with the empirical data on the ratio of bad loans to total loans and on the default rates: both these indicators are higher in the *Mezzogiorno* than in the rest of Italy (see section 1, above). Hence, the Southern banking system seems to have taken on remarkably high risks [see also: Fazio (1985), Faini et al. (1992)].

A second explanation considers not only the banking costs due to borrowers' default risk, but also the costs due to banks' liquidity constraint[94]. To elaborate this further, it is sufficient to refer to the models which connect asset and liability management[95]. In these models, the liquidity cost of a given bank is determined by the probability distribution that the ratio of bank reserves to bank deposits goes below a given binding value exogenously decided by the central bank. This probability distribution, in its turn, is the more unfavourable the lower the expected deposit reflux of the bank under examination, i.e. the higher the portion of the loans which this bank expects not to flow back as its deposits but to be transformed into legal tender and/or into deposits of competing banks. Zazzaro (1993) rightly stresses that, for a local Southern bank which is only able to attract deposits in its segment of the Southern credit market, the rate of deposits' reflux is inversely related to the rate of its loans granted to borrowers with economic links

---

[93] This assumption is theoretically suggested by several authors [see for example: Dow (1987), Marzano et al. (1983)] apply it to the Southern Italian banking system.

[94] The importance of liquidity constraint for an interpretation of the behavior of the Southern Italian banking system is emphasized by Zazzaro (1993). I should acknowledge that, until now, I have neglected this aspect. The reasons for which will become clear below.

[95] A review of the literature on this subject, published before (1980), is offered by: Baltensperger (1980). For some of the more recent models, see: Tobin (1982), Prisman et al. (1986), Messori–Tamborini (1993).

outside the local sub–area [see also Sharpe (1992)]. Taking for granted that the most efficient and innovative local borrowers of a marginal area such as the *Mezzogiorno* have to build up strong non–local economic relationship[96], it could follow in principle that local Southern banks choose to finance traditional and inefficient activities of local borrowers. This case occurs when, in order to finance more efficient non–local borrowers or innovative local borrowers, these banks would have to increase liquidity costs in larger measure than to decrease the costs due to borrowers' default risk.

This second explanation is limited because, even in this case, the theoretical conclusions do not fit with the empirical data. As shown in section 2, a specific feature of the Southern banks balance sheets is the high ratio of deposits to total liabilities. Moreover, even if the Southern banks have seen a fast decrease in the net balance towards other banks during recent years, their balance is still largely positive. It follows that, despite the Southern banking system's low ratio of own capital to total liabilities, the expected level of banks' liquidity cost in the *Mezzogiorno* does not yet represent a binding constraint on their loans policy.

There remains a third explanation which starts from the theoretical and experimental point of view that individuals as well as economic organizations, face limits in their ability to foresee and compute. In an uncertain world these limits lead to "bounded rationality" which, according to Simon (1959, 1979), implies that "satisfying" replaces maximizing choices. One of the main features of satisfying behavior is that individuals inside an economic organization can be involved in learning processes, i.e. in the search for and in the screening of new information and knowledge, and in the exploitation of the most useful information and knowledge found in order to frame and solve economic problems.

The point to be emphasized here is that problem framing and problem solving apply to economic problems of a different nature [Simon (1979)]. Adopting Nelson and Winter's (1982) definitions drawn from the Schumpeterian model[97], I distinguish between the "routine decisions" and the "innovative decisions". In routine decisions, individuals inside an organization face situations which are customary and repetitive. In this case, the problem framing and the problem solving are quite simple being based, at most, on the adaptation of a few marginal novelties to the past experience. This means that learning process is unimportant. On the other hand, in the case of innovative decisions, individuals inside an organization must face and solve new problems and, hence, they must build up new and appropriate forms of co–operation and co–ordination. The latter require not only an efficient management, but also a flexible strategy which can adapt the internal organization to new problem

---

[96] Take for example a firm which is located in a marginal area but is specialized in high-tech production. It is likely that the most important suppliers of this firm are non–local.

[97] These definitions are slightly different with respect to those elaborated by Simon (1979) to whom until now I have explicitly referred.

framing and problem solving. In this sense, the learning process is quite important but difficult: it becomes an attempt to put into practice organizational innovations in order to answer in an innovative way the new external problems.

It is useful to apply the distinction between routine and innovative decisions to banks behavior. In principle, the latter decisions would involve any bank which is screening non–traditional or unknown prospective borrowers to design a debt contract and to define its credit supply. The role of *social accountants* for credit allocation, which Stiglitz and Weiss (1988) attribute to the banks, and the *establishment* of long–term relationships is part of the innovative cases. In such cases, the bank management would face new problems and his/her learning process could involve innovative organizational changes and decisions. On the other hand, in the case where a bank only finances well–known borrowers who undertake traditional activities, I would consider its loan policy a routine decision. Banks behavior, which follows passively the market demand of traditional and inefficient borrowers, is part of the routine cases. This suggests that the passive loan policies, adopted by banks in the Southern Italian credit market (see above), represent an extreme case of routine decisions.

The previous observations point also out that individuals and organizations have to overcome delicate difficulties in order to handle problem framing and problem solving in the innovative cases. The market and the organizational failures, which characterize the financial relationships in Southern Italian economy (see sections 2 and 3), can therefore be the causes as well as the consequences of Southern banks' inability to replace routines with innovative decisions. This last explanation of the adaptive behavior followed by the Southern Italian banks, has the advantage of fitting with the empirical data. It is credible that inefficient local banks, which can survive only thanks to the segmentation of the Southern credit markets, follow routine behavior and are unwilling or unable to substitute new debt contracts with unknown or innovative borrowers for their long–term relationships with local and traditional borrowers. On the other hand, the limited economic size of the market's niches and these same long–term relationships impede the outside banks or the few Southern banks with innovative lending policies from entering into competition.

I am then able to state that the behavior of the banking system in the *Mezzogiorno* is characterized by an inability or unwillingness to perform the function of *social accountants* which would be necessary for the economic development of this area. The banks have not pursued an active strategy to decrease the default risk of Southern borrowers or to increase the competitiveness of the productive and financial systems of the South. The result has been heavy inefficiency in credit allocation (mainly, in the allocation of long–term loans), and in the internal relations of the Southern banking system. It is sufficient to recall the permanent large number of inefficient local banks in highly segmented credit markets. This indicates the extent to which the lack of dynamic and efficient allocation has crippled the operative efficiency of the banks.

VIII. CONCLUSIONS

The conclusions of the previous section point to how the high default risk of Southern borrowers and the related economic indicators (i.e. the rate of bad loans to total loans in the region) can be ascribed, at least partially, to the inability or unwillingness of the banking system to steer the economic initiatives of the *Mezzogiorno*. This suggests that the accepted interpretation of the empirical framework, given in the first three sections, and of the geographic relationship between banks and firms is superficial. It is accurate on a descriptive level to maintain that the high interest rates on credit supply and the strong impact of credit rationing, which characterize banking loans in the *Mezzogiorno*, can be attributed to the high default risk of the local borrowers, to the weakness of local banks' balance sheet, and to the segmentation of the regional credit market. But the causal links between the behavior of the banking system and borrowers' default risk in the *Mezzogiorno* show that the demand and the supply sides are not independent in the local credit market; and that this interdependency makes the determinants of the credit supply so intricate that the previous description is rendered meaningless.

Given the high default risk of prospective Southern borrowers, the passive behavior of the banking system in this area has flowed on high interest rates on loans supply and on credit rationing. On the other hand, these features of the credit supply side have hindered the economic development of the *Mezzogiorno*, and hence the potential improvement in the loans' 'quality`.

This vicious circle has strengthened, and been fed by, other inefficient links. The reproduction of a high–risk credit demand has had a negative impact on the profitability of Southern banks. On the other hand, the realization of moderate but non–negative operating profits has reinforced the willingness of these banks to take up a passive and defensive credit policy. This attitude, in its turn, has emphasized the relative weakness in the balance sheet of Southern banks (in particular, the low ratios of bank own capital to total liabilities) and has protected the strong segmentation of the Southern credit market. These inefficient elements of the Southern banking system have led to an upward tendency in the average interest rates on credit supply and, due to the adverse selection and incentive effects, an upward tendency in the recourse to credit rationing. These negative tendencies have been greatly facilitated by the strong segmentation of the Southern credit market. The result has been the worsening quality of Southern credit demand. Hence, the vicious circle is complete.

The policy implications of this analysis are quite clear. It is not sufficient for monetary policy to stimulate Northern Italian banks to open new branches in the *Mezzogiorno*. Given the high sunk costs in the Southern Italian credit market, this opening would not lead to the elimination of inefficient local banks, but instead could stimulate a collusion between too many credit suppliers, and would not decrease market segmentation. A more promising policy would be to stimulate mergers between big or medium–sized Northern banks and Southern local banks. In the last years the merging processes have redoubled. It is still too early to judge whether these mergers can improve the workings of Southern capital markets. The risk is that the stronger Northern banks would use the opportunity of the merger to enlarge but also to strengthen the segments of the Southern credit market; the hope is that this merging process would flow into an increase in the degree of contestability and, thus, in a more efficient organization of this market.

REFERENCE

BALTENSPERGER, E. (1980), "Alternative approaches to the theory of the banking firm", **Journal of Monetary Economics**, 6, 1–37.

BANCA D'ITALIA (1989), **Relazione annuale sul 1988**, Roma.

BANCA D'ITALIA (1990), **Il sistema finanziario nel Mezzogiorno**, G. Galli (ed.), Contributi all'analisi economica, numero speciale, Roma.

BAUMOL, W.J.; PANZAR, J.C.; WILLIG, R.D. (1982), **Contestable Markets and the Theory of Industry Structure**, New York.

BAUMOL, W.J.; PANZAR, J.C.; WILLIG, R.D. (1983), "Contestable Markets: An Uprising in the Theory of Industry Structure: Reply", **American Economic Review**, 73, 491–6.

CAMERON, R. (1967), **Banking in the early stages of industrialization**, Oxford.

CHICK, V.; DOW, S.C. (1988), "A post–Keynesian perspective on the relation between banking and regional development", **Post–Keynesian monetary economics**, P. Arestis (ed.), Aldershot.

CURRIE, M.; MESSORI, M. (1993), "New institutional economics and new Keynesian economics", **mimeo.**

D'AMICO, N.; PARIGI, G.; TRIFILIDIS, M. (1990), "I tassi d'interesse e la rischiosità degli impieghi bancari nel Mezzogiorno" in Banca d'Italia (1990).

DE MEZA, D.; WEBB, D.C. (1987), "Too Much Investment: A Problem of Asymmetric Information", **Quarterly Journal of Economics**, 101, 282–92.

DE MEZA, D.; WEBB, D.C. (1988), "Credit market efficiency and tax policy in the presence of screening costs", **Journal of Public Economics**, 36, 1–22.

DI BATTISTA, M.L; GRILLO, M. (1988), "La concorrenza nell'industria bancaria italiana" in Banca Commerciale Italiana, **Banca e mercato**, Bologna.

D'ONOFRIO, P.; PEPE, R. (1990), "Le strutture creditizie nel Mezzogiorno" in Banca d'Italia (1990).

DOW, S.C. (1987), "The treatment of money in regional economics", **Journal of Regional Science**, 27.

FAINI, R.; GALLI, G.; GIANNINI, C. (1992), "Finance and development: The case of Southern Italy", **Banca d'Italia – Temi di discussione**, no. 170, giugno.

FAZIO, A. (1985), "Credito e attività produttiva nel Mezzogiorno", **Banca d'Italia – Bollettino economico**, Ottobre, 27–42.

FUDENBERG, D.; TIROLE, J. (1986), "A Theory of Exit in Duopoly", **Econometrica**, 54, 943–60.

GALE, D.; HELLWIG, M. (1985), "Incentive–compatible debt contracts: The one-period problem", **Review of Economic Studies**, 52, 647–63.

GALLI, G.; ONADO, M. (1990), "Dualismo territoriale e sistema finanziario" in Banca d'Italia (1990).

GAMBETTA, D. (1988), **Trust. Making and breaking cooperative relations**, Oxford.

GIANNOLA, A.; MARANI, U. (1991), "The financial structure of Mezzogiorno's industrial firms", **Quaderni del Dipartimento di Scienze economiche e sociali**, Universitá di Napoli, no. 2.

GOLDSMITH, R.W. (1969), **Financial structure and development**, New Haven.

GREENWALD, B.C.; STIGLITZ, J.E. (1987), "Keynesian, new Keynesian and new classical economics", **Oxford Economic Papers**, 39, 119–32.

GROSSMAN, S.J.; HART, O.D. (1986), "The costs and benefits of ownership: A theory of vertical and lateral integration", **Journal of Political Economy**, 94, 691-719.

HART, O.; HOLMSTROM, B. (1987), "The Theory of Contracts" in **Advances in Economic Theory**, T.F. Bewley (ed.), Cambridge.

HART, O.D.; MOORE, J. (1988), "Incomplete contracts and renegotiation", **Econometrica**, 56, 755–85.

HART, O.D.; MOORE, J. (1990), "Property rights and the nature of the firm", **Journal of Political Economy**, 98, 1119–58.

KING, R.G.; LEVINE, R. (1993), "Finance and growth: Schumpeter might be right", **Quarterly Journal of Economics**, 107, 733–37.

KREPS, D.M. (1990), **A Course in Microeconomic Theory**, New York.

MARULLO REEDTZ, P. (1990), "La redditività delle aziende di credito. Un'analisi per aree geografiche" in Banca d'Italia (1990).

MARZANO, F.; DEL MONTE, A.; FABBRONI, M.; MARTINA, R. (1983), "Il sistema bancario meridionale, lo sviluppo del Mezzogiorno e l'ingresso di banche estere in Italia", **Economia italiana**, 61–107.

MCKINNON, R.I. (1973), **Money and capital in economic development**, Washington.

MESSORI, M. (1989), "Sistemi di imprese e sviluppo meridionale. Un confronto fra due aree industriali" in **Modelli di sviluppo**, G. Becattini (ed.), Bologna.

MESSORI, M.; SILIPO, D. (1991), "Un'analisi empirica delle differenze territoriali del sistema bancario italiano", **Cespe Paper**, no. 6/90.

MESSORI, M.; TAMBORINI, R. (1993), "Money, credit and finance in a sequence economy", **Working Papers**, Dipartimento di Scienze Economiche, Università di Roma "La Sapienza"

MILDE, H.; RILEY, J.G. (1988), "Signaling in Credit Markets", **Quarterly Journal of Economics**, 103, 101–29.

MILGROM, P.; ROBERTS, J. (1990), "Bargaining costs, influence costs and the organization of economic activity" in **Perspectives on Positive Political Economy**, J.E. Alt and K.A. Shepsle (eds.), Cambridge.

NELSON, R.R.; WINTER, S.G. (1982), **An Evolutionary Theory of Economic Change**, Cambridge Mass..

ONADO, M.; SALVO, G.; VILLANI, M. (1990), "Flussi finanziari e allocazione del risparmio nel Mezzogiorno" in Banca d'Italia (1990).

PITTALUGA, G.B. (1987), "Il razionamento del credito bancario in Italia: una verifica empirica", *Moneta e credito*. Now in G.B. Pittaluga (1989), **Il razionamento del credito: aspetti teorici e verifiche empiriche**, Genova.

PRISMAN, E.Z.; SLOVIN, M.B.; SUSHKA, M.E. (1986), "A general model of the banking firm under conditions of monopoly, uncertainty, and recourse", **Journal of Monetary Economics**, 17, 293–304.

SCHUMPETER, J.A. (1912), **Theorie der wirtschaftlichen Entwicklung; engl. trans. The Theory of Economic Development**, New York, 1961 (first edition: 1934).

SCHUMPETER, J.A. (1939), **Business Cycles. A Theoretical, Historical and Statistical Analysis of the Capitalist Process**, 2 vol., New York.

SCHUMPETER, J.A. (1970), **Das Wesen des Geldes**, F.K. Mann (ed.), Goettingen.

SHARPE, S.A. (1990), "Asymmetric information, bank lending, and implicit contracts: A stylized model of customer relationships", **Journal of Finance**, 45, 1069–87.

SHARPE, S.A. (1992), "Consumer switching costs, market structure and prices: The theory and its application in the bank deposit market", **Finance and Economics Discussion Series**, 183, Federal Reserve Board, Washington, D.C..

SIMON, H.A. (1959), "Theories of decision making in economics", **American Economic Review**, 49, 253–283.

SIMON, H.A. (1979), "Rational decision making in business organizations", **American Economic Review**, 69, 493–513.

SIRACUSANO, F.; TRESOLDI, C. (1990), "Le piccole imprese manifatturiere nel Mezzogiorno: Diseconomie esterne, incentivi, equilibri gestionali e finanziari" in Banca d'Italia (1990).

STIGLITZ, J.E. (1987a), "The Causes and Consequences of the Dependence of Quality on Price", **Journal of Economic Literature**, 25, 1–48.

STIGLITZ, J.E. (1987b), "The new Keynesian economics: Money and credit", **mimeo**.

STIGLITZ, J.E.; WEISS, A. (1981), "Credit Rationing in Markets with Imperfect Information", **American Economic Review**, 71, 393–410.

STIGLITZ, J.E.; WEISS, A. (1988), "Banks as Social Accountants and Screening Devices for the Allocation of Credit", **NBER Working Paper**, no. 2710.

STIGLITZ, J.E.; WEISS, A. (1992), "Macro–Economic Equilibrium and Credit Rationing", **Oxford Economic Papers**, 44, 694–724.

TOWNSEND, R.M. (1979), "Optimal contracts and competitive markets with costly state verification". **Journal of Economic Theory**, 21, 265–93.

WILLIAMSON, O.E. (1975), **Markets and Hierarchies**. New York.

WILLIAMSON, O.E. (1985), **The Economic Institutions of Capitalism**, New York.

WILLIAMSON, O.E. (1989), "Transaction cost economics" in **Handbook of Industrial Organization**, R. Schmalensee and R. Willig (eds.), Amsterdam.

ZAZZARO, A. (1992), "Banche locali e sviluppo economico regionale: costi di liquidità e costi di solvibilità", **Rivista di politica economica**, 83, 107–52.

15 THE REGIONAL IMPACT OF THE INTERNAL MARKET: AN ANALYSIS FOR LAGGING REGIONS

Michel Quévit

RIDER, Université Catholique de Louvain.

Louvain–la–Neuve, Belgium

I. INTRODUCTION: BEYOND THE CLASSICAL APPROACH TO THE REGIONAL IMPACT OF THE COMPLETION OF THE EUROPEAN INTERNAL MARKET

Studies of the application of the Single European Act have revealed two types of consideration, one concerning the macro–economic effects of the Internal Market and the other concerning its spatial effects. On the first, the Cecchini Report [Commission of the European Communities (1988)] favoured a global, amplified growth of the national economies of the Community. This growth is justified by scale savings achieved with the abolition of frontiers and, above all, by the harmonization of production and marketing standards. Surprisingly, the regional dimension appears to have been given on consideration in this approach to the problem, except for the postulation that general growth will benefit the whole of the European Community irrespective of existing spatial imbalances.

The studies carried out, particulary those on behalf of the European Commnunity, concerning the regional impact of completion of the Internal Market generally conclude with a pessimistic vision from a spatial perspective. In a simplified version, they are summed up in the following statement: "the main effect of completion of the European Market will be to concentrate economic activity in a smaller number of places where cost reductions and scale savings are used to the best advantage" [PA Cambridge Economic Consultants (1988)]. The authors reach the conclusion that increased aid to the less–favoured regions will be necessary, whilst advising that they be better adjusted to local realities than in the past. It is from this perspective of 1992 that the decision was taken by the European Community to double the Structural Funds. Such an atitude is relatively common. Consequently, it is not a matter of questioning the decision, which derives from the rule of equity, but knowing whether the appreciable quantitative effect of that measure will have a qualitative effect through more favourable economic positioning of the regions concerned, bearing in mind the economic and technological changes engendered not only by the dynamics of completion of the Internal Market, but also by techno–industrial changes in the highly industrialized countries.

X. Vence-Deza and J. S. Metcalfe (eds.), Wealth from Diversity, 343–360.

The approach of this paper, therefore, is to focus on the current relations between spatial development and industrial changes. It objects to the neo–classical analysis, because it does not really take into account the predominant integration of science and technology in productive systems in future decades of development of the industrial economy. For proper perception of regional developments, it is necessary to link the impact of the Internal Market to techno–industrial development. In this context, the Single Market can present new development opportunities for the regions of the Community.

Within this conceptual framework, the purpose of this paper is to examine some theoretical, empirical and policy issues raised by the completion of the Internal Market in relation to the regions of the Community. The first section examines the results of costs of non–Europe studies, the emphasis being on the theoretical approach to the Internal Market and on some lessons raised by the studies from a regional perspective. The second section presents an analysis of the sectoral impact of the Single Market focusing on issues of policy adjustment. The third and fourth sections examine the sectoral impact of the completion of the Single Market for a specific categorie of regions, the lagging regions of the community. The last section considers possible innovative policy adjustment scenarios for the orientation of regional policy.

II. REGIONAL IMPACT OF THE COMPLETION OF THE INTERNAL MARKET: SOME LESSONS FROM THE COST OF NON–EUROPE STUDIES[98]

II.1 The Direct Effects of the Lifting of NTBs: a Positive and Inmediate Impact on the Global Economy

In general, the theoretical approach of studies relating to the assessment of the impact of the lifting of non–tariff barriers (NTBs) is based on the existence of economic advantages linked to the removal of trade barriers and market distortions which can occur in two particular situations: one where existing comparative advantage may be exploited; the other where there is an absence of comparative advantage for products and the advantages result from an intensification of competition. The approach is based principally on the more fundamental studies on the theory of international trade [e.g. Dixit and Norman (1980), Corden (1984), Baldwin (1984)].

---

[98] The *direct* effects of the lifting of non–tariff barriers (NTBs) are based on the assumption that this measure will lead to a lowering of costs which will have repercussions on costs and margins and will spread throughout the economy via a lowering of intermediary consumption costs. With respect to the assumed *indirect* effects, the creation of the Internal Market will have a determining influence on the mechanisms linked to the economic environment by creating a more competitive climate (e.g. size of markets, effects of pricing and non–pricing, strategic company behaviour, and innovation).

A workable approach for examining the impact of the NTBs is provided in the study by Nerb (1987) regarding the perception by business managers of the effects of the SEM. The author described eight categories of barriers to note in order of decreasing importance: technical standards and regulations; administrative barriers; border formalities; regulation of goods transport; disparities in VAT matters; national regulations of capital markets; restrictions in government–market matters; and the application of Community law. On the basis of this typology, the study demonstrates that perceptions of the barriers linked to NTBs vary appreciably according to sectors, countries and size of firms.

The study made a number of major observations. First, the technical regulations (and standard and administrative barriers) constitute the most important barrier for the industrialists of the countries studied. Second, the disparity between technical standards and regulations is particulary important for the mechanical engineering industry. Third, for a large number of industries, government market restrictions have only a limited effect, but they are important for the transport, office equipment and electrical sectors. Lastly, small and medium–sized enterprises (SMEs) are less concerned than large businesses by restrictions in government markets, and more concerned with matters relating to customs procedures.

From a *regional* point of view, it can be assumed that the direct effects of the lifting of NTBs will be positive, either for the traditional industrial regions (RETI) or the lagging regions (Objective 1 regions) of the Community, in that firms will benefit from the lowering of production costs. However, differences could occur relating to the structural components of the productive fabric of these regions.

A series of studies under the supervision of the Commission to assess the economic costs for these diverse barriers seems to confirm this hypothesis. The following findings are particularly noteworthy. The costs of customs procedures are of major importance for SMEs (between 30% and 45% higher than for the large businesses) and in countries where the administrative procedures are relatively more extensive [Ernst and Whinney (1987)]. Also, the direct effect of opening government markets on capital equipment apply, above all, in the following areas: industrial boilers (reduction of 20%); turbine generators (12%); eletric engines (20%); and the telephone industry (30–40%) [Atkins (1987)].

More specifically, studies of several industrial sectors (food products, pharmaceutical products, car manufacturing, textiles and clothing, construction material and telecomunications equipment) highlight the importance of the sensitivity of the national economies to the probable effects of the SEM according to their sectoral specialization. These studies indicate that the direct effects of the SEM are beneficial for most regions. Of course, a more detailed analysis is needed to confirm this hypothesis, in as much as these sectoral findings are too global, and they do not integrate the very great differences which may exist at the sub–sectoral level along the entire industrial production chain [see Quévit et al. (1991), Buiges et al. (1990)].

345

II.2 The Indirect Effects of the Completion of the Internal Market

Studies on the cost of non–Europe have highlighted the importance of the indirect effects of the SEM following the creation of a more favourable competitive climate linked to two major factors: market size and reinforcement of competition. According to completed studies, the creation of the SEM will have integration effects of three kinds. First, a reduction of costs will be achieved by better explotation of production units through technical economies of scale or range and comparative advantages through increases in volume and inter/intra–industry restructuring processes. Second, price competition pressure (above all in protected sectors) by way of diverse mechanisms will lead to: encouragement to increase efficiency (reduction of the X–inefficiency, i.e. of internal inefficiency); rationalization of industrial structures; construction of a more adequate price system in relation to real production costs (price/cost margin); and adjustment between industries affected by the interplay of comparative advantages. Third, the non–price effects of increased competition will encourage businesses to improve their organization, to increase the quality and variety of their products, to innovate production processes and, on the macroeconomic level, to intensify innovative flow relating to products and processes. This triple effect should lead to a set of economic links around two basic mechanisms which reinforce one another –the phenomena of size and competition.

From a regional perspective, the results of these studies suggest differential regional effects of the SEM. In general, there are only limited opportunities for regions to generate technical economies of scale in industry because of the incidence of sectors where demand is stagnating or sectors of low technology products. Many SMEs will find it difficult to pass the minimum efficiency size threshold. There are also few possibilities for many regions to generate technical economies of scale in services, except for those which have good urban infrastructure, and the peripheral geographical position of many regions is an obstacle to non–technical economies of scale aiming at the reduction of transport costs.

More important from a regional point of view is the global statement of these studies which conclude that, in the long term, there will be a predominance of indirect effects over direct effects of the completion of the Single Market according to the nature of the structural environment of the firm. On the basis of Porter's (1985) methodology on comparative advantages linked to types of structural environment, i.e. structural environment of a fragmented type, structural environment of a specialist type, and volume environment, the authors rightly note that the impact of the completion of the Internal Market will be of a different nature according to the structural characteristic of a firm's sector of activity. The completion of the Single Market represents an *opportunity* for firms situated in a specialist or volume environment, but the impact in an environment of a fragmented type would be limited, except to change the rules of the competition "game" and to move to industrial production for standardized products [Emerson et al. (1990)].

346

It is a primary requirement that efforts be made to increase knowledge of the structural environment of the regions.

III. THE SECTORAL IMPACT TO THE COMPLETION OF THE INTERNAL MARKET

The study carried out by Buiges et al. (1990), on the sectoral impact of the Single Market on the Community countries provides useful insights on the nature of the indirect effects of the SEM according to their structural components. Of 120 industrial sectors (NACE 3 digits), the authors identified forty as likely to be the subject of NTBs affecting intra–Community trade. This concerns industries which are currently protected by NTBs and where these barriers prevent the explotation of economies of scale or permit the retention of wide price discrepancies between the Member States. These sectors represent 50% of the value added to the Community (see Table 1).

III.1 Relative Importance of Countries in Sectors Sensitive to the Impact of the SEM

In general, the importance of the less–favoured countries in the sectors in question is as significant as that of the most–favoured countries but is different in nature. In fact, in the less–developed countries, there are few high–technology industries linked to government markets or traditional or regulated markets.

Table 1.– Weight of the Forty Identified Sectors at Community Level

Country	Share in industrial value added	Share in industrial employment
Ireland	59.6	43.3
Germany	54.6	54.5
France	53.4	50.8
United Kingdom	52.5	50.0
Belgium	48.8	50.1
Italy	47.7	48.6
Netherlands	47.0	44.9
Portugal	43.3	48.1
Greece	44.8	45.4
Spain	40.9	39.1
Denmark	39.6	39.4

Source: Commission of the European Communities.

It is above all in the more traditional and labour–intensive industries where specific barriers exist, notably in the textile and farm product industries (e.g. export subsidies in Greece, quotas and tariff barriers in Portugal).

III.2 Structural Impacts of Lifting NTBs Differentiation between Northern and Sourthern Countries

A static approach of performance assessment of the sectors affected by the SEM shows the great drawback of over–simplifying. It does not take account of the structural effects of lifting NTBs which, as shown by the studies on the cost of non–Europe, are more important in the medium and long–term than the static effects of the decrease in costs and prices. The effects of lifting NTBs will, at the same time, affect comparative advantages, economies of scale and product differentiation. The combination of these factors can cause different behaviour in the production strategy of economic agents and create different developments in the regional productive systems.

With respect to structural achievements and comparative advantages of productive systems, a new differentiation is appearing between the economies of the sourthern and northern countries in the EC, based on an examination of their comparative advantages in relation to three structural characteristics–capital intensity, labour intensity and R&D. On the basis of forty sectors sensitive to removing NTBs and a redistribution of groups of countries with comparable achievements in all sectors (i.e. those which are mainly strong or weak), five groups of sectors may be identified: capital and R&D intensive sectors; capital but not R&D intensive sectors; labour–intensive sectors; skilled labour–intensive sectors; and less labour and capital intensive. sectors (see Table 2). According to this typology, a differentiated redistribution between the countries of North and South is again apparent with comparative advantage for the northern countries in R&D and/or capital–intensive sectors and comparative advantages for the southern countries in labour–intensive sectors.

Inter–industrial anda intra–industrial scenarios are distinguished in the relationship between the diversity of productive systems and the dynamic effects of the SEM suggested by Buiges et al. (1990), in that they separate out two intra-Community trade scenarios in the linkage between the idea of comparative advantage –that of economies of scale and product differentiation. An *inter–industrial scenario* is where integration in the Internal Market would favour specialization in the relatively efficient sectors or those which intensively use abundant resources. If similar products are involved, the lifting of NTBs will favour the geographic concentration of production according to the comparative advantages of the country or region. An *intra–industrial scenario* is where integration could, where products are differentiated, favour specialization in different kinds of products on an intra-community scale. According to the authors, the lifting of NTBs will affect countries and sectors in different ways depending upon whether inter–industrial or intra-industrial type trade dominates.

This distinction is important for a regional perspective to the extent that, before Greece, Spain and Portugal joined the EC, intra–Community trade vas principally of an intra–industrial type, with the exception of southern Italy [Padoa-Schioppa (1987)]. However, with the arrival of southern countries in the Community, trade by these countries with the rest of the Community has been mainly of a inter–industrial type. More precisely, four sectors have been identified which are almost exclusively of the inter–industrial type: wine (Portugal, Spain, Italy and France); the footwear industry (Portugal, Spain, Italy and Greece); clothing (Portugal, Italy and Greece); and machine–tools (Italy, Germany and Spain). It is important to note that more extensive analysis of the situation of the southern countries shows a coexistence within the same country of two scenarios, for example in Spain.

The comparative analysis of the sectoral impact of the SEM on Community countries suggests the existence of at least three types of region:

. Regions of high achievement exclusively in inter–industrial type trade. This is a category characterized by a significant degree of non–skilled and skilled labour–intensive sectors.

. Mixed regions with high achievements in inter–industrial type trade but which have intra–industrial type trade sectors. These are regions which consist of labour–intensive sector and capital–intensive sectors.

. Regions of weak structural achievement in inter–industrial type trade which excludes them from the benefics of the completion of the Internal Market.

Table 2.– Comparative Advantages of Each Member State in the Forty Sectors: Relative Export/Import Ratio in the Five Groups

Country-Group	B+L	DK	D	GR	E	F	IRL	I	NL	P	UK
Capital and R&D-intensive sectors	110	61	118	25	85	105	112	65	103	49	109
Capital-intensive sectors	147	216	83	83	118	105	143	120	310	41	85
Skilled labour-intensive sectors	83	101	197	27	68	152	41	120	88	31	130
Labour-intensive sectors	75	88	37	196	379	71	52	345	52	416	87
Less capital and labour-intensive sectors	14	108	186	29	53	103	48	189	89	36	137
Comparative Advantages											
Capital Intensity	X	X			X		X		X		
Labour Intensity				X	X			X		X	
R&D Intensity			X			X	X				X

Source: Commission of the European Communities.

IV. THE IMPACT OF THE COMPLETION OF THE INTERNAL MARKET ON THE COMMUNITY'S LAGGING REGIONS[99]

IV.1 The High Sensitivity of the Lagging Regions (Objective 1 regions) to the Completion of the Internal Market

The analysis of the presence of sensitive sectors in the productive fabric of Objective 1 regions shows an important degree of sensitivity to the SEM, with the excepcion of Italy. Whereas the weight of the sensitive sectors is, in general, relatively high in the entire Objective 1 regions compared to the national average, the Italian regions are a well–known exception, where the sensitivity to the SEM impact is very weak (5.1% of the regional employment versus 12.2% for the whole country), and nearly insignificant in regions such as Calabria, Sardinia and Sicilia. The degree of sensitivity to the SEM is highest in Greece (more than 75% of the regional manufacturing employment), with the exception of the Oriental Macedonia region (33%). In terms of added value, responsiveness to the SEM can be a little weaker.

A great sensitivity to the SEM is also evident in the Portuguese regions where, except for the region of Lisboa and Vale Tejo, the impact on sensitive sector is higher than the national average in terms of added value. The high sensitivity of the Centro region can be explained by the presence of sectors protected by the partitioning of the public markets and industry being of little importance with regard to added value.

In the Objective 1 regions in Spain, the impact of the sensitive sectors is clearly further below the national average in terms of added value (37.9% versus 55.2%) than in terms of employment (48.1% versus 56%). Regions very sensitive to the SEM such as Comunidad Valenciana, Castilla–La Mancha, Andalucia and Galicia co–exist with others clearly less sensitive, such as the Canaries and Extramadura. Finally, the Irish regions also show a degree of sensitivity to the SEM superior to the national average (60.9% in added value and 46.7% in terms of employment) with the exception of the South East (34%), the East (41%) and the Midlands (43%). Two regions are particularly affected –the West (67%) and the North East (57%).

---

[99] This chapter is part of the study carried out by RIDER on behalf of GREMI on **Development Prospects of the Community's Lagging Regions and the Socio–Economic Consequences of the Completion of the Internal Market**, commissioned by DG XVI of the EC, 1991.

IV.2 The Main Impacts of the Completion of the Internal Market on the Objective 1 Regions

With regard to the direct effect of the lifting of NTBs on trade, the regions will benefit from the general movement towards lower costs but they will be more directly concerned by the lifting of specific NTBs. The Objetctive 1 regions will be more sensitive to the lifting of barriers increasing costs than to the lifting of restrictions to market access [Emerson (1990)] for several reasons: the importance of SMEs in their productive fabric; less inmediate opportunities for economies of scale; and a low degree of sectoral specialization dependent on government markets (with the exception of certain export subsidies or quota systems and tariff barriers in some countries). This tendency is confirmed by the evolution of the effects of specific NTBs.

As indicated in the study of Buigues et al. (1990), concerning the southern countries, the Objective 1 regions are mainly concerned by the impact of 1992 on sectors with medium NTBs (Group 4)[100]. For most of these regions, 1992 will have only a small impact on sectors with high technology linked to public markets. However, familiar differences appear when the national positions are examined.

The Italian regions are clearly more sensitive to the 1992 impact on the sectors with high NTBs —especially those with high technology linked to public markets, both traditional and regulated— than on sectors with medium NTBs. In Spain, the sectors with medium NTBs dominate, but the relative importance of sectors linked to the regulated public markets in Andalucia and the Canaries is notable. In the Portuguese regions, the 1992 impact will also be stronger for those sectors with medium NTBs. However, there will be a sensitive impact of high-technology sector linked to public and regulated markets in Centro and Alentejo regions and sectors linked to traditional public markets and regulated markets in the Algarve, Lastly, Ireland shows a dual configuration between regions essentially characterized by sectors with medium NTBs and regions highly influenced by the joint presence of these sectors with high technology sectors and sectors linked to traditional public markets or regulated ones like the Midwest, the West and, to a lesser degree, Donegal and the North West. The North East is a special case, where the presence of sectors linked to traditional public markets or regulated markets is

---

[100] For the authors, the sensitive sectors are re–grouped in four sub–sets according to two criteria: their degree of openness to intra–Community trade: and the price differences between the Member States for identical products.

Group 1.– *High–technology public–procurement markets*: this concerns sectors where demand is experiencing substantial growth and where technological content is high (e.g. information, telecomunications and medical instruments).

Group 2 and 3. – *Traditional or regulated public–procurement markets*: two groups are united this title, one group where intra–Community trade is low but price differences high (e.g. material for energy production and railway material), a second group where intra–Community trade is low but whose price differences are less significant than the fact of intra–Community imports.

Group 4.– *Sectors with moderate NTBs*: this concerns activities where technical, administrative or fiscal barriers still exist which allow the retention of wide price differences.

significant.

The main *indirect* effect of the Internal Market on the Objective 1 regions is related to the effects of phenomena of size. In particular, there is generally little scope for many Objective 1 regions to generate technical economies of scale in industry with sectors where demand is stagnating or sectors of low–technology products. Modest economies of scale are possible in the food, drink and tobacco sectors, in textiles and clothing, leather and wood articles, but there may be greater opportunities for economies of scale in the distribution and marketing of certain subsectors of farm products. Many SMEs operating in the Objective 1 regions may find it difficult to pass the minimum efficiency size threshold. In the service sector, there are no opportunities for most Objective 1 regions to generate technical economies of scale except for those which have good urban *infrastructure* because of the predominance of non–market services over market services, and structural weakness in the supply of services to businesses. Lastly, there is the problem of the peripheral geographical location of many Objective 1 regions with regard to non–technical economies of scale aiming at the reduction of transport costs.

IV.3 Structural Performance of Sensitive Sector and Vulnerability of Objective 1 Regions to the Effects of the SME

The productive fabric of the Objective 1 regions in general is vulnerable with respect to the completion of the Internal Market. Amongst the strong regions, only one –the Midwest– has a configuration of intra–industrial type and is thus well-positioned to face the impact of 1992. The other strong regions are inter–industrial in nature: Norte, Puglia or, with a dual configuration, Murcia, Castilla–La Mancha, South West and Campania. Most of the other regions (twenty–one out of thirty-eight) are in a vulnerable or threatened position (see Table 3).

However, some of the mixed regions (ten) have vulnerable and strong sectors co–existing: Comunidad Valenciana, Abruzzi, Centro, West, Sicilia and Calabria, regions with a dual configuration of intra and inter–industrial type; and Extramadura, of essentially inter–industrial type.

The result of these different statements is that most of the Objective 1 regions must, as a matter of urgency, develop adjustment policies that go beyond the classical inter–industrial exchange scenarios. To stay in an inter–industrial scheme by making profits from the comparative transition cost advantages would make them even more vulnerable to intra–community competition, developing countries or competition from Eastern Europe. The development in these regions of the intra-industrial scenario, centred essentially on the exogenous contribution of foreign firms with high technological potential, will *maintain* and/or reinforce the dualism of regional economies by isolating the endogenous potential of these regions from the growth process.

Table 3.– Level of Vulnerability of the Productive Fabric of Objective 1 Regions facing the SEM

	Vulnerable or strong but threatened	Strong	Mixed	Insignificant
Inter-industrial ouline	Algarve Central Macedonia East Macedonia Thessalie West Greece Islands and others Greek regions	Norte Puglia		Extremadura
Intra-industrial outline	Asturias Castilla-Leon Molise Lisbon	Mid-West	Sardenia South-East	Alentejo
Dual	Andalucia Galicia Basilicata Peloponnese Nord-Est Donegal NW Midlands East Attica Central Greece	South West Campania	C. Valenciana Castilla La Mancha Abruzzi Centro Murcia West Sicilia Calabria	Canaries

Accordingly, two more innovative adjustment scenarios are proposed. One is of the inter–industrial type, the other of the intra–sectoral type; both scenarios avoid the dangers described above. In this context, the Objective 1 regions have room for improvement in the enforcement of their policy insofar as (for most) they could combine innovative adjustment scenarios of inter–industrial and intra–industrial types and try to establish complementarities and synergies between these two approaches. For the regions dominated by inter–industrial exchanges, it is important to note that an inter–industrial adjustment that upgrades regional specialization is perfectly accessible in the perspective of Internal Market completion.

V. PLURALITY OF ADJUSTMENT SCENARIOS ADAPTED TO THE REGIONS

The analysis of the impact of achievement of the Internal Market in the Objective 1 regions reveal the existence of three types of regions distinguished by their sensitivity to 1992 and their structural performance.

. Regions characterized mainly by intra–sectoral exchanges, but where intensive capital and/or R&D sector dominate. These regions will tend to position themselves more positively than others towards the SEM insofar as they contain industrial sectors where scale economies can still be mobilized. These are mainly the Objective 1 regions such as Asturias, Castilla–Leon, Lisbon and Vale Tejo, Alentejo, Molise and Mid–West.

. Regions characterized mainly by inter–industrial exchanges where labour intensive sectors dominate. These regions will mainly aim at profiting from the comparative advantages linked to their industrial specialization. They are essentially the regions of Extramadura, Norte, Algarve, Central and Oriental Macedonia, Thessalia, Puglia, Western Greece, the Islands and other Greek regions.

. Regions with a dual configuration, which are the most numerous, where inter–industrial and intra–industrial type exchanges are in balance. They include the following regions: Andalucia, Comunidad Valenciana, Murcia, Castilla–La Mancha, Galicia, Centro, Abbruzzi, Campania, Basilicata, Sicilia, Calabria, Peloponnese, Attica, Central Greece, West, Southwest, East, North East, Donegal, North West; and Midlands.

V.1 Probable Adjustment Strategy for the Regions to the Inter–Industrial and Intra–Industrial Scenarios

The adjustment strategy of most regions with follow the classical inter-

355

industrial or intra–industrial scenarios if dynamic regions policies are implemented. The *classic inter–industrial scenario* is that of a region whose structural achievements are characterized by low–skilled, labour–intensive sectors. In this scenario, the removal of NTBs will encourage industries to increase their production in areas where they have a comparative advantage and they will, therefore, aim to increase their inter–industrial specialization. This policy is reinforced by the existence of low wage costs. From this point of view, following the removal of NTBs, the behaviour of economic agents will favour the improvement of the internal capacity of the business but principally the adaptation of production processes and not intra–Community co–operation. They will remain oriented mainly towards the domestic market and/or towards intra–Community exports of simplified and traditionally well–made products.

Such a scenario has a certain number of short–term advantages to the extent that companies draw maximum inmediate profit from the lifting of the NTBs and north–south relocation movements are favoured. Nevertheless, this type of scenario is not without risk in the medium–term. As is already the case with industries in certain Objective 1 regions, these regions must eventually contend with competition from developing countries and changes in Eastern Europe.

The *classic intra–industrial scenario* is where regions choose to support development of capital and R&D intensive sectors. The lifting of NTBs allows more concentrated exploitation of economies of scale and an increase in competition which will increase pressure for product differentiation. The behaviour of economic agents aims jointly at an intensive investment policy in R&D (technological creation type), highly–qualified human resources and the development of intra–Community technological partnerships at the level of pre–competition research. This long–term scenario has its dangers, especially for the Objective 1 regions which must rely on the contribution of foreign investment in advanced industries and the attraction of multinacional companies. There is a risk, therefore, of the scenario leading to a dualization of the regional productive fabric between high–performance technology sectors and less–competitive local industries whose adjustment costs can be very heavy for regional economies such as Ireland.

V.2 Adjustment Policy Aimed at Innovative Inter–Industrial and Intra-Industrial Scenarios

The vulnerability of the productive fabric to regions facing the SEM suggests that it is imperative for most of the regions to develop adjustment policies which allow them to avoid the negative effects of industrial strategies associated with the two classical scenarios described above. To face the necessity of by–passing the classical adjustment schemes, two alternative scenarios are proposed and labelled as innovators because they belong to approaches that produce a double dynamic: the valorization of the existing regional economic potential; and the taking into account of structural logics determined by the creation of an internal market. One belongs to the inter–industrial logic, the other to the intra–industrial logic (see Table 4). Those two scenarios aim to improve the competitiveness of regional economic structures, and the focus especially on four areas: regional structural performance; the industrial development mode corresponding to comparative advantages of regional productive fabrics; the strategic behaviour of firms facing the growth of the NTBs; and the market orientation of firms.

The inter–industrial innovative scenario is based on classic inter–industrial trade, in which the regions profit from the opportunity of lifting NTBs to integrate with products of a higher value added in the context to homogeneous and traditional production (upgrading of the range). The behaviour of economic agents will remain centred principally upon the improvement of internal company capacity but favour policies of investment in production innovation and technology transfer. In view of this, the business manager will simultaneously develop partnership co–operation with other businesses and technology transfer institutions on the local and intra-community plane, notably in the area of technological trade and distribution. The objective will be to increase the share of the intra and extra–community market. In view of this, the regions will be able progressively to improve their specialization in more capital intensive sectors. Such a scenario is more particularly suited to Objective 1 regions which are, from the outset, more dependent on qualified and non–qualified labour–intensive sectors.

The intra–industrial innovative scenario is one in which the regions will try to profit from the lifting of the NTBs to increase the competitiveness of capital and qualified labour–intensive sectors progressively to integrate them in intra–industrial economic trade, based on existing human and technological resources. The final objective sought is product differentiation. To achieve this, the economic agents will favour agreements of intra–Community co–operation with the aim of increasing the technological content of their products. They aim to establish themselves in technological miches and/or in tailored products. They will try progressively to enter into pre–competitive European research partnerships with advanced technology businesses and research institutions. Particular attention is given to the training of highly–qualified employees and local labour to meet the new technological requirements of the business by developing co–operation with institutes of higher education. In view of this, the regions will be able to improve the achievements of their productive fabric in the capital and technology–intensive sectors. This scenario is more particularly suited to the RETI or the Objective 1 regions which possess capital–intensive sectors with a minimum of technological capacity (such as research centres and universities).

Table 4.– Proposal of Dynamic Adjustment Scenarios Adapted to the Problem of Objective 1 Regions

	Classic inter-industrial scenario	Inter-industrial innovative scenario	Intra-industrial innovative scenario	Classic intra-industrial scenario
Regional structural achievements	Dominance of non-qualified labour intensive sectors	Dominance of qualified labour intensive sectors	Dominance of sectors intensive in capital and in average technological content	Dominance of capital intensive and high technology sectors
Method of industrial development linked to the comparative advantages and IM impact	Increased industrial specialization	Upgrading of the range of products, specialist products in great demand	Technological niches Tailored products	Product differentiation
Strategy of businesses in view of NTB	Improvement of internal capacity wich emphasis on the innovation of production processes	Improvement of internal capacity with emphasis on product innovation, technological transfers	Modern management: technological expertise, technological product adaptation, technological partnerships and inter-company co-operation	Increase of R&D investments, technological production, international technological partnership
Marketing strategy	Priority orientation towards the domestic market	Initiation of Intra-Community exchanges and growth of market sectors	Initiation of intra-Community trade for target products and tailored products	Technical and non-technical economies of scale, intra-industrial strengthening, oligopolistic control
Risks and advantages	Short-term advantages: costs/margin. Medium and long-term risks: competition of less developed countries and Eastern countries	Medium-term advantages: control of intra and extra-Community market sectors. Long-term danger: competition of new less developed countries	Medium-term and long-term advantages of growing population	Dependence on foreign investments and increase in the division of the regional productive fabric high adjustment costs

Anticipated evolution: change from an inter-industrial type of economy to an intra-industrial type of economy

Source: The author.

REFERENCES

ATKINS, W.S. (1987), **The Cost of Non–Europe in Public Sector Procurement**. Atkins Management Consultants. Reading.

AYDALOT, P.; KEEBLE, D. (eds.) (1988), **High Technology Industry and Innovative Environments: The European Experience**. Routledge, London.

BALDWIN, R.A. (1984), "Trade policies in developed countries" in Jones, R.W. and Kenen, P.B. (eds.), **Handbook of International Economies**, Vol. 1, pp. 571-619. North–Holland, Amsterdam.

BLISS, C.; BRAGA DE MACEDO, J. (eds.) (1990), **Unity with Diversity in the European Economy: The Community's Southern Frontier**. Centre for Economic Policy Research, Cambridge University Press, Cambridge.

BUIGES, P.; ILKOVITZ, F.; LEDRUN, J.–F. (1990), "The sectoral impact to the Internal Market", **European Economy** (Social Europe, special issue), CEC, Brussels.

CAMAGNI, R. (ed.) (1991), **Innovation Networks**, GREMI/Belhaven Press, London and New York.

CEC (1988), **Recherche sur le cout de la non–Europe** (under the direction of P. Cecchini), Office of Official Publications for the European Communities, Luxembourg.

CORDEN, M. (1984), "The normative theory of international trade" in R.W. Jones and P.B. Kenen (eds.), **Handbook of International Economies**, Vol. I, North–Holland, Amsterdam.

DIXIT, A.; NORMAN, V. (1980), **Theory of International Trade**. Cambridge University Press, Cambridge.

EMERSON, M. (1990) 1992, **La nouvelle économie européenne: une evaluation par la Commission de la CE des effects économiques de l'achèvement du March Interieur**, De Boeck Université, Bruxelles.

ERNST & WHINNEY (1987), **The Cost of Non–Europe: Border–related Controls and Administrative Formalities**. CEC, Brussels.

MAILLAT, D.; QUEVIT, M.; SENN, L.–F. (eds.) (forthcoming), **Milieux innovateurs et reseaux d'innovation: un defi pour le développement régional**, GREMI III (ed), Universitaire de Neuchatel.

NERB, G. (1987), **A Survey of European Industry's Perception of the Likely Effects**. Directorate General of Economic and Financial Affairs. CEE, Brussels.

PA CAMBRIDGE ECONOMIC CONSULTANTS (1988), **The regional impact of policies implemented in the context of completing the Community's Internal Market by 1992, Final Report**, DG XVI, Commission of the European Communities, Brussels.

PADOA–SCHIOPPA, T. (1987), **Efficiency, stability and equity: a strategy for the evolution of the economic system of the European Community**. Report of a study group appointed by the European Communities, Brussels.

PORTER, M. (1985), **Competitive Advantage**, Free Press, New York.

QUEVIT, M.; HOUARD, J.; BODSON, S.; DANGOISSE, A. (1991), **Impact regional 1992: les Regions de tradition industrielle**. De Boeck Université, Bruxelles.

QUEVIT, M. (1990), "Regional technology trajectories and European research and technology development policies" in E. Crociatti, M. Alderman and A. Thwaites (eds.), **Technology Change in a Spatial Context: Theory, Empirical Evidence and Policy**, pp. 317–38. Springer–Verlag, Berlin/Heidelberg.

16 BOOSTING EUROPEAN GROWTH: STRATEGIES FOR INTEGRATION AND STRATEGIES FOR COMPETITION

Sergio Bruno
U. La Sapienza, Roma

I. INTRODUCTION

The purpose of this paper is that of outlining a European set of policies for boosting a new phase of development. The policy proposals make the maximum (and hopefully the best) possible use of the still limited knowledge we begin to have on how economic systems behave when changing. The reasons why the proposals can only be "outlined" are dealt with in the last section.

Our main arguments may be summarised as follows. We are not experiencing the downswing of an ordinary business cycle, but rather a "phase" of a more profound behavioural/institutional transformation. The situation may thus last and even worsen. No single country alone can likely lead Europe (and the rest of the world) towards a way out from a crisis, which risks thus to be of the same weight and hardship of the recession of the thirties.

The present phase of recession is basically induced by wrong policies and attitudes, guided by models which are inadequate to interpret the basically dynamic features of our economic systems. The problems of change and of growth have been overlooked, and an excess of emphasis has been put on short run and on monetary/financial problems. An equal and connected emphasis has been put on international competitiveness, with the result of triggering a self–defeating cathing-up game, both between companies and between public policies, at an infra–European level.

We need a new phase of development to take off. Only in such a framework the problems of the public debt and of unemployment would have a chance to be properly solved. We do believe that the activation of a growth process, though difficult, is possible. The policies which are needed to achieve such a result should be managed at the European level, should establish a credible climate of expectations about a lengthy expansionary phase, and should be able to keep under control the time–sequence of distortions, concerning the flow and stock aggregate magnitudes of the economy, which takes necessarily place during the transition to higher growth rates.

X. Vence-Deza and J. S. Metcalfe (eds.), Wealth from Diversity, 361–387.

The instruments to be activated in order to achieve such goals are then outlined. Most of them are not new. However their coordinated use makes a rather innovative policy. The advocated policies are conceived as genuinely European, aiming at softening internal European conflicts and at building up, instead, a full fledged competitive pole at the planetary level. On the all, this should strengthen the tendencies towards a European political and economic union. The option for the European Monetary Unification (EMU) is maintained, but the emphasis on it is down–graded with respect to the prior need of building up an institutional and technical framework able to deal with the issues of the real side of our economies.

II. THE INADEQUACY OF THE PREVAILING APPROACHES

The hypothesis that the present recession is not an ordinary business cycle has important implications. The chances for a spontaneous way out from the present recession are poor, and it would be risky to rely, as in the past, on the hope that some of the larger or more economically powerful countries may lead the remaining ones towards an expansionary cycle[101]. The recession can last well beyond the length of a normal business cycle; it can also worsen because of wrong reactions to the length of the crisis[102].

We hold that the major evil which is now affecting Europe is an excess of aggressive, destructive, un–regulated competition. However, this should be regarded as connected to, and as the end–result of a long and complex sequence of "errors", mainly concerning the management of macroeconomic policies and their connections with "meso" and micro policies and agents' behaviours[103].

The policies are wrong because the underlying models according to which the functioning of the economic systems is interpreted, though shared by the overwhelming majority of the economists, are inadequate for understanding and managing the present phase of rapid change[104]. The existing models had been conceived instead for facing the problems of a relatively stable environment, and only slightly re–adapted since then[105].

---

[101] Although such a possibility cannot be completely ruled out, the likelihood of its occurrence is highly restricted.

[102] As it will be better explained later on, the attempts to react to the recession by the means of short breath expansionary spells are likely to fail or to result exceedingly weak. On the other hand, their reinforcement, out of a long run program of the kind proposed here, would likely elicit inflationary pressures; the need to react to this by the means of traditional restrictive policies would then result as disastrous, making the shift to a development phase even more remote.

[103] Bruno (1986, 1987, 1989).

[104] Amendola–Gaffard (1988), Bruno (1987, 1993), Bruno–De Lellis (1992).

[105] This raises problems which, though having a basically political and epistemological nature, present practical and concrete implications. The discussion of this issue would be of the utmost importance, since, if the dominating paradigm is inadequate, what is required to the political leader wanting to change the present state of affairs is to take decisions which go against the more commonly shared economic beliefs. And even more will this require courage and wisdom, the farther is the present state of un–orthodox approaches from the possibility of communicating "messages" susceptible

In particular, the models which inspire governments' conduct, presently as well at the times when the so called Keynesian models were more popular, are basically focused on the behaviour of short run variables. They overlook the complexity of capital accumulation and of the dynamic paths associated to it. The attention they pay on expectations is essentially limited to prices, while the role which the expectations play in determining the investments and the overall strategies, adopted by the firms and by the institutions in order to affect the real side variables of the economy, is almost completely neglected.

Even more specifically, we can assert that no attention whatsoever has been paid to the effects of public actions on the climate of expectations induced among the economic agents. We need growth, and this requires investments, but no wise company would venture in any ambitious investment plan so long as all the leading authorities and influential characters keep on promising them gloomy horizons. In the context of such a mood, even some spells of traditional demand management would fail, since the companies would interpret them as transitory phenomena, undertaken along the traditional approach of "stop and go policies", which would not deserve to be responded to through an increase in the investment rates. It would not be false, then, to say that the present phase is basically produced and induced by a series of repeated negative messages conveyed by public authorities to the economic agents[106].

III. NATIONS' AND COMPANIES' RIVALRY AS A SELF-DEFEATING GAME

The policies which have been widely practiced in Europe since the early seventies are basically restrictive and have produced devastating effects on expectations. These effects have been substantially overlooked, though they have to be regarded, not surprisingly, as the major responsible for the fall of growth trends and for the substantial employment stagnation[107] since the seventies.

Such policies were used by the single European countries not only for keeping the inflationary pressures under control, but also, together with wages and labour market policies, in order to make their domestic production more competitive with respect to that of the other countries; and since most of the exports of the European countries go to other European countries, this has resulted in an a boost

---

of being easily grasped and rapidly implemented. In any case see section 10 below.

[106] This helps explaining why the lowering of interest rates should not be regarded as a sufficient response. If the firms expect a stagnant demand and prices which are unrewarding because of the competitive aggressiveness, they have no reason to invest, independently of the interest rate level. The lowering of interest rate would then ease the financial strain which is faced by most of the firms, and would occasionally allow some additional investments in few specific sectors, but could not elicit by itself a reversal to growth.

[107] Apart from short run movements and, above all, structural reshuffling in favour of the tertiary activities, which helped sustaining the overall employment level, because of the increasing weight of the activities with a lower dynamics of productivity.

of infra–European rivalry, initially played mainly at the level of public strategies.

However, since such strategies were played by all the countries, the resulting catching–up game was basically self–defeating[108], with a substantial stability of the shares of infra–European trade[109].

The above mentioned inter–countries competition concerning policy strategies accompanied,induced or reinforced an increasing rivalry among firms, both at the international level and at the national levels. The end of a highly growing environment pushed in the seventies the European firms to undertake processes of reorganisation and restructuring and reinforced their competitive attitudes. The initial adjustment processes worked in many activities as a learning–inducing experience: many firms, in certain cases involving entire sectors, learnt how to innovate and pursue innovation. The eighties became thus a period of high and increasing rivalry between the European companies, initially played essentially by the means of qualitative/innovative competition, and since the end of the eighties increasingly in terms also of price competition. Such a boosting rivalry added itself to the inter-country rivalry recalled above.

However, also inter–firm rivalry is a catching–up game, which can result, in the long run, not only self–defeating but even destructive, as the more recent European tendencies to profit squeeze and financial strains are likely to prove.

In order to understand how it is possible that competition may become a destructive process, one has to keep the distance from the traditional theory, which depicts competition essentially in terms of equilibrium; that is, a state of quiet, only occasionally disturbed by lacks of efficiency and/or by attempts to collude and to raise prices above the marginal costs. The more recent generalisation of competitive theory in terms of contestable markets, which widely inspired (or at least reinforced) the de–regulatory wave of the eighties, strengthened such a static connotation of competition[110].

The competition of the real world is a different thing. *It is an ever–moving catching–up game, where each of the players attempts to gain market shares at the expenses of the other players*. If the game is pushed too far, it may become a deadly game for at least part of the gamblers. The more this tends to happen the more the firms have only a tentative and partial control over their costs, and the more some of their aggressive, cost–cutting actions produce negative effects which become perceivable only after relevant time lags.

---

[108] Also because of the possibility (and sometimes the practice) of competitive devaluations.

[109] Bruno–Sardoni (1989).

[110] Baumol–Panzar–Willig (1982).

Our feeling is that, in the last year or so, many firms have found themselves with costs in excess of the expected ones, as a consequence of restricted sales and the impossibility of repaying past expenditures for investment and Research and Development, and with prices too low, with respect both to expected prices and to actual costs, because of an exceedingly high rivalry. Many of them are now likely to be pushed by their own aggressiveness, or compelled by the aggressiveness of other firms, to cut expenditures having the nature of investments, impairing thus the possibility of future recovery and efficiency[111].

Such a destructive trend might be interrupted only through tacit or explicit collusion, aiming at self–restraining rivalry[112]. However, the rivalry game is easily started, but hard to be interrupted, if not after some important business failure, above all in a situation of recession.

When the situation reaches such a stage, in any case, the effects of the excess of competition become themselves macroeconomically relevant, reinforcing the restrictionary effects of macro–policies, and putting into motion a vicious circle.

IV. THE INSTITUTIONAL REINFORCEMENT OF THE TENDENCY TOWARDS EXCESS COMPETITION

The tendency towards aggressive rivalry finds its origins in the conjunction of the overall decline of growth and the increasing macroeconomic uncertainty since the early seventies[113], and of the innovative attitudes flourished among the firms (since the end of the seventies), which brought them to compete increasingly through the quality of their products/services and the efficiency of the underlying productions processes.

In our view, thus, the tendency was simply reinforced by the deregulatory attitudes which acquired strength within most of the European Member States during the eighties[114].

More recently, however, the role of institutions, and in particular of European institutions, is becoming of overwhelming importance in pushing towards

---

[111] We are referring here to any kind of expenditures which affect the production and/or the productivity in future periods, and not, or not only, in the period during which they are met. Such a feature concerns not only the expenditures corresponding to the acquisition of physical capital, but usually also those met for Research and Development, for training, and even those for paying current wages and salaries as long as there are relevant processes of learning by doing [Amendola–Gaffard (1988)].

[112] The same type of collusion which prevailed in the thirty years following World War II.

[113] The higher is the rate of growth, the lesser the firms need to compete for market shares in order to grow.

[114] Such attitudes produced instead dismantling effects above all for what concerns the sphere of social policies. The reinforcement of aggressive competition acted thus mainly through the erosion of industrial relations and labour market standards.

a state of unconstrained (and thus possibly destructive) competitiveness. This is due to the undesired implications of the rules which had been set up for the unified European market, in conjunction with the change of attitudes which has been produced, at the European policy level, by the advent of the benchmark of European Monetary Union (EMU).

Just a bit of ideologically un–biased historical reflection is needed to argue that, within single national states, industrial policy authorities and anti–trust authorities (whatever their labels) behave as countervailing powers in an ever-moving dialectical game. None of them, indeed, has historically acted on the basis of a stable and uniform set of criteria, which have been instead changed over time and across different historical environments, in response to contingent pressures and options[115].

The political U–turn which has led the European partners and the EC to focus on the issue of monetary union, despite the dangers of EMU in terms of the activation of anti–cohesive forces[116], has shifted the attention away from the real–side economic problems and interrupted the long–lasting construction of European strategies in the fields of industrial and social policies. This contributed to overlook the fact that, in the absence of a real European authority in the domain of industrial policy, the rules aimed at allowing the implementation of the European unified market would have been interpreted and implemented by the European High Court as rules for anti–trust actions. This is what is happening more recently, on the basis of the initiative either of firms or of the Member States against other firms belonging to a different foreign European country. However, in the absence of a European authority in the domain of industrial policy, the Court has become the only arbiter in the definition of which behaviours might be considered as acceptable in terms of market strategies, with the immediate result of rapidly dismantling what was left over of the previous national industrial policies.

Competition is thus becoming an end in itself, as the result of sentences dictated by a mere juridical viewpoint; that is, something which has never been advocated even by the more enthusiast liberal economists nor has been pursued within the most established institutional tradition in antitrust activities, such as that of U.S.A. Competition was in fact mainly regarded by them as a mere instrument for reaching efficiency under specific environmental conditions, and never an end in itself.

Furthermore, according to the more recent approaches inspired by Coase's early contributions, which appear now to be broadly accepted even by neoclassical

---

[115] This is corroborated by the contingent scope of most of the publications about these issues, despite the professional tendency of the economists to treat them at the most theoretical and general level.

[116] Which are well visible now but had been clearly forecasted [Bruno (1992)].

scientific environments, the markets appear to be as institutionally framed spaces for transactions, governed by rules which are only in part defined by the countries' general commercial legislation, while largely depend on a regulatory framework, which is defined and continuously reshaped by a set of customary or legally accepted actors, either de facto or by the means of legally binding but contingent statements[117]. Both the authorities which are in charge of industrial and social policy, and the authorities which are in charge of antitrust control (when they exist), belong to such a set, with their roles being framed within a reciprocal "check and balance" play context.

The conduct of economic issues, both industrial and social, cannot be left only to the legal culture such as it can be manifested in the Courts, that is, as the formally derived implementation of a statically "given" legislation[118]. It must be much more flexible and it must respond to much wider political, technological, industrial and social concerns.

V. THE NEED TO SEEK FOR OVERALL EUROPEAN COMPETITIVENESS

An exceedingly low degree of competition is an evil; an excess of aggressive and unregulated competition may result disastrous.

Europe has presently more the problem of strengthening its overall competitive ability with respect to the other planetary competitive poles than that of inducing an exceedingly high competition among the European firms. In most of the strategic sectors the dimensions of European companies are on average too small to compete effectively with the largest companies involved in global competition; therefore, a reshuffling of strategies leading to more concentration and/or to more cooperation among European companies would highly help, above all for what concerns Research and Development and marketing activities.

On the other hand, the European leaders should keep in their minds that the other planetary competitive poles do have industrial and trade policies. Europe not only has not such policies, but presently there is, as we saw earlier, a real "effort" to dismantle the corresponding national policies. However correct such a behaviour might be from a purely legal viewpoint, such actions cannot but have negative effects on European competitiveness at a planetary level.

The establishment of a strong European Authority for industrial and trade policy is likely to be the most effective reaction to the above recalled dangers. Such

---

[117] Coase (1937, 1988), Williamson (1985).

[118] In our case such a legislation, being the result of complex international agreements, could not be easily and rapidly up-dated.

an Authority, however, should be properly endowed from the viewpoint of the juridical, technical and financial instruments. Above all we regard it as the institutional agency which is necessary to establish a link with a set of more systemic actions aimed at eliciting a new phase of development.

We believe we do need a new phase of development to take off. Only in such a framework the problems of the public debt and of unemployment –which affect, though with different intensity, all the European Countries– would have a chance to be properly solved. On the desirability of this there could not be more agreement. The problems arise, however, when discussing *whether it is possible, or, for the less pessimistic, when and how it is better to induce the advocated expansion.*

VI. THE NEED FOR A LENGTHY GROWTH

We take for granted that a development is possible. Our systems, though made quite rigid by the more recent tendencies of business behaviours and of public policies, have potentially the margins and the culture for putting again into motion processes of capital widening, that is, processes aimed at enlarging the productive capacity together with the employment.

However, most of the members of the economists' community think that it is not the right time to push for growth; before doing so, indeed, they assert that the public budgets and the inflationary pressures should be kept under better control.

What makes this attitude rather odd, however, is that most of the economists have been keeping on thinking in such a way, and producing consistent advising to the leading authorities, for the last decade or so.

The right moment was never the present one.

The doubt that the budgetary and the inflationary problems might have been someway correlated to an insufficient rate of growth –though obviously through a system of causal links more complex than those considered by the standard short–run models– apparently never crossed the minds of the economic establishment.

It is our view that behind inflation there is a multiplicity of causes[119]; and that among such causes an important one, which has been gaining momentum during the eighties, is a dynamic shortage of supply, due to an insufficient accumulation of

---

[119] The hypothesis of multi–causality –which is widely embodied in the present attitude of many disciplines, above all in the biological sphere– is rarely used in standard economic models.

productive capacity [Bruno (1986, 1987)][120]. Should such a hypothesis be correct, even only partially, the reinforcement of inflationary pressures in our economic systems might be attributed to the lack of development, and thus indirectly to the attitude of those policy–makers (and their economic advisors) for whom it is never the proper moment for eliciting growth.

VII. HOW TO PROCEED TOWARDS A DEVELOPMENT PROCESS

The real problem is instead *how* to induce growth; more precisely

a) which are the conditions which are necessary to disrupt the forces which keep the economy in a stagnation trap (or more in general around a certain unsatisfactory rate of growth) and to put the system into motion towards an expansion;

b) which are the features of the dynamic path towards a higher growth rate, namely, which types of "distortions" should necessarily be associated to such a transition.

This appears to be an analytical problem nearly forgotten by the majority of the economists since the end of the great literature on economic development, led by authors like Kaldor (1957a, 1957b), Hirschman (1958), Streeten (1959), Lewis, Rosenstein Rodan (1943), Nurkse (1953), Scitovsky (1959), and since the precious, but neglected contribution by Richardson (1960)[121]. The best indication that can be drawn from the above literature and from further reflections and elaborations may be summarised as follows.

In order to put a process of growth into motion, a credible climate of expectations of a lengthy expansionary phase should be established throughout Europe, while removing those obstacles, above all the financial ones, which might discourage or constrain the preliminary stage of investments expansion[122].

However simple the task might seem from the above statement, it will come out that it is not so. What is implied is in fact *a rupture of extrapolative expectations in the minds of the agents, and a radical change of models in the minds of the policy makers*.

---

[120] We are equally convinced that, during the same period, the countries have attempted to sustain their economies by gaining export shares, but that this has resulted in a fundamentally useless catching up game [Bruno–Sardoni (1989)].

[121] Bruno–De Lellis (1992) provide a critical survey and argue the need for re–examining such contributions and further exploiting the hints which are opened by them.

[122] Bruno–De Lellis (1992).

The agents, which are presently trapped by negative prospects for what concerns expected sales, due to the stagnant shape of the macroeconomic environment, and for what concerns profits, due to the low margins on costs brought about by the exceedingly high degree of competition, have to be convinced that at certain conditions, which involve their own beliefs about the future of the economy, and their own behaviours in terms of investments, a new phase of expansion will occur. Moreover, such a view must become widely shared among the agents[123].

The policy–makers, on their turn, must be the first ones to be convinced of such a possibility, since this is the condition for undertaking the actions which are needed to convince the agents.

To start a process of expansion, however, is simply the first step.

The process should then be kept alive, and this is an even more difficult task. During such a process, in fact, the economy will unavoidably run into a series of distortions.

If the ensuing unbalances are not properly interpreted and managed, but *are regarded* instead *according the spectacles provided by the old models, a violent interruption of the development process and a return to recession would be the more likely outcome*.

On the policy side all this requires on the one to motivate the companies to invest in capital widening (and obviously to remove any obstacles to investment), on the other one to keep under control the disposable income, though avoiding to discourage the motivations to invest.

VIII. THE DYNAMIC PROBLEMS BROUGHT ABOUT BY THE TRANSITION TO EXPANSION: THE RATIONALE FOR MONETARY AND FISCAL POLICIES

A process of change –and the taking off of a development process is a change– *cannot but be a process of "distortion"*. To put into motion a change amounts to induce a feasible sequence, where additional inputs come *before* additional outputs, where additional productive capacity should be set up *before* it might be used for producing more consumption goods[124], where t*he activities which are performed by different subjects should be coordinated*[125].

---

[123] In other words what is suggested is that at certain conditions economic systems may behave as self–fulfilling expectations systems. The extraordinary phase of development after World War II may be interpreted in such a way.

[124] As it has been argued by path–breaking contributions of Hicks and Amendola–Gaffard.

[125] Bruno–De Lellis (1992).

370

VIII.1 The Induction of Expansionary Expectations

In the present mood, the firms are not only unlikely to invest in additional capacity, but even to hire additional workers, should they expect a purely transitory upswing. They need to rely —as it has been observed already— on a credible promise of a lengthy development phase.

This implies that in the first stages of the change the investments should be motivated by expectations which by no means can be the extrapolation of previous trends: *not thus actual demand*, but instead reliable prospects for *future demand* should be the motivating engine of the envisaged change; a task which is not at all easy to achieve. The bulk of a growth strategy is to build a strong and shared belief —among the economic agents— that a durable period of expansion is not only possible, but that it will constitute an undefeatable priority of European governments for the next decade.

The ideal would be a solemn statement that the expansionary policy should not be abandoned independently of the unbalances. However, this would not be credible, while, in order to convince people to risk, the statements of governments must be credible. A more complex and articulated strategy should thus be defined; namely, an appropriate deal centred around a development plan, where not only the actions to be undertaken by the relevant subjects should be scheduled and agreed upon, but

– the expected state of the main economic indicators along the transition, conditional to the action concerted between the involved partners (private as well as public), should be spelled out;

– should the expected values not correspond to the actual ones, the actions, or the procedures to decide the actions to be undertaken, ought to be clearly defined in advance;

– the governments should make clear that the expansionary program may be either relented or strengthened, according to the state of the indicators, but not abandoned;

– the subjects —be them companies or workers— who accept to comply with the indications of the program should be someway "insured" for the losses they might incur because of the compliance to the program.

VIII.2 Micro and Macro Economic Dynamic Distortions

Let us delay any operational discussion about how to reach this goal, and let us assume that the governments have someway succeeded in producing such a

371

change of attitudes among the firms, and that the firms are willing to embark themselves in a sharp increase of their investment rates. The fact that they want to invest is not enough. Two further problems would arise, the first one at the micro, and the second one at the macro economic level.

Financial resources should be made available to the companies willing to embark in what cannot but be considered as an over–risky decision (no matter how strong the public reassurances might be); no ordinary credit mechanisms would do. Special credit conditions, connected with the system of mechanisms and conditional forecasting of the concerted plan referred to above, should be devised and implemented.

At the macroeconomic level, a displacement in real terms of the relative weight of the productive activities, those for consumption and those for investments, in favour of the latter ones, must occur.

In monetary economies the two events –namely, the provision of larger liquidity to companies willing to invest and the reallocation of the relative weight of productive factors– are strictly connected, and cannot but consist in a sequence of dynamic distortions affecting prices and relative incomes.

VIII.3 The Liquidity Cycle Brought about by an Acceleration of Growth

Any process of change affecting the productive capacity implies a time-sequence of distortions concerning the flow and stock aggregate magnitudes of the economy; in particular an additional inflow of liquidity is required in order to finance the additional investments. Such a liquidity, however, ultimately transforms itself into payments to the additional factors of production engaged in the building up of capital equipments, and produces thus the expansion of disposable income, and possibly of consumption demand. This, however, is likely to happen before the production flow made possible by the additional investments might be activated. A supply/demand gap is thus likely to occur[126].

---

[126] Let us suppose, for the sake of simplicity, to start from a situation of purely reproductive stationary state, and that the firms, which had ordered until then only the replacement investments I, begin to order additional investments at a positive rate of increase of g per period. Since the initial cashflow of firms (in excess of current inputs payments) is I by definition, the firms must borrow gI the first period, $g(1+g)I$ the second period, and so forth, until the moment –say N periods after the first increase has occurred– when the additional productive capacity is accomplished and the corresponding additional production may start to yield additional proceeds.

However, the sums which are borrowed are used to pay the additional factors of production which build the capital equipments, and at least a part of them will increase the income which is disposable for demanding the same goods which ought to be produced with the additional capacity whose construction is still on the way. This explains the possible supply/demand gap.

Such a result strictly depends on the fact of assuming that the production of currently consumed goods takes a fraction of the time which is necessary, instead, to build up the corresponding productive capacity. Such an assumption –completely overlooked by standard dynamic models, which assume that investments transform themselves into productive capacity the period after having being ordered– is widely corroborated by factual observation.

Such a gap may result either in a tension over prices or in a tendency to increase imports.

It could be argued that such unbalances are someway physiological in the taking off of a development process and that they have nothing in common with the —only descriptively similar— unbalances associated to certain phases of stagflation managed through stop–and–go type of policies.

However, the policy–makers and the public opinion would be hardly convinced of this, in face of the bad and superficial cognitive attitudes acquired in the past.

Since, once the expectations in a lengthy period of expansion had been activated, their frustration would be the worst of the possible evils, it is advisable not to rely only on persuasiveness. It stems from here the opportunity to attempt to avoid, or anyway to minimise the impact of such unbalances.

At this purpose we will suggest a set of coordinated actions in order to pursue such a goal. They will concern the time scheduling of the take off process, the opportunity to control selectively the speed of growth of different sectors according to their different flexibility to expansion, the international coordination of the expansionary effort, the need to control disposable income through time and other complementary measures. We will delay their illustration (see section IX.6 below), since most of them are connected with the strategies aiming at changing the expectations mood and at letting a wide plan of investments to take off, so that they will become more meaningful once such issues will have been dealt with.

IX. THE POLICY ACTIONS

In the previous sections we have made explicit some of the complexities and some of the problems of coordination and interdependencies which have to be faced in order to activate and to keep alive a development process. They appear to be so sharp and strong that strategic simplifications are needed.

IX.1 The Need of Special Relationships with European Strategic Companies

The first simplification consists in the acknowledgement of the differences which exist between the companies which have, or may potentially acquire, a strategic relevance at the European level, and those which have a basically local relevance.

The first ones have to be involved in the concerted definition and in the cooperative implementation of strategic European programs. Such firms must become the "European champions" for future planetary competition. At the same time these firms have to become the crucial actors which have to mature a real change of expectations and reverberate them towards the rest of the economy.

The second set of firms has instead to be involved more indirectly, either as subcontractors of the first group, or by the means of automatic fiscal and monetary incentives, or through regional programs.

IX.2 Short–Medium Term Programs and Strategic Programs

In order to minimise unavoidable unbalances, where and when to induce the necessary acceleration of production and of investments should be carefully scheduled. A second strategic simplification consists thus in the distinction between short–medium term programs and long–term, strategic programs.

The present phase of the recession is such that some of the sectors have margins of unused capacity, while in other ones such margins are restricted, either because the fall of demand concerning their products has been lower than the average or because the previous state of expectations had already led them to slow down their pace of capital accumulation.

In the first group of sectors the demand should be reactivated immediately, with an intensity proportional to the margin of unused capacity. Therefore, targeted programs should be devised such as to put immediately into motion the sectors which have a higher degree of presently unused capacity or appear to be anyway more flexible to expansion[127]. Such programs should have a higher temporal priority, independently of the priority they would have because of the importance of the needs which would be fulfilled[128]. In other words, such activities and their products need not to be important in themselves, but simply because their activation may "put things into motion" and favour the formation and the diffusion of a climate of positive expectations[129].

On the other hand, in the sectors with a low margin of unused capacity, the expansion should be scheduled in such a way as to start with the investments aimed at building up additional productive capacity. This implies the ability of inducing

---

[127] That is, sectors which need, for the simplicity of their productive activities, a shorter time for becoming ready to supply additional output in response to a higher level of demand without exerting pressures upon prices.

[128] What is proposed here is basically a classical indication of demand management. Not only selectivity is important, however, but the smoothness of the time scheduling: even if the margin of unused capacity were high, an exceedingly fast acceleration should be avoided, in order to keep under control initially the indirect effects through disposable income. It would be important, on the other hand, to make clear to the concerned firms that the expansionary trends might be accelerated in the future, should they invest enough.

[129] The expansion of such activities should be likely relented in a second phase, in order to release the financial and human resources which are more needed in the expansion of more strategic activities.

enough trust on a durable phase of expansion. Therefore programs conceived as more long–term ones should be set up, independently of unused capacity, on the basis of perceptions of the priorities concerning technological opportunities and emerging needs. In other words, the long run programs have to be strategic programs, whose referees should obviously be the *European champions* referred to above and their subcontractors, and they should be defined with *the aim of pursuing innovative opportunities for acquiring planetary competitive advantages.*

IX.3 The Institutional Set–up

Though it is not our task to draw the exact contours of the institutional arrangements which ought to be set up in order to produce the programs referred to in the previous sections, a few words have to be spent on the necessary, and by no means sufficient conditions that have to be met.

An Authority for industrial policies must be ultimately set up at the European level, with a status of prestige and power similar to that which had been conceived for the European Monetary Institute, and with competences and skills similar to those of MITI in Japan.

Since the establishment of such an Authority will require time and it will highly benefit from previous pilot experiences, transitory arrangements may (should) be started earlier, also because, while the legal instruments of which the Authority might be empowered is certainly important, what is even more important is the establishment of a *network of cognitive abilities and of positive attitudes towards informational exchanges*, both among firms and among entrepreneurial associations on the one side, and between such subjects and formal governmental and European institutions an the other side[130].

The transitory arrangements could consist in the establishment of a task force at the European level, which can use the support of the existing offices of the Commission and proceed through negotiations. The partners of such negotiations are the industrial associations at sectoral levels, the largest European companies in

---

[130] This stems from the need to abandon the previous explicit as well as implicit schemes of economic policy. These are in fact based on two assumptions which have to be abandoned. The first one is that the public authorities have a larger and better knowledge than the companies. The second one is that the companies passively respond to changes of constraints and stimuli established by the public authorities through specific changes of their behaviours, whose patterns are univocally known in advance by the policy–makers. Both assumptions are false.

The policy–makers and the firms have partially different cognitive sets, and both can gain from targeted and selective exchanges of information, above all when the aim is that of constructing a commonly shared system of expectations. Furthermore, the firms, far from being the mechanical automata implicitly defined by the standard economic literature, are able to develop strategic thoughts and behaviours, that is, are endowed with systemic consciousness, even if such cognitive abilities are not so extreme and evenly distributed as the rational expectations hypothesis asserts.

general and, in the transitory phase, the industry Ministries of the Member States[131], the social partners, occasionally the leaders of the Member States governments. Such actors should have different roles, in general and in relation with the different types of programs mentioned above (and their coordination); also the procedures and the "products" might meaningfully differ. Let us discuss these issues separately in rough terms in the light of the simplifications introduced earlier, starting from the strategic programs.

IX.4 Strategic Programs/European Champions

A preliminary problem is that of establishing which are the strategic activities for which to establish the strategic European programs. The solution –or at least the initial and preliminary solution– is easier to find than it might appear. We propose here to "paraphrase" the contestability principle, made so popular by Baumol, Panzar and Willig (1982) for regulating competition: Europe should consider as "strategic" for present and future competition those activities which are considered as such by the other competitive poles at the planetary level, first of all USA and Japan[132], and because of this are subject to special attention and support on the side of the governments and/or of broad industrial alliances.

The contestability principle suggests also the directions where to proceed to in the establishment of specific programs. The aim is in fact that of avoiding that the activities which are carried on in Europe in any given strategic domain might be damaged or might not have the same chances of being successful than the corresponding activities undertaken by companies belonging to the other global competitive poles, because of better structural or environmental conditions, or because of the actions of their governments. In a first phase, thus, the role of the institutional network (including thus the whole of the involved actors), should aim at the compensation of competitive disadvantages. The ways and the instruments through which to pursue such a task cannot be defined a priori, since they highly differ for the different activities. In many cases there is a problem of lacking appropriate captive markets. In other cases there is a problem of insufficient companies' scale, which does not allow a competitive critical mass of R&D efforts[133]. In other cases yet there are problems of (quantitatively or qualitatively) insufficient public infrastructuring. In most of the cases, finally, there are problems

---

[131] We are thinking at the Ministries of all the European Member States. At the beginning, however, it would be worthwhile to proceed even with some absence, that is, gathering the Ministries of the states which are interested to establish a relatively transparent coordination and willing to cooperate in such a direction.

[132] Notice that this is a merely defensive criterion, which ought to be adopted and maintained only in a first phase, during which the European partners –both companies and Member States– must overtake their reciprocal mistrust and their ideological biases in favour of unconstrained competition. The criterion should be dropped later on.

[133] This is the case, e.g., of the sector which produces for telecommunication, which is characterised by an excess of (comparative) fragmentation. At this purpose we talked of the need to "oligopolarise" the competition [Amendola et al. (1991)]. By no means this implies the need of companies' fusion, but simply of strategic cooperative alliances.

of insufficient coordination, partly due to insufficient attitudes towards informational exchanges and partly due to residual nationalistic attitudes.

The definition of tasks and procedures is of the utmost importance.

Companies are regarded as being diffident because they fear that the information they make available can be exploited against them by actual or potential competitors. This belief is less true and important than the civil servants tend to think. Firms, above all the more innovative ones, are more accustomed to "trade" pieces of information than it is usually thought (as our field investigations show), and above all more trained to do so avoiding risks of opportunistic behaviours of the other dealing firms. The problem is thus that of enlarging the occasions and the "desks" where to deal, that of supporting the transaction costs which are met for these purposes, and only finally to offer more guarantees for what concerns the opportunistic exploitation of shared information. At this purpose, anyway, what has to be kept in mind is that the risk of opportunistic behaviours largely depends on the context which exists or which is created; if the aim and the occasion of negotiations is that of constructing an expansionary environment, where all the participating partners have to gain more with respect to the expected ongoing environment, the forces pushing for aggressive and exploiting behaviours would slacken, while those pushing for cooperation would be fostered.

Member States are regarded as being diffident, since they are brought to think that the agreed actions may result of (more) advantage to other countries and not (than) for themselves. This perception is correct, though the arguments about the difference which is brought about by the bet to construct an expansionary environment should be of help also against such a tendency. There are ways of fighting against such an attitude or at least for establishing fair systems of compensation[134].

The ultimate purpose of the institutional activities should be that of establishing one or several programs in each of the considered strategic activities. Such programs have to be regarded as temporally specified schedules of agenda of actions and of actors implementing the actions, accompanied by a system of deals among the involved actors, be them private as well as institutions. Firms may establish deals for various forms of cooperation and/or for binding the spheres of hegemony concerning the markets; the public actor may deal for preserving a certain

---

[134] One of the best strategies which has been experienced in the past has been that of compelling firms involved in European contracts to constitute pools involving firms of different member countries. Given this constraint, which appears to be inefficient in itself according to pure static criteria, it was of interest of the firms of the initial pool belonging to most efficient environments to select the most promising firms belonging to the more backward environments and promote, through technical and organisational assistance, their fastest upgrading.

More in general, the forms of compensation which might be envisaged and experienced are various and are limited only by the lack of imagination. If several strategical programs are set up, and not only one, the possibilities of balancing the opportunities and the actual benefits at the infra–European level are highly enlarged.

degree of competition, for setting limits on prices, for prescribing standards, etc., using its power, in terms of normative and financial instruments (including public demand), as a means of suasion. He may also play the role of guarantee and arbiter with respect to the inter–firms deals.

These results have to be attained through a round–about process, aiming at letting gradually emerge the problems, the options and the possible actors. A system of conferences to be held at the European level, focused upon sectoral problems or strategic problems, and involving both the largest companies and specialised representatives of the social partners and of the concerned ministries of the Member States may be the starting point[135]. Beginning from this, however, the further stages of the negotial procedure should aim at involving only the companies which are potentially interested to go up to the end in the definition of the strategic programs. The provisional definitions of the programs might be further discussed with the social partners before they are eventually decided.

The public authority, thus, on the one side has the role of promoting or easing the definition of deals among companies, but on the other one may become himself a dealer, either as a third party of inter–firms deal, with the purpose of favouring the decision to deal, or instead as a real counterpart of pools of companies, with the aim of conferring certain features to their behaviours. This explains why the envisaged industrial European Authority should be empowered with autonomous normative as well as financial instruments and a high level of prestige should be granted to it. The tasks which have been defined often imply set–ups which cannot but be derogatory with respect to a more general system of rules concerning the competitive behaviours and criteria; and it is not conceivable that these derogatory practices and decisions should conform to the ordinary procedure which constrains the European legislation. It would be contradictory, indeed, with the basically negotial nature of the procedures according to which the programs are defined, and with the needs of flexibility, but also of timeliness and reliability they require in order to work. However anomalous the envisaged political and legal regime which is envisaged might appear at a first glance, it does not differ actually from what had been envisaged about the power and above all the autonomy attributed to the European Monetary Institute.

IX.5 Short–Medium Term Programs

Here the aim of the programs is normally not that of producing deals among specific partners, but on the one side that of providing reliable signals to the multitude of firms belonging to given sectors, and on the other one to the

---

[135] This stage might be in common with the procedures aiming at setting up the short–run and more traditional programs.

governments, both at national and (European) regional levels.

The role of the Authority (or of its provisional equivalent) is thus, first of all, that of an active and reliable monitoring of the sectors. This can be obtained through the improvement of data collection concerning the industrial activities. The information which is needed concerns timely data on costs, prices, demand, actual and potential output, orders and accomplishment of investments, expectations and plans. The aim of the monitoring activities is that of forecasting the margins the different sectors have for responding in a non inflationary way to demand increases, and that of finding out ways for addressing the demand towards those productions which appear to be more flexible to expansion.

The activity of data collection and data elaboration could be conveniently integrated by sectoral conferences, of the same kind of those which have been proposed for starting the processes aimed at the elaboration of strategic programs. Also here it exists a problem of construction of a climate of shared expectations favourable to expansion, and that of the promotion of consistent investment increases.

The monitoring and coordination activity might carry to the suggestion to use fiscal and credit incentives, and to the advising –addressed to national and regional governments– concerning their expenditure plans; such advising might concern when to spend in certain directions, and thus also to which plans attribute a higher priority. Two complementary tasks would result as useful. The first one would be to monitor the expenditure plans of the governments (national and regional), helping thus the process of formation of plans and coordination within the involved sectors. Conversely, the Authority might provide back to the governments information about prices and contractual conditions, in such a way as to help keeping prices and public expenditures the lowest possible[136].

IX.6 Actions Aiming at the Control of Disposable Income and Prices

In section VIII.4 we mentioned the need to attempt to minimise the unbalances brought about by the transition towards a higher growth path. In the present section we will illustrate the suggestions in a greater detail, which will result more meaningful in the light of the issues which have been treated in the previous sections IX.1–IX.5.

---

[136] The purpose to relaunch an expansionary phase concerns Europe, but the same purpose would be better served should Europe be able to relaunch expansion at a more planetary level. An important factor would thus be that of the opening of new markets, both in the East of Europe and in the South of the world. This, however, requires more optimistic and generous attitudes towards the financing of development processes in those areas, of the same type –as it has been often mentioned– of the post–war Marshall Plan. Given this, anyway, an European Industrial Authority would be essential for a correct and non–inflationary administration of the equivalent Marshall Plan.

(i) **The time scheduling of the take off process**. The pace of the take off process is of overwhelming importance. In general terms it can be said that the start off must be kept very slow, and then there has to be a progressive acceleration; the acceleration should be neat and steady, and should be maintained independently of an acceleration of inflation, to be regarded as physiological and transitory, at least so long as the program of expansion is maintained. What should be avoided at any cost is the reactivation of stop–and–go policies. Finally, should the situation actually reach the desired level of growth rate –which might well be around the 5%– a soft landing should be assured[137]. This suggests also that, in order to maintain a proper expectations climate, a sound use of intertemporal coordination and thus of clear announcement effects should be pursued. Since the bulk of the outlined strategy is based on private behaviours, elicited to play according to shared expectations of expansion, *the use of announcement effects is highly recommendable*. Private agents should be put in the conditions of knowing, clearly and in advance, that the governments will do certain things at clearly stipulated calendar times and/or environmental conditions.

(ii) **Sectors selectivity**. The major risk of inflation is due to a dynamic supply shortage. This suggests to differentiate the speed of expansion someway proportionally to the degree of unused capacity and likely speed of response. This implies two different tasks: the first one is to keep under selective control, as far as possible, the speed of demand expansion; the second one is to support the speed of capacity construction in the sectors which have less unused capacity and lower flexibility to expansion[138].

(iii) **International coordination**. It is essential that a multiplicity of nations adopt the same expansionary strategies and that they make efforts of coordinating such strategies. This will ease the problem of external unbalances; furthermore it will contribute in slowing down the pressures for competitive rivalry, both between firms and between countries. An appropriate international coordination at the sectoral level, aiming at exploiting in a compensatory way the different sectoral flexibilities to expansion and at "administering" the export–import flows in an overall compensatory way might also contribute to keep down the inflationary pressures, avoiding supply shortages. Such arrangements could result in lower pressures on currency exchange rates.

(iv) **Disposable income policy**. There is the need to plan and to Implement

---

[137] The provisional simulations conducted on a two–sectors model able to reproduce the main interpretative features outlined in the first sections of this paper show that the dynamic path of transition to a higher growth rate is very sensitive to the smoothness of the transition.

[138] The simulations referred to in the previous footnote put into evidence that the existence of a certain degree of excess capacity, above all in the sectors which produce capital goods, is a strategic prerequisite for the transition. Conversely, the reaching of capacity limits might have devastating effects, not only because it forbids the envisaged acceleration but because it risks to exert a backlash upon expectations, not to mention the possibility of triggering tensions upon prices and imports.

an appropriate disposable income policy, while avoiding to frustrate the motivations to invest. As we saw above, the demand–supply gap fundamentally depends on the fact that the expenditures which are met in order to build up the productive capacity tend to transform themselves in increases of the disposable income, and thus possibly of demand, earlier than the moment when the additional productive capacity might be used for producing additional consumption output. The disposable income is usually controlled by the means of personal taxation. In our case, however, it is important to convey a credible message that the contraction of disposable income will be a strictly temporary measure. There are several technical arrangements through which governments (or better the European Commission) may engage in such a credible promise[139].

(v) **Residual measures**. As residual and/or complementary measures, the Commission and the national governments may resort to other instruments, such as agreements with the major companies, eased credit conditions made conditional to price controls, etc.[140]. More in general, the expansionary thrust activated by the large European programs should be integrated by provisions directed to smaller firms. Partly this might be achieved through automatic fiscal incentives aimed at rewarding the increase of the investments in real terms. Partly this should pursued through a large collective effort aimed at providing technical and scientific assistance to the smaller enterprises.

X. CONCLUDING REMARKS

A good deal of the analytical background of this paper will appear as heretical to an orthodox reader, and probably (s)he would be ready to accuse it not to have solid theoretical foundations. We would be ready to accept the first comment. As for the second one, we will limit ourselves to few observations.

The traditional theories connected to the neoclassical paradigm are little concerned with long run and with processes of change. As for the concrete functioning of markets and of institutions, the traditional approaches are obliged to rely either upon adhockeries or upon applied (a–theoretical) good sense. We hold that the problems which the world economy, and the European one in particular, is presently facing, have to do with an insufficient rate of growth; that a higher rate of growth is both desirable and possible, as any not short–sighted historical perspective would suggest; that the presently prevailing trends have to do with the prevailing

---

[139] One way is to make clear that part of the taxes have the nature of a special loan, which will be given back in the next years, with a maximum lag of 3–4 years. Another way would be to give generous fiscal allowances proportional to the subscription of European securities, with the proviso that such securities will be not negotiable earlier than certain calendar dates, which the Commission may decide to anticipate.

[140] It is to be acknowledged that very little imagination and "intelligence" (that is ability to gather and use industrial information) has been manifested by public policies with respect to the issue of price control.

conduct of both, markets and public institutions –a conduct which highly differs from that which was considered as sound and was observable during the historical phases when a high rate of growth succeeded in establishing itself.

Two implications stem from this. The first one is to draw analytical inspiration from the relatively few scholars who have attempted to face the problems connected with the dynamics of economic change, first of all Sir John Hicks[141]. The second one is to look for inspiration from historical experience[142], under the assumption that the real world functioning of our economic systems is too complex and culturally biased to be entirely captured by the prevailing over–sophisticated, over–abstract, but also basically simple minded models. We are the first to acknowledge how weak is our knowledge of the processes of change; but this depends on the fact that the bulk of economic theory overlooked such an issue. We are among the few who have attempted to face the problem in modern terms. We are the first to acknowledge how our results are preliminary and provisional.

But, if our diagnosis –namely, that the origin of the present evils has to do with an exceedingly low growth– has an even small chance of being correct, then even our allegedly poor pieces of knowledge about how a growth process may be started might be of the greatest help. In fact the scholars are free to choose their focus of interest. The policy–makers are (or should) not. So the latter ones should be compelled to make the best use of the available knowledge, however poor and unsatisfactory this might be.

A part of the policy proposals developed in this paper might appear as vague, the remaining part as made of traditional instruments. Both perceptions are correct, but they need not to have a negative connotation.

Even if single instruments are traditional, the way they might be jointly used, the context in which are they used, and finally the time specification of their use can make of a given mix of traditional instruments something different and some way new. In any case it would be pretentious on our side to hold that we are proposing something completely new. Our economic systems have experienced, in the past, periods of outstanding development, and what we are proposing is aimed at reproducing the more essential features which belonged to those periods.

The factors which produced the development processes in the past were partly the result of specific historical contingencies which cannot be reproduced in themselves. Furthermore, our systems have become much more complex –and above all much more internationally conditioned– since then. It stems from this the need

---

[141] The inspiring contributions are too many to mention them all; but see Hicks (1939, 1965, 1973, 1974, 1979).

[142] Including more recent empirical investigations on real business behaviours and the patterns of innovative competition.

to use a bundle of instruments which are at least partially different from those which were used or anyway at work in previous development phases. What the development phases have in common is the fact that an important core of agents is moved –that is, motivated and pushed to act– by a common faith that the system may change and evolve in certain directions, whose contours are at the beginning only vaguely defined, but which become more precise along the change process itself, as long as further stages are constructed and a process of learning takes place, concerning what is possible and desirable to do next, and which are the more important constraints and the likely tasks to be performed.

And the alleged "vagueness" stems exactly from this. If there has to be a learning process, affecting all the involved actors, and thus also the institutions and the so called "policy–makers", not all the aspects of the development process can be specified since the beginning, but only the initial hints and the directions to be chosen may be provided. And this by definition.

Is this too risky?. Should really our boat abandon the secure banks of traditional policies to venture itself in an apparently vaguely defined direction of navigation?.

The point is, however, whether the present banks are really reliable, or are they simply asserted to be so. In the last 10–15 years the governments have increasingly conformed their conduct to the prescriptions of the economic orthodoxy, in the ways they are interpreted and transformed in operational indications by specific influential environments, such as the IMF, the economic and financial sections of the OECD and of the EEC, the Central Banks of the national states[143]. Things do not appear to have improved significantly since then; and the more the situation was worsening, the more the same influential environments were prescribing to intensify the same therapy.

The doubt that there might be something wrong either in the diagnosis or in the therapy seems to have never occurred to those who had the responsibility of economic affairs[144]. On the contrary, when the systemic rate of growth was roughly able to keep close to that of productivity, this was regarded as a success, despite the high unemployment rates[145] and despite the even larger manpower reserve potential

---

The majority of the influential officers of such agencies are trained in the same universities in which the orthodoxy forms itself and is preserved.

[144] An event like this is simply inconceivable in disciplines which have a more robust experimental basis than the economic one. In economics, which McCloskey has rightly enough indicated as a rhetoric discipline, often the empirical failures tend to reinforce the models.

[145] The rhetorical device has been that of letting shift upward, along the last quarter of century, the unemployment rate regarded as the equilibrium one.

that our systems hide[146]. May thus such a banks really be thought as being more reliable?. And may really the traditional prescriptions be regarded as less vague, once one takes into account how many times, in the course of the last 10–15 years, the scenarios and the forecasting provided by the same influential environments have been falsified by ex post experience and the prescribed therapies have been acknowledged as being insufficiently strong?.

, A final observation. Most of the attention is presently focused on the crisis of EMS, and on the best way for re–establishing it and proceed towards the EMU and the European Monetary Institute. Such issues have been instead overlooked in this paper. We have a twofold answer to such a possible objection.

While we believe that a regime of wildly fluctuating exchange rates constitutes a serious obstacle to trade and to the establishment of stable cooperation among companies belonging to different countries, we are also strongly convinced that occasional and concerted realignments of the currency parities are physiological and useful in the long run for infra–European cooperation. It is a corollary of such a view the fact that we consider the push towards the EMU, such as it has been decided at Maastricht, as a premature step which could not but produce, as it actually did, anti–cohesive pressures[147]. There is no reason why the European countries, which are so different on so many economic and institutional grounds, should evolve at the same pace, from the productivity and nominal price–level viewpoint, so as to have the possibility of maintaining given currency parities (while approaching the date of EMU) and the same kind of stable economic performances (after the EMU).

The counter–argument of the EMU supporters to this is that the European economies should converge. Our objection is that the type of convergency the EMU supporters have in mind is neither clear nor rapidly achievable, and that some aspects of *diversity* are a richness, and not a weakness, of Europe. The forces brought about by a situation of aggressive market competition may certainly produce some tendency towards a certain homogeneity of the modes of production, because of the short run pressure this might exert towards the adoption of the (presently) best available techniques. However, this could be achieved only at high social and economic costs, in terms of business failures and unemployment, and in terms of the emergence of new dualisms; that is, events which could not fail to produce (or reinforce) anti–cohesive reactions at the European level. Furthermore, what is not clear is whether long–term competitiveness, which nowadays depends more upon the ability to innovate than upon the ability to curtail short–run costs, might be better favoured by a *selective* preservation of diversities –managed thus through the

---

[146] It is well known that the rate of activity of potentially active population is not a constant, but a variable which is sensitive to the probability of finding jobs. We are brought to guess that, on the average, the potential reserve of labour which exists in our industrialised systems is at least double of what is manifested by the unemployment rates.

[147] Bruno (1992).

culturally endowed actions of institutions– rather than by a politically forced tendency towards a pure productive convergency, guided by the blind forces of the invisible hand.

It is a second corollary of our argument that in our view the EMS was substantially proceeding in the best of the *possible* ways before the idea of an early EMU became the winning horse. The Central Banks had learnt quite well how to manage the ordinary fluctuations of the exchange rates and how to discourage too precocious realignments of the stipulated parities and/or to condition them[148]. The situation was far from being the more desirable one, but it was likely to be the best compromise in the given circumstances. And these were circumstances, in our view, which not only allowed a sufficiently reliable context for the development of trades, but also a relatively efficient environment for the establishment of infra–European cooperative relationships among firms belonging to different member countries and between them and the European institutions. The latter ones, in particular, were beginning to be active and increasingly effective in the induction of processes of integration between institutions, firms and persons belonging to different Member States. Such things, which necessarily take a long time, are the only ones able to carry to a suitable, and not forced integration of cultures, values and eventually of economic activities. This ongoing process has been, in our view, delayed, if not impaired by the anti–cohesive forces put into motion by the choice of attaining the EMU too early.

There is a second connected answer. The attention which has been brought in more recent periods upon the financial and monetary performances, above all by the governments but also by private companies, has brought all the concerned actors to overlook the state of the real–side variables of our economies. What the present paper suggests is that we ought to reverse such a system of priorities of concern. Should we succeed in re–establish a positive climate of expectations, should we be able to induce a new phase of capital widening and a higher employment level, then our financial and monetary problems, not to mention those concerning inflation, would come very close to being solved.

---

[148] Today the management of the same policy would probably be harder, after the unrestricted liberalisation of capital movements.

REFERENCES

AMENDOLA, M.; BRUNO, S. (1990), "The behaviour of the innovative firm: Relations to the environment", **Research Policy**, no. 5.

AMENDOLA, M.; BRUNO, S.; DE LAZZARI, S.; INGRAO, B.; PIACENTINI, P. (1991), "The Case Italy" in Modes of Usage and Diffusion of New Technologies and New Knowledge (MUST), MONITOR/FAST, March.

AMENDOLA, M.; GAFFARD, J.L. (1988), **The Innovative Choice. An Economic Analysis of the Dynamics of Technology**, Basil Blackwell, Oxford.

BAUMOL, W.J.; PANZAR, J.C.; WILLIG, R.D. (1982, 1988), **Contestable Markets and the The Theory of Industry Structure**. Harcourt Brace Jovanovich Publishers.

BRUNO, S. (1986), "Incertezza, complessita' e crisi della 'Economia del controllo'", **Economia e Lavoro**, no. 3, December.

BRUNO, S. (1987), "Micro–Flexibility and Macro–Rigidity: Some Notes on Expectations and the Dynamics of Aggregate Supply", **Labour**.

BRUNO, S. (1989), "The Secret Story of the Rediscovery of Classical Unemployment and of its Consequences on Economic Advisors", **Studi Economici**.

BRUNO, S. (1992), "The dangers of EMU: An Industrial Policy Viewpoint", **Revue d'Economie Industrielle**.

BRUNO, S. (1993), "The Limits of Laissez Faire", Dipartimento di Scienze Economiche, mimeo.

BRUNO, S.; SARDONI C. (1989), "Productivity Competitiveness among Non-Cooperating Integrated Economies: A Negative–Sum Game", **Labour**, Vol. 3, no. 1.

BRUNO, S.; DE LELLIS, A. (1992), "The Innovative Systems: The Economics of ex–ante Coordination".

COASE, R.H. (1937), "The Nature of the Firm", **Economica**.

COASE, R.H. (1988), **The Firm, the Market and the Law**, The University of Chicago Press

HICKS, J.R. (1939), **Value and Capital**, Clarendon Press, Oxford.

HICKS, J.R. (1965), **Capital and Growth**, Clarendon Press, Oxford.

HICKS, J.R. (1973), **Capital and Time; a neo–Austrian Theory**, Clarendon Press, Oxford.

HICKS, J.R. (1974), **The Crisis in Keynesian Economics**, Basil Blackwell, Oxford.

HICKS, J.R. (1979), **Causality in Economics**, Basil Blackwell, Oxford.

HIRSCHMAN, A.O. (1958), **The Strategy of Economic Development**, Yale University Press, New Haven.

KALDOR, N. (1957a), "Inflaçao e Desenvolvimento Economico", **Revista Brasileira de Economia**, March.

KALDOR, N. (1957b), "O Problema do Crescimento Accelerado", **Revista Brasileira de Economia**, March.

NURKSE, R. (1953), **Problems of Capital Formation in Underdeveloped Countries**, Oxford.

RICHARDSON, G.B. (1960), **Information and Investment**, Oxford University Press.

ROSENSTEIN–RODAN, P.N. (1943), "Problems of Industrialization of Eastern and Southern–Eastern Europe", **Economic Journal**, june–sept.

SCITOVSKY, T. (1959), "Growth: Balanced or Unbalanced", **The Allocation of Economic Resources**, Stanford University Press.

STREETEN, P. (1959), "Unbalanced Growth", **Oxford Economic Papers**, june.

WILLIAMSON, O.E. (1985), **The Economic Institutions of Capitalism**, The Free Press, New York.

AUTHOR INDEX

A
Abernathy, 63
Abramowitz, 92-4
Acs, 104
Amendola,ix,8,11,20,112,119,120,122,146,147,159,
160,192,249,362,365,370,376
Amin, 126,153,164,269,274
Andersen, 21
Archibugi, 41,93-4
Arnold, 200,221
Arora, 88
Arrow, 79,82,90
Arthur, 20-1
Ashcroft, 114,267-70,278,281,283,285,286
Atkins, 345
Audretsch, 104,148,170
Aydalot, 117-8,147-8,150,159,165,187

B
Baldwin, 344
Baltensperger, 334
Banerjee, 27
Baumol, 321-2,364,376
Becattini, 104
Beckouche, 154
Bell, 40
Benko, 152
Berthomieu, 256
Bertin, 43
Bianchi, 126
Bingham, 124
Bisignano, 295-6,299
Blakely, 118
Boddy, 115
Boeckhout, 104
Boisgontier, 112
Boudeville, 151, 231
Bradshaw, 110
Branson, 295
Braun, 112
Breheny, 115
Brooks, 81
Bruno, 121, 249,364, 366,369-70,384
Buckley, 246

Buiges, 345,347-8,352

C
Cantwell, 66
Camagni, 103,117,121
Cameron, 307
Carlsson, 64
Casson, 246
Castells, 118
CICYT, 182
Clark, 67
Coase, 245,367
Cohendet, 261
Colombo, 236
CEC, ii,184,188-9,293-4,
301,343
Contractor, 243-4,246,
251,343,347
Cooke, 117
Coppins, 267
Corado, 94
Corbett, 63
Corden, 344
Cowan, 83,84,91
Cressman, 27
Crow, 22,33
Currie, 329
Charbit, 112, 261-2
Chenery, 210
Chesnais, 120, 243,247
Chick, 294, 297,300,307,327

D
D'Ambrogio,117
D'Amico, 311,321
D'Onofrio, 325
Dahlman, 45,200
Dasgupta, 82
David, 20,78,81-3,88
Davis, 298
De Bernardy, 112
De Meza, 308
De Lellis, 121,362,369-70
De Smidt, 228
Decoster, 116

389

X. Vence-Deza and J. S. Metcalfe (eds.), Wealth from Diversity, 389–393.
© 1996 *Kluwer Academic Publishers. Printed in the Netherlands.*

Economics of Science, Technology and Innovation

KLUWER ACADEMIC PUBLISHERS – DORDRECHT / BOSTON / LONDON